ORGANIC FARMING: METHODS, ECONOMICS AND STRUCTURE

Organic Farming: Methods, Economics and Structure

Michael Nelson
and Ivan Artamova
Editors

Nova Science Publishers, Inc.
New York

For permission to use material from this book please contact us:
Telephone 631-231-7269; Fax 631-231-8175
Web Site: http://www.novapublishers.com

NOTICE TO THE READER

LIBRARY OF CONGRESS CATALOGING-IN-PUBLICATION DATA

Nelson, Michael, 1962-
Organic farming : methods, economics and structure / Michael Nelson and Ivan Artamova.
p. cm.
Includes index.
ISBN 978-1-60692-864-6 (hardcover)
1. Organic farming. 2. Humus. 3. Soil microbial ecology 4. Natural foods. I. Artamova, Ivan. II. Title.
S605.5.N45 2009
631.5'84--dc22

2009000136

Published by Nova Science Publishers, Inc. ✢ New York

CONTENTS

PREFACE

Organic farming is a form of agriculture relying on such processes like crop rotation, green manure and biological pest control. Organic farming is gaining worldwide acceptance and has been expanding at an annual rate of 20% in the last decade, accounting for over 24 million hectares worldwide. This book discusses the agricultural practices in organic farming which cause changes in soil organic matter and microbial biomass content. This book also explains the concept of scientific paradigms and draws parallels with paradigms in agriculture, and in particular focuses upon organic farming and genetic engineering. An overview of the evolution of global market demand for organic meat and factors affecting it are also described. Future prospects for development of organic meat consumption and possible problems concerning trade are analyzed as well.

Chapter 1 - The olive tree is of great socio-economic and ecological importance in the Mediterranean region. In several countries, namely in Portugal, olive production is a significant land use, and so, the olive agro-ecosystem is an important reservoir for biodiversity. During the last years we are assisting to a growing interest for organic olive grove cultivation, to fulfil the requirements of consumers who are more and more interested in ecologically clean products due to health and environmental reasons as well as the increasing concern for safe and quality food. In Portugal organic olive surface has steadily increased in the last few years and from only 3 781 ha in 1994, it grew to 19 341 ha in 2006. Almost all this surface (about 99.3%) was distributed by the districts of the interior along the Spanish border (Trás-os-Montes, Beira Interior and Alentejo) where climatic conditions are suitable for the crop and pest and disease pressure is relatively low. Moreover traditional farming practised in these regions approximates organic farming methods, which eases conversion. The financial support provided by the European Union under the agri-environmental measures for organic production through the Common Agricultural Policy, which largely have stimulated farmers´ interest towards organic agriculture, have also expanded the development of scientific and technical research regarding this type of crop management. In this chapter, results from organic olive research carried on during the last ten years in the northeast of Portugal will be discussed. Emphasis will be given on activities having as objective the enhancement of soil fertility through the application of composting of plant and animal materials, as well as those focusing on arthropod pest management. Progresses attained and areas which need to be explored will be analyzed.

Chapter 2 - This paper discusses the concept of scientific paradigms and draws parallels with paradigms in agriculture, and in particular focuses upon organic farming and genetic engineering. There is anecdotal evidence that they are both potential new agricultural paradigms. Different writers have referred to both as new 'paradigms' in agriculture, though the qualitative evidence is stronger for organic farming than genetic engineering. The attitudes of agricultural professionals are important as they are responsible for creating, developing and diffusing innovations as well as informing and educating farmers and the public. This paper summarizes past research on the attitudes of Australian agricultural professionals (n=185) towards organic farming (Wheeler 2008) and genetic engineering (Wheeler pending). These two sets of results are compared to assess if professional attitudes to the two innovations are driven by similar influences or are quantitatively significantly different. In addition, analysis using bivariate ordered probit regression is conducted in this study on the nature of professionals' answers to the net benefits of the two innovations themselves. Do professionals tend to be negative or positive about both innovations, or positive about one and negative about the other? From a comparison of the results, it is shown there is some quantitative evidence that organic farming represents a potential new sub-paradigm in agriculture, but only very weak evidence that genetic engineering is a potential new paradigm.

Chapter 3 - Global estimate of described millipedes is about 12,000 species out of a total estimate 80,000 species in all the continents except for Antarctica. Millipede collection is available in 268 institutions of 143 countries but understudied in Africa, Central and South America and Asia. The family Sphaerotheriidae (Order: Sphaerotheriida) consists of giant millipedes possessing 13 body segments with 21 pairs of walking legs called 'pill millipedes' as they attain sphere shape on conglobation. The subfamilies and tribes have been separated based on the shape of female vulva and stridulating organ. They predominantly hail from Southern Hemisphere (Indo-Australia, Madagascar and South Africa) and wiped out in the intermediate regions due to unsuitable environment. Limited dispersal ability led to endemism of the Malagasy giant pill millipedes (*Sphaeromimus* and *Zoosphaerium*), which exhibits female gigantism and Batesian mimicry. Pill millipedes are hosts for a variety of microbes and from food to faecal matter the bacterial load increases and fungal load decreases. Presence of ergosterol in the faecal pellets of *Glomeris* indicates intensive digestion of fungi in the gut passage. *Glomeris* consume up to 1.7-10% of annual litter production, which amounts to 10 times the body mass. *Arthrosphaera* prefers mixed litter than monolitter and the mixed litter ingestion and faecal pellet production is higher than other tropical millipedes. Increased nitrogen, phosphorus and potassium, narrow C/N ratio and shift of pH towards neutral in mixed litter compost produced by *Arthrosphaera* reveals its value as alternative to vermicompost. Millipedes are influenced by the forest under storey and act as buffer stock for significant increase of calcium and magnesium and augment mineral recycling. Clear evidence has been emerged on the invasion of *Arthrosphaera* from evergreen and semi-evergreen forests to plantations in Western Ghats in India. Long-term intra-site heterogeneity in organic matter input in mixed plantations might have facilitated such invasion. *Arthrosphaera* is known for a narrow range of distribution and exhibits single site endemism in the Peninsular India. They are sensitive to litter chemistry at narrow spatial scales than the earthworms. In spite of conducive ecological conditions, millipedes are highly sensitive and fail to overcome the limitation of even a single edaphic factor (e.g. water deficit, soil texture, litter thickness). Such sensitivity of *Arthrosphaera* qualifies its suitability to serve as indicator of soil quality loss or restoration in managed plantations or forests.

Chapter 4 - Farm structure and production traits are numerous, and it is a real difficulty to integrate such a large bulk of knowledge into simplified categorisations. One common tool is discriminant analysis, which requires an *a priori* categorisation. Conversely, clustering methods are available to identify homogeneous groups *a posteriori*. The difficulty of clustering methods is in the very large number of options. The discriminant analysis and the clustering methods intend to propose categories on traits which are evaluated at one moment without any attempt to relate and to suggest any hierarchy in the order or time of appearance of the traits. They are comparable to a snapshot at one moment. Phylogenetic analyses are intended to propose a sequence of appearances of the traits; they intend to reconstruct a history of events. In the present paper we used data from meat sheep farms in the centre of France. Some farms were organic and others were conventional; all farms studied were subsidised by the state. Firstly, we used cluster and phylogenetic analyses for categorisation of conventional (9) and organic (7) meat sheep farms. Secondly, the evolution on several years in seven organic farms was studied. The phylogenetic analysis (Hennig86 programme implemented with the freeware Tree Gardener) is performed on re-coded data into evoluted (derived) *versus* non-evoluted (ancestral) trait or level of the trait. The phylogenetic analysis we propose is a flexible tool that reconstructs the sequence of traits, and probably better corresponds to the strategy of the farmer in its environment.

Chapter 5 - This chapter provides an overview on the evolution of global market demand for organic meat and factors affecting it. Future prospects for development of organic meat consumptions and possible problems concerning trading are analysed. The price that people are willing to pay is the major determinant of the market share of organic meat. Although intent to purchase depends upon the interactions of quality attributes such as appearance and colour, it has been hypothesized a consistent effect of organic labelling as compared to some sensory characteristics on the price offered by consumers for organic meat. Previous experiments have shown that expectations induced by the information can affect the quality perception. Therefore, consumers' behaviour and attitudes toward organic meat are tested through the study of the effect of information about organic farming practices on product liking and consumer willingness to buy organic beef and pay the extra production costs. Our results indicate that this information, if given to the consumers, can be a major determinant of beef liking, thus providing a potential tool for meat differentiation to traditional farms where husbandry is based on extensive rearing systems and high animal welfare standards. Accordingly, consumers are prepared to spend more for organic beef, thus indicating that reliable information about the organic farming system may markedly increase consumer willingness to pay.

Chapter 6 - The objective of this study was to optimize organic steer production by analyzing the effect of production factors as feed level, grazing strategy, and finishing and welfare strategies on production objectives as liveweight gain, carcass conformation and health. Six Danish organic farms were part of the experiment that endured 2½ years. Registrations included liveweight, feed consumption, welfare-related observations and slaughter results with special emphasis on the finishing period. Daily gain of steers varied between the farms from 533 to 823 g and from 562 to 867 g on pasture and during the stable period respectively. Average feed consumption per steer per year was 1770 Scandinavian Feed Units (SFU) ± 186 based on 4 farms. Feed consumption in summer accounted 0.47 to 0.57 of total feed consumption per year. Finishing strategies showed great variations between farms with liveweight at slaughter from 458 to 595 kg, age at slaughter from 20 to 25 months and length of finishing period from 2.3 to 3.4 months. The total amount of cereals consumed in the finishing period was 100-357 kg/steer. Slaughter results show a variation between farms in carcass conformation from 3.5 to 4.9 and fatness from 2.5 to 3.2. Liveweight at slaughter was the only variable with general effect on carcass conformation (0.6 point on the EUROP-scale with increase in 100 kg liveweight at slaughter). The welfare registration scheme indicates good conditions both in the barn and on pasture, low disease rate and minor problems concerning welfare aspects in organic steer production. It is concluded that finishing strategy should focus on liveweight at slaughter. Length of finishing period as well as daily gain during the finishing period in the range practiced on the farms seemed to be of minor importance with regard to carcass conformation.

Chapter 7 - All kinds of fertilizers available on the Austrian market have been analyzed for their nutrient element und micro-element contents. From a total of 2759 samples, taken in 2002 - 2006, 2391 samples were assigned as solid and nutritional, and classified due to their main element characteristics N-P-K-C-S. Commercial organic fertilizers, manures of any kind, and composts have been summarized in one category each. The elements (in alphabetical order) Al, As, B, Ba, Be, Ca, Cd, Co, Cr, Cu, Fe, K, Li, Mg, Mn, Mo, Na, Ni, P, Pb, Sr, V, and Zn were determined by ICP-OES after acid digestion. Because all datasets are asymmetric, with many low and some high values, medians and 5-95% concentration ranges are given, for each micro-element in 17 tables. Concentrations in dry mass are compared with mean soil and mean crust values. Loads for a fertilization rate of 100 kg N and 100 kg P were compared with atmospheric deposition data and presumable crop uptake. At usual fertilization rates for N and P, the input of most micro elements was maximum for composts. Manures and commercial organics also supplied micro elements at higher levels than mineral fertilizers, except Cd-As-Cr. In addition, manures, some K-salts, and PK-fertilizers supply a lot of Na. On a long term, phosphate fertilizers will contribute to the enrichment of arable soils with Ba, Be, Cd, Cr, and V, because they supply more than atmospheric perrecipitation and crop outputs. Composts will probably enrich soils with As, Ba, Be, Cu, Cr, Li, Mn, Ni, Pb, V, and Zn.

Chapter 8 - Agricultural production systems have to combine management practices in order to sustain soil quality and also profitability. Organic farming is gaining worldwide acceptance and has been expanding at an annual rate of 20% in the last decade, accounting for over 24 million hectares worldwide. Organic practices avoid applications of synthetic fertilizers and pesticides, rely on organic inputs and recycling for nutrient supply, and emphasize cropping system design and biological processes for pest management, as defined by organic farming regulation in the world. In comparison with conventional farming, organic farming has potential benefits in improving food quality and safety. Plant production in organic farming mainly depends on nutrient release as a function of mineralization processes in soils. The build-up of a large and active soil microbial biomass is an important pool of accessible nutrients, and therefore is an important priority in organic farming. Nutrient mineralization and immobilization are soil microbial processes governed by C availability and are linked closely to soil organic matter (SOM) content. Furthermore, soil microbial biomass respond rapidly to changes resulting from agronomic practices and other disturbances and have been used to assess early changes due to different soil management practices. Therefore, agricultural practices in organic farming can cause changes in soil organic matter and microbial biomass content.

Chapter 9 - U.S. farmland under organic management has grown steadily for the last decade as farmers strive to meet consumer demand. By 2003, the United States had over 2 million acres of certified organic crops and pasture. USDA implemented an organic regulatory program in 2002, and Federal research and education activities have also emerged.

Chapter 10 - Congress passed the Organic Foods Production Act (OFPA) in 1990 as part of a larger law governing U.S. Department of Agriculture (USDA) programs from 1990 through 1996 (P.L. 101-624, the Food, Agriculture, Conservation, and Trade Act of 1990). The act authorized the creation of a National Organic Program within USDA to establish standards for producers and processors of organic foods, and permit such operations to label their products with a "USDA Organic" seal after being officially certified by USDA-accredited agents. The purpose of the program, which was implemented in October 2002, is to give consumers confidence in the legitimacy of products sold as organic, permit legal action against those who use the term fraudulently, increase the supply and variety of available organic products, and facilitate international trade in organic products.

Due to the newness of the National Organic Program, it is difficult to gauge its overall impact on the organic industry. However, a recent USDA report states that the number of people seeking USDA accreditation to become certification agents increased by more than 130% between 2002 and 2005, suggesting significant growth in the number of farm and processing operations going into organic production. USDA estimates that sales of organic foods rose from $6 billion in 2000 to $10.4 billion in 2003. The annual rate of market growth since 1990 has remained steady at about 20%.

Policy issues affecting the National Organic Program since implementation largely reflect the differences in interpretation among stakeholders of the language and intent of OFPA and the actual operation of the program under the final rule. The most recent issues concern, first, USDA's efforts to write a new regulation governing livestock access to pasture, and second, a lawsuit brought by a plaintiff who argued that USDA's administration of the NOP has been too lenient concerning the use of certain synthetic ingredients in foods bearing the "USDA Organic" label, and too lenient concerning the process for converting a dairy herd from conventional to organic production. The final judgment in the case, which was issued in June 2005, requires the Department to develop new rules based on the original statutory language. The revised rules would become effective two years after the date of the judgement, up until which time products made under the old rules would be allowed into commerce.

Subsequently, however, conferees on the FY2006 USDA appropriations bill attached a provision that amends the OFPA in such a way as to largely maintain the NOP regulations as they were before the court decision. This action is likely to worsen the rift between those who welcomed the court ruling because they felt it would force USDA to write stricter regulations that would more closely reflect the statute, and those (a sizeable consortium of major food companies) who predicted that new rules written to satisfy the court decision might severely limit current and future lines of organic products and economically harm all sectors of the industry.

Chapter 11 - This compilation cites only documents published directly by the U.S. Department of Agriculture's primary agencies during the years 1977 through August 2005. The full range of published work by USDA scientists, analysts, advisors and other employees at stations throughout the world appears in many other publishing venues including peer-reviewed journals, research reports, newsletters, conference presentations, Web pages, bulletins and factsheets, databases and more. USDA funded projects and organizations also publish documents that appear in many guises and places.

In terms of subject, only publications pertinent to organic production, food and marketing are represented. There are many others that are relevant to organic systems that are not cited. In addition, this listing does not address the large body of research published before 1940 when reliance on synthetic fertilizers and pesticides in agriculture began to become widely adopted. Much of this historic information, which predates the common use of the term "organic agriculture," is directly applicable to current organic systems. For information on other organic-related resources from USDA and elsewhere, or for information about accessing materials from the National Agricultural Library collection, contact the Alternative Farming Systems Information Center (AFSIC).

In: Organic Farming: Methods, Economics and Structure
ISBN 978-1-60692-864-6
Editors: Michael Nelson and Ivan Artamova

Chapter 1

ORGANIC OLIVE GROWING IN NORTHEASTERN PORTUGAL: PRESENT STATE AND FUTURE PROSPECTS

Sónia A.P. Santos[1], Cristina Cunha-Queda[2], Fátima Gonçalves[3], José Alberto Pereira[1], Albino Bento[1] and Laura M. Torres[3]

[1]CIMO & Escola Superior Agrária, Instituto Politécnico de Bragança, Bragança, Portugal

[2]UIQA & Instituto Superior de Agronomia, Tapada da Ajuda, Lisboa, Portugal

[3]CITAB & Universidade de Trás-os-Montes e Alto Douro, Vila Real, Portugal

ABSTRACT

The olive tree is of great socio-economic and ecological importance in the Mediterranean region. In several countries, namely in Portugal, olive production is a significant land use, and so, the olive agro-ecosystem is an important reservoir for biodiversity. During the last years we are assisting to a growing interest for organic olive grove cultivation, to fulfil the requirements of consumers who are more and more interested in ecologically clean products due to health and environmental reasons as well as the increasing concern for safe and quality food. In Portugal organic olive surface has steadily increased in the last few years and from only 3 781 ha in 1994, it grew to 19 341 ha in 2006. Almost all this surface (about 99.3%) was distributed by the districts of the interior along the Spanish border (Trás-os-Montes, Beira Interior and Alentejo) where climatic conditions are suitable for the crop and pest and disease pressure is relatively low. Moreover traditional farming practised in these regions approximates organic farming methods, which eases conversion. The financial support provided by the European Union under the agri-environmental measures for organic production through the Common Agricultural Policy, which largely have stimulated farmers′ interest towards organic agriculture, have also expanded the development of scientific and technical research regarding this type of crop management. In this chapter, results from organic olive research carried on during the last ten years in the northeast of Portugal will be discussed. Emphasis will be given on activities having as objective the enhancement of soil fertility through the application of composting of plant and animal materials, as well as those focusing on arthropod pest management. Progresses attained and areas which need to be explored will be analyzed.

1. Introduction

The olive tree (*Olea europaea* L.) is a typical and emblematic tree in the Mediterranean countries (Loumou & Giourga, 2003) where it has a multiple importance particularly in that concern to economic, social, cultural and landscape aspects.

In Portugal, the olive sector was considered by the government as a priority area and the traditional characteristics of the olive cultivation have been used to develop Product Denominations of Origin (PDO's) (Gouveia & Soeiro, 1997). Olive cultivation represents a significant land use in Portugal, where olive trees are grown in an area of 335 028 ha from which 19 341 ha are managed according organic farming guidelines (Anonymous, 2007; INE, 2007). Trás-os-Montes region (Northeastern Portugal) is an important olive production area, representing 21.6% of the total Portuguese olive cultivation area (Anonymous, 2007) and 22.5% of the organic managed area (INE, 2007). In this region, organic farming can generate high-quality products and can be seen as an alternative since olive groves occur in sloping and mountainous areas and yield is not competitive. If marketed as products with PDO, these high-value products can be sold at a substantially higher price, improving profitability of traditional farmers (Fleskens, 2007).

The success of the organic agriculture is largely dependent on the understanding of the mechanisms operating in the agroecosystem. In this scope, Trás-os-Montes offers an interesting case study for addressing organic olive farming, as a result of different research projects developed in the region throughout the last years. Research was focused on two different approaches: a) composting of biodegradable organic matter, and b) the development of an arthropod pest management for organic olive growing.

Composting as a bioenergetic process of transformation of biodegradable organic matter, allows to its treatment/valorization, and is one of the solutions recommended in the guidelines of the European Community, in order to minimize the deposition of biodegradable wastes into landfills. This way it will be possible the conciliation of environment protection and valorization of these biowastes as organic fertilizers, constituting a basic resource for organic farming, as it is indicated in the communitarian legislation and in the Portuguese Handbook for Organic Farming (Ferreira *et al.*, 2002).

Arthropod pest management is an alternative to the use of chemicals and integrates pest management activities with other farming practices that maintain soil productivity and crop health, while ensuring food quality and economic viability. Concerning this issue, research was centred mainly on the effects of different agricultural management practices in the control of the three most important pest species in the region: the olive moth, *Prays oleae* (Bernard) (Lepidoptera: Yponomeutidae), the olive fruit fly, *Bactrocera (Daculus) oleae* (Gmelin) (Diptera: Tephridae) and the black scale, *Saissetia oleae* (Olivier) (Hemiptera: Coccidae). Due to weather conditions, damages caused by these pests are highly variable between years and geographical regions, which render difficult their control. In addition, high temperatures (>30-35°C) that normally occur during summer can contribute significantly for the decline of pests' populations. This is particularly noteworthy in the case of the olive fruit fly and the black scale.

Considering the olive moth, the phenology of this pest, which develops three generations per year, is well synchronized with that of the olive tree. Larvae of the leaf (or phyllophagous) generation overwinter as leaf-miners and adults emerge early spring. Eggs are laid on the calyx of the flower buds and the emerged larva of this generation (anthophagous) develops on flower buds, flowers and, at later instars, on whole flower clusters. Larvae of the following generation (carpophagous) penetrate the developing fruit and feed inside the endocarp causing the detachment and, consequently, the early fall of the fruit (Arambourg, 1986, Bento, 1999). *P. oleae* is considered the key-pest of olive groves in northeastern Portugal, where it causes yield losses up to 50% every year (Bento, 1999).

The olive fruit fly adults lay eggs singly in the mesocarp of the olive fruit, after the hardening of olive fruit pit, which coincides with the end of July. Larvae develop in the pulp destroying it and then pupate in the fruit or in the soil where it overwinters (Arambourg, 1986, Civantos, 1998). Olive fruit fly causes both direct damages by destroying the fruit pulp and indirect damages by both causing an early fall of the fruits and reducing the quality of the olive oil.

The black scale is an oviparous species with parthenogenetic reproduction since males are very rare or absent (Passos-Carvalho *et al.*, 2003). Females have three nymphal stages before the adult stage that presents two distinct phases, the immature female or fourth instar nymph stage and the mature or ovipositioning female. In Trás-os-Montes, *S. oleae* is an univoltine species which winters on immature stages, mainly on second and third-instar nymphs. However, under favourable conditions, a partial or even a complete second generation can take place (Pereira, 2004). Third-instar nymphs usually migrate from leaves to younger sprout in spring. During April and May, the third-instar nymphs give origin to young females, which mature and lay eggs from May to the end of July (Passos-Carvalho *et al.*, 2003; Pereira, 2004). In the olive tree, *S. oleae* feeding causes both direct damages by sucking plant sap and indirect damages by favoring the development of fungi such as the sooty mold.

The goal of this chapter is to provide an overview of research work developed in olive groves from northeastern Portugal considering issues like composting of organic matter and arthropod pest management.

2. Composting of Biodegradable Organic Matter

In general Mediterranean soils have low organic matter content, which is highly correlated with their potential productivity and fertility due to its direct influence on physical, chemical and biological soil properties (Albuquerque *et al.*, 2007). The application of organic wastes through soil is one practice that improves soil organic content. However, when biodegradable organic matter (without any previous stabilization treatment) is applied to soil it may be responsible for depressive effects on plant growth, and can also induces quick development of biochemical transformations, which may lead to harmful effects like odors and gaseous emissions during transport, storage and land application (Guardia *et al.*, 2004). Another important aspect is the quality of organic matter, namely the presence of heavy metals and other contaminants that can be harmful when organic matter is applied to soils. In this case the source separation and collection of organic wastes are essential to guarantee the quality of organic matter, especially when applied in organic farming. In this context the use of by-products or wastes from food industries, like olive oil and wine production is advantageous because usually they have high quality and are not contaminated.

On the other hand, the stabilization of by-products and waste organic matter can be achieved through composting, which is a controlled aerobic process of bio-oxidation of heterogeneous biodegradable substrates, resulting of the concerted action of microorganisms (bacteria, actinomycetes and fungi) naturally associated to the substrates, during which occurs a thermophilic phase and the biomass suffers deep transformations (partial mineralization and humification). The final product, the compost, must be homogeneous, humified and sanitized and its application to the soil should not produce any adverse effect to the environment and crops (Cunha-Queda, 1999).

Because nitrogen is one of the most expensive inputs in organic farming (Raviv *et al.*, 2005), losses of nitrogen during composting may be reduced by using high C:N mixtures that enhance N immobilization, or by lowering the pH of the compost solution (Raviv *et al.*, 2005). The characterization of -on and -off farm organic residues and by-products is very important in order to establish the adequate mixtures for composting process. For example, on Table 1 is displayed the physical and chemical characterization of several waste and by-products from olive oil (Two-Phase Olive-Mill Waste - 2POMW, Leaves of Olive Trees - LOT) and wine (Grape Stalks – GS, Grape Marc - GM) production as well carnaz (C) a residue produced from tanneries without treatment with heavy metals.

Table 1. Physical and chemical characterization of several waste and by-products from olive oil and wine production and carnaz (mean ± standard deviation; d.m. – dry matter) (Cunha-Queda *et al.*, 2006).

Waste and by-products	Parameters					
	Moisture [%]	Organic Matter [% d.m.]	Total Nitrogen [% d.m.]	C/N ratio	pH	Electrical Conductivity [mS cm^{-1}]
Two-Phase Olive Mill Waste (2POMW)	65.83 ± 0.58	95.92 ± 0.34	0.72 ± 0.02	66.75 ± 0.23	5.03 ± 0.01	2.68 ± 0.03
Leaves of Olive Trees (LOT)	52.13 ± 0.87	94.60 ± 1.32	1.28 ± 0.18	36.99 ± 0.48	5.96 ± 0.07	0.06 ± 0.01
Grape Stalks (GS)	37.81 ± 2.25	86.46 ± 1.29	0.76 ± 0.04	56.98 ± 0.85	7.65 ± 0.11	1.74 ± 0.09
Grape Marc (GM)	56.78 ± 0.83	88.85 ± 2.52	1.93 ± 0.14	23.00 ± 0.65	3.78 ± 0.13	1.72 ± 0.09
Carnaz (C)	57.84 ± 0.18	96.34 ± 0.28	13.78 ± 0.08	3.50 ± 0.01	3.71 ± 0.01	2.13 ± 0.04

The Two-Phase Olive Mill Waste presents a high moisture and C/N ratio, but low nitrogen content and pH, for this reason it is important to mix other materials in order to have an initial mixture with appropriate characteristics (such as moisture, C/N ratio and porosity) for composting process. Grape Marc and Carnaz are interesting materials because of their high nitrogen content.

Results of some composting piles (with mechanical turning), which were carried out with different proportions of each raw material showed that the utilization of Grape Stalks had a beneficial effect, especially due to their bulking agent properties that enhancing the porosity of composting biomass (Gea *et al.*, 2005). In fact, composts produced with the waste and by-products described above showed good quality (Table 2). Copper and Zinc contents were below the limits reported in the Community Rules for Organic Farming, Annex II of Regulation (EEC) N° 2092/91 (Copper: 70 mg kg^{-1} dry matter and Zinc: 200 mg kg^{-1} dry matter).

Table 2. Physical, chemical and biological characterization of produced composts [C1 – 63.4% of 2POMW + 33.3% of GS + 3.3% of LOT; C2 – 90.9% of GS + 9.1% of C; C3 – 54.2% of 2POMW + 33.3% of GS + 8.3% of LOT + 4.2 of C; C4 – 47% of GS + 26.5% of GM + 26.5% of C; composting mixtures expressed by weight on a net basis] (d.m. – dry matter; nd – not detected) (Cunha-Queda *et al.*, 2006).

Parameters	Composts			
	C1	C2	C3	C4
Moisture [%]	26.27	37.79	39.47	37.92
Organic Matter [% d.m.]	53.70	38.56	58.40	55.80
Total Nitrogen [% d.m.]	1.04	1.67	2.01	3.28
Ammonium Nitrogen [mg kg^{-1} d.m.]	26.50	45.45	267.31	1057.85
C/N ratio	25.82	11.57	14.50	8.50
pH	7.46	9.03	6.56	5.45
Electrical conductivity [mS cm^{-1}]	0.77	2.75	2.36	4.65
Cu [mg kg^{-1} d.m.]	28.96	39.61	50.97	28.80
Zn [mg kg^{-1} d.m.]	50.37	63.40	45.90	42.26
Oxygen Consumption - AT4 [mg O_2 g^{-1} d.m.]	2.98	0.81	0.88	0.54
Stability Class (self-heating test)	V	V	V	V
GI [%]	82.07	68.17	102.50	96.54
Salmonella spp. [g^{-1}]	nd	nd	nd	nd

Other important aspects are stability and maturity of composting products. All the composts produced (Table 2) from the waste and by-products listed on Table 1 presented good stability level. In fact, the oxygen consumption (AT4 Test) was lower than 10 mg O_2 g^{-1}

of dry matter, which is the threshold value indicated on the "2nd Draft Working Document Biological Treatment of Biowaste" (DG Env.A.2., 2001), for a residue to be considered not biodegradable and be stabilized in accordance with the definition presented in the Article 2 (m) of Directive 1999/31/EC. This stability level is also confirmed by self-heating test; according to the classification presented on the reference document Laga-Merkblatt M 10 (1995) composts presented the highest class of stability - class V (evaluated by self-heating test). Regarding evaluation of maturity degree, all composts had Germination Index values higher than 60%, threshold value considered by Pera (1991) in order to classify composts as not phytotoxic. This aspect is very important especially when composts are produced from materials, such as Olive Mill Waste (OMW), in which the main characteristic is the presence of organic compounds such as organic acids, lipids, alcohols and polyphenols that turn OMW into phytotoxic materials, representing a great environmental hazard when it is not managed properly (Roig *et al.*, 2006).

As for the nitrogen conservation, the use of Grape Stalks seems to be important, for example compost C3 (Table 2), in which Carnaz (with high total nitrogen content) was added to Two-Phase Olive Mill Waste (with high C/N ratio), the porosity of biomass given by Grape Stalks and the improvement of aerobic conditions of biomass, prevented the eventual losses of nitrogen. Values of pH for composts C3 and C4 are lower than those observed for composts C1 and C2; these results may be due to the formation of NO_3^- by nitrification process (Sánchez-Monedero, 2001).

The presence of Salmonella spp. was not detected, result that evidences the good conditions for the development of composting thermophilic phase necessary to achieve an adequate sanitation level of composts for agricultural use.

Further research is needed to confirm these results namely using the composts produced in long-term field experiments.

3. Arthropod Pest Management

The organization of this issue follows the pyramidal model proposed by Wyss *et al.* (2005) where the concept of insect pest management program for organic systems can be developed in four phases. The first phase comprises the adoption of cultural practices like the inclusion of resistant or tolerant cultivars that help to prevent pest outbreaks and the enhancement of soil quality by incorporation of specific cover crops and/or the addition of soil amendments (e. g., manure, slurry or composts). In the second phase, the habitat is managed to preserve or enhance survival, fecundity, longevity, and behaviour of natural enemies in order to prevent pest outbreaks (Landis *et al.*, 2000). The diversity and abundance of these species can be improved by the incorporation of suitable natural habitats into or nearby the crop (Landis *et al.*, 2000, Fiedler *et al.*, 2008, Jonsson *et al.*, 2008). In the third phase, direct measures like mass-reared live agents are used in inundation or inoculation releases. Unlike first- and second-phase strategies that involve advance planning, biocontrol agents give organic farmers the opportunity to react rapidly when pest populations reach action thresholds. Finally, disruption techniques and the application of approved insecticides of biological and mineral origin constitute the fourth phase. In organic agriculture these are used as a last option for the control of pests when all methods used in preceding phases have failed. In a plant protection program, indirect measures should be favoured over direct regulation measures that can be employed when indirect measures were not sufficiently efficacious.

3.1. First-Phase Strategies

This phase includes different techniques used for the reduction of the likelihood of pest infestation and damage. Cultural practices should be implemented in the initial stages of a long-term organic farm-plan.

3.1.1. Olive Cultivars Resistance

Arthropod resistance may at times be a primary consideration in cultivar selection when new organic farms are planed. When compared with other species that have been the centre of intense research and genetic improvement programs, the olive tree has the considerable advantage of being a species of very ancient domestication within the Mediterranean area, which has allowed the over time co-evolution of pests and natural enemies (Fabbri & Ganino, 2002). Thus, different genetical differentiations occurred within each geographical region, with those genotypes today called cultivars being well adapted to its environment and, consequently, the cultivation of new olive groves can be strongly improved by the correct choice of the less susceptible cultivars (Malavolta *et al.*, 2002). But, it is important to keep in mind that the selection of cultivars may have opposing effects on different pests and that must be based on an overall risk assessment (Zehnder *et al*, 2007).

Studies developed in northeastern Portugal by Bento *et al.* (2003) and Cardoso *et al.* (2006) aimed to find differences for the degree of *B. oleae* fruit attack among the three most important cultivars in the region, Cobrançosa, Madural and Verdeal Transmontana. The risk attack caused by this pest was traditionally low when compared with other regions in the South of the country. However, for several years during the last decade, a great percentage of fruits were attacked by the olive fly at harvest. Those authors found that, in general, cv. Madural showed higher percentage of *B. oleae* fruit attack than cv. Cobrançosa, differing 15-30% between each other, and also that fruit attack strongly increased during autumn, when temperatures are mild and relative humidity is high. Furthermore, damaging infestations, caused mainly by L3 larval stages, increased with the degree of ripeness and an early harvest will contribute for reducing detrimental effects of larvae. Thus, factors like the definition of a timely harvest and the minimisation of the interval between harvest and milling should be important considerations in order to control the level of fruit infestation and, consequently, to get high quality olive products. In course, is a study conducted by Gonçalves *et al.* (unpublished data) aiming to find compounds and physical characteristics like epiderm and pulp hardness/firmness that could be implicated in processes determining the low incidence of olive fruit fly attack.

3.1.2. Tillage Practices

Soil organisms provide many ecological functions such as decomposition of organic matter, mineralization of nutrients and also as agents of biological control of the olive pests that spend a period of their life cycle in the soil (e.g. the olive fruit fly, *B. oleae*) (Herrick,

2000). Furthermore, they are actively involved in the build-up of soil structure, have positive effects on water infiltration, drainage, water-holding capacity and soil aeration (Holland, 2004). The high level of diversity of organic farms significantly enhance resilience (Bengtsson *et al.*, 2005; Hole *et al.*, 2005), resistance to pests (Zehnder *et al.*, 2007; Wyss *et al.*, 2005; Pfiffner *et al.*, 2003) and diseases and a better utilization of soil nutrients and water (Altieri *et al.*, 2005). In northeastern Portugal, soil tillage is a traditional practice in the olive groves, with olive-growers drastically suppressing the growth of non-crop ("weeds") vegetation between olive trees by raking the soil. This vegetation management is carried out to conserve water and nutrients to the olive tree, to facilitate harvesting of the olive fruits and also to reduce the likelihood of wildfire risk (Fleskens, 2007) but it conducts to a high degree of habitat disturbance, affecting the populations of natural enemies (Jervis *et al.*, 1992). In a study where the effect of soil tillage was investigated in northeastern Portugal, traditional tillage affected the abundance and diversity of soil arthropods. In particular, Formicidae family was the most responsive group suffering a decrease in species richness and diversity in the tilled plot (Lousão *et al.*, 2006). In frequently tilled olive groves, sensitive species like *Messor barbarus* (L.), *M. lusitanicus* Tinaut, *Cataglyphis hispanicus* (Emery) and *C. ibericus* (Emery) were progressively affected (Lousão *et al.*, 2006) because they are associated to open areas between trees (Redolfi *et al.*, 1999). Consequently, the structure of the community changes and resistant and resilient species, like *Tapinoma nigerrimum* (Nylander), will dominate the assemblage (Lousão *et al.*, 2006). Several authors like Peck *et al.* (1998), Lobry de Bruyn (1999) and Redolfi *et al.* (1999) suggested that Formicidae are potentially important biological indicators of soil condition in agro-ecosystems and, for that reason, this group can be easily and reliably used by farmers to monitor soil quality (Lobry de Bruyn, 1999). Within the scope of the agri-environmental measures, environmentally friendly agricultural practices like the maintenance of strips of meadows between tree lines should be gradually implemented by olive farmers in order to promote the conservation of habitats with high ecological value. In some cases, irrigation can be applied according to need and with the best methods to avoid losses (e.g. micro-irrigation) (Malavolta *et al.*, 2002). Field strips can have important functions like: a) to conserve soil moisture for a longer period of time (Niggli *et al.*, 2007), b) to prevent soil erosion (Fleskens, 2007), c) to provide shelter that might sustain natural enemies during the period of olive pest scarcity (Jervis *et al.*, 1992) and d) to favour a rich soil biota that can improve nutrient recycling and plant health (Kladivko, 2001).

3.2. Second-Phase Strategies

After the implementation of structural strategies derived from the first phase, ecological engineering advances, namely in conservation biological control, are explored. Importantly, second-phase strategies can be applied independently of the first stage to remedy, at least partially, situations in which soil conditions, geographical area, or cultivar is suboptimal. The second phase can help the process of conversion to organic production by weaning conventional production systems away from synthetic pesticide inputs.

3.2.1. Conservation Biological Control

Conservation biological control (CBC) represents a sustainable practice of enhancing the potential of natural enemies to control pests (Jonsson *et al.*, 2008). Thus, the reduction of pest problems involves the manipulation of agricultural habitats for enhancing abundance and diversity of the natural enemy community (e.g. predators, parasitoids, and pathogens) and, consequently, the possibility of natural biological control (Malavolta *et al.*, 2002, Böller *et al.*, 2004, Griffiths *et al.*, 2008, Jonsson *et al.*, 2008, Straub *et al.*, 2008).

The olive tree is characterized by an abundant and diverse community of natural enemies with economical interest, comprising predators (ants, coccinellids, spiders, chrysopids, mirids and anthocorids) and parasitoids (hymenopterans and dipterans) (e.g. Morris *et al.*, 1999b, Santos *et al.*, 2007a) that can develop functions as natural control agents of olive pests.

3.2.1.1. Parasitoids

Parasitoids belonging to order Hymenoptera were included in seven families: Aphelinidae, Encyrtidae, Pteromalidae, Eulophidae, Ichneumonidae, Braconidae and Trichogramatidae.

Nine species of parasitoids were identified from larval and pupal stages of the olive moth. From those, four species (*Chelonus elaeaphilus* Silvestri, a specific solitary endoparasitoid, *Apanteles xanthostigma* (Haliday), a polyphagous endoparasitoid, *Ageniaspis fuscicollis* Dalman, a specific polyembrionic endoparasitoid and *Elasmus flabellatus* (Fonscolombe), a polyphagous ectoparasitoid or hyperparasitoid) were the most representative for this pest (Bento, 1999, Gonçalves, 2002, Bento *et al.*, 2004, Gonçalves *et al.*, 2006a). The minimum and maximum abundances observed in olive groves from northeastern Portugal are summarized on Table 3.

Table 3. Minimum and maximum relative abundances of parasitoid species emerged from larvae and pupae of the three generations of *Prays oleae*, in northeastern Portugal (Bento, 1999, Gonçalves, 2002, Bento *et al.*, 2004, Gonçalves *et al.*, 2006a).

Family/Species	Abundance (%)		
	Leaf	Flower	Fruit
Encyrtidae			
Ageniaspis fuscicollis Dalman	2.4-53.5	9.0-67.0	1.1-4.6
Eulophidae			
Pnigalio agraules Walker	0.0-0.3	0.0	0.0
Dicladocerus westwoodii Westwood	0.0	0.0-0.2	0.0
Elasmus flabellatus Fonscolombe	0.0-4.4	5.4-38.9	0.0-0.6
Baryscapus nigroviolaceus Nees	0.0-0.3	0.0	0.3
Ichneumonidae			
Diadegma armillatum (Gravenhorst)	0.0	0.3-1.0	0.0
Braconidae			
Chelonus elaeaphilus Silvestri	2.0-67.7	1.0-24.1	20.5-88.6
Apanteles xanthostigma Haliday	2.0-16.3	1.0-21.0	6.0-19.1
Bracon crassicornis Thomson	0-0.6	0-8.9	0.7-31.0

On average, higher parasitization rates were observed on larval and pupal stages of the carpophagous generation compared with the other two generations. Unidentified species of *Trichogramma* were obtained from the parasitized eggs with parasitization rate ranging from 0.0% to 1.3% (Bento, 1999).

For the black scale, eight species of parasitoids were recovered from nymphal and mature stages of this pest. *Metaphycus lounsburyi* (Howard), *Coccophagus lycimnia* (Walker) and *Scutellista nigra* Mercet were, in general, the most representative parasitoids of *S. oleae* (Pereira, 2004). The minimum and maximum abundances registered are summarized on Table 4.

Table 4. Minimum and maximum relative abundance of parasitoid species recovered from the different developmental stages of *Saissetia oleae*, in northeastern Portugal, 1997-1999 (Pereira, 2004).

Family/Species	Abundance (%)
Aphelinidae	
Coccophagus lycimnia (Walker)	10.2-27.4
Coccophagus semicircularis (Förster)	3.6-21.9
Encyrtidae	
Metaphycus flavus (Howard)	4.0-9.8
Metaphycus helvolus (Compere)	2.5-7.2
Metaphycus lounsburyi (Howard)	2.0-62.4
Pteromalidae	
Moranila californica (Howard)	0.0-0.2
Scutellista caerulae (Fonscolombe)	0.0-9.9
Scutellista nigra Mercet	2.1-26.7

Black scale parasitoids are usually specific to one or two consecutive scale stages (Daane & Caltagirone, 1989, Pereira, 2004) with ovipositing females reaching the highest parasitism rate, where up to 57% of the specimens were found to be parasitized mainly by three parasitoid species, *M. lounsburyi*, *Scutellista caerulae* (Fonscolombe) and *S. nigra* (Pereira, 2004). *Coccophagus semicircularis* (Förster) was the main parasitoid emerging from young females with the parasitization rate reaching 16.7% whereas *Metaphycus flavus* (Howard) preferred the second and third instar nymphs (Pereira, 2004).

Considering the olive fruit fly, the biological control of this pest can be considered of difficult achievement because it develops most of its life cycle in the pulp fruit. Furthermore, differences of pulp thickness among cultivars may hamper the role of parasitoids with relatively short ovipositors (Hoelmer *et al.*, 2007). As a result, natural enemies, i.e. parasitoids and predators, may exert some control mostly when the insect leaves the fruit in search of a place for overwinter, or during the overwintering period in the soil. In a study conducted in 2006, only two parasitized larvae were found (0.13%), and on one of them, the parasitoid *Pnigalio agraules* (Walker) was identified (Gonçalves *et al.*, 2007a).

3.2.1.2. Predators

The guild of insect predators mainly belonged to five families: Formicidae, Chrysopidae, Coccinellidae, Miridae and Anthocoridae.

Considering ants, these are one of the most abundant groups of arthropods in the olive agro-ecosystem (Arambourg, 1986, Morris *et al.*, 1999b, Lousão, 2006, Santos *et al.*, 2007a,

Santos *et al.* 2007b). Even if some species reduce parasitoid efficiency (Panis, 1981), the beneficial role of the group may outweigh potential disadvantages, as several species are recognized as predators of olive pests, namely the olive moth (Morris *et al.*, 1998; 1999a; Lozano *et al.*, 2002), the olive fly (Neuenschwander *et al.*, 1986) and the olive bark beetle (Jarraya, 1986; González & Campos, 1990). In northeastern Portugal, species identified belonged to three subfamilies and 22 species (Table 5). *Tapinoma nigerrimum* (Nylander), *Cataglyphis hispanicus* (Emery), *Crematogaster scutellaris* (Olivier) and *Camponotus piceus* (Leach) have been indicated important predators of the olive moth with several specimens carrying larvae (Pereira *et al.*, 2002).

Table 5. Minimum and maximum relative abundance for ant species caught from soil and olive canopy, northeastern Portugal (Pereira *et al.*, 2002, Lousão, 2006, Santos *et al.*, 2007b).

Subfamily/Species	Abundance (%)
Dolichoderinae	
Tapinoma nigerrimum (Nylander)	24.0-55.4
Formicinae	
Camponotus aethiopus (Latreille)	0.1-5.1
Camponotus cruentatus (Latreille)	0.2-0.8
Camponotus lateralis (Olivier)	0.0-4.3
Camponotus fallax (Nylander)	0.0-0.1
Camponotus foreli Emery	0.0-0.1
Camponotus piceus (Leach)	0.1-12.1
Cataglyphis hispanicus (Emery)	2.0-6.4
Cataglyphis ibericus (Emery)	0.0-3.7
Camponotus truncatus (Spinola)	0.0-0.2
Formica cunicularia Latreille	0.1-0.6
Formica subrufa (Roger)	0.3-1.2
Lasius niger (Linnaeus)	0.0-0.1
Plagiolepis pygmaea (Latreille)	0.1-2.4
Myrmicinae	
Aphaenogaster iberica Emery	0.3-0.4
Crematogaster auberti Emery	0.1-14.1
Crematogaster scutellaris (Olivier)	5.7-20.0
Messor bouvieri Bondroit	2.2-9.6
Messor barbarus (Linnaeus)	0.4-23.3
Messor lusitanicus Tinaut	0.0-0.1
Tetramorium semilaeve André	3.8-30.8
Leptothorax angustulus (Nylander)	0-0.4

Coccinellids are also abundant predators in the olive grove and, in northeastern Portugal, twenty species were identified in organic olive groves (Table 6). From those, *Chilocorus bipustulatus* L., *Scymnus (Pullus) subvillosus* (Goeze), *Scymnus (Mimopullus) mediterraneus* Iablokoff-Khnzorian, *Scymnus (Scymnus) interruptus* (Goeze) and *Rhyzobius chrysomeloides* (Herbst) were the most abundant species completing several life cycles in this agro-ecosystem (Santos, 2007).

Table 6. Minimum and maximum relative abundance for coccinellid species caught from olive canopy, northeastern Portugal (Santos, 2007).

Subfamily/species	Abundance (%)
Scymninae	
Nephus (Bipunctatus) bisignatus	0.3-0.6
Nephus (Sidis) helgae	0.0-0.1
Scymnus (Scymnus) apetzi	0.9-1.2
Scymnus (Scymnus) interruptus	21.4-63.1
Scymnus (Mimopullus) mediterraneus	2.5-7.4
Scymnus (Scymnus) rufipes	0.0-0.1
Scymnus (Pullus) subvillosus	7.4-11.8
Stethorus punctillum	0.0-0.2
Chilocorinae	
Chilocorus bipustulatus	14.1-30.9
Exochomus quadripustulatus	0.1-0.4
Exochomus nigromaculatus	0.1-0.9
Platynaspis luteorubra	0.1-0.4
Coccidulinae	
Rhyzobius chrysomeloides	9.5-24.3
Rhyzobius litura	0.3-0.9
Rhyzobius lophantae	0.0-0.1
Coccinelinae	
Adalia (Adalia) decempunctata	0.0-0.1
Coccinela septempunctata	0.0-0.1
Oenopia conglobata	0.0-0.1
Oenopia lyncia	0.0-0.1
Propylaea 14-punctata	0.0-0.1

Chrysopids are other group of predators, known for their voracity, and that have been mass-reared to use as control agents of several pests (Stelzl & Devetak, 1999). In northeastern Portugal, six chrysopid species were identified (Table 7) being *Chrysoperla carnea* (Stephens) the most abundant species, with relative abundances ranging between 49.0% and 63.4% (Bento *et al.*, 1999a).

Table 7. Minimum and maximum relative abundances for chrysopid species caught in McPhail traps in northeastern Portugal, 1994-1996 (Bento *et al.*, 1999a).

Species	Abundance (%)
Chrysoperla carnea (Stephens)	49.0-63.4
Dichochrysa flavifrons (Brauer)	23.9-29.6
Dichochrysa picteti (Malachlan)	12.9-13.3
Dichochrysa prasina (Burmeister)	9.3-11.3

Species	Abundance (%)
Nineta vittata (Wesmael)	0.5-0.9
Rexa lordina Nevás	0.0-0.5

Furthermore, the peak of abundance of adult chrysopids occurred between May and August which coincided with the occurrence of the larval stages of the anthophagous generation and the initial larval stages of the carpophagous generation of the olive moth (Bento *et al.*, 1999a). It was also found that an important mortality factor of the carpophagous generation was egg predation, since 33% to 67% of the eggs deposited on the fruits were destroyed. Most of this predation is thought to have been caused by chrysopids, as these insects are considered to be responsible for up to 80% to 90% of *P. oleae* egg predation in some regions and years (Ramos *et al.*, 1983)

Among mirids, the most abundant species were *Deraeocoris lutencens* (Schilling), *Heterotoma merioptera* (Scopoli) and *Campyloneura virgula* (Herrich-Schaeffer). On the other hand, *Anthocoris nemoralis* (Fabricius 1794) was the most relevant species identified among anthocorids. Morris *et al.* (1999c) considered these two groups, potential predators of the olive moth in Spanish olive groves, but more studies are needed to infer about their feed preferences and, consequently, their importance as biocontrol agents in the olive grove.

Unlike parasitism or disease, insect predation is usually difficult to observe directly in the field. Consequently, analytical methods such as serological tests (ID-ELISA - indirect enzyme-linked immunosorbent assay) were developed and successfully used to detect remains of *P. oleae* and *S. oleae* in the gut of field collected predators. Results obtained by Lozano *et al.* (2002) showed that Formicidae, Coleoptera, Araneae and Heteroptera were the most important predators of the olive moth in northeastern Portugal. Predation likely occurred over eggs and larvae of the anthophagous generation, when the highest number of predators tested positive. Morris *et al.* (1999a) found similar results in Spanish olive groves. For *S. oleae*, only coccinellid species were investigated and serological tests detected that 21.2% of the overall field collected coccinellids reacted positively with the *S. oleae* antiserum. *C. bipustulatus* and coccinellid larvae obtained the highest percentages of positives. Moreover, in laboratory experiments, *C. bipustulatus* consumed different phenological stages of *S. oleae* and completed its life cycle when fed exclusively with eggs or with first instar nymphs of *S. oleae*. Temporal synchrony established between each coccinellid species and the different phenological stages of *S. oleae* showed significant associations between *S. interruptus*, *S. subvillosus*, *S. mediterraneus* and the first (principally those recently emerged) and second instars nymphs and between *C. bipustulatus* and the second and third instars nymphs of *S. oleae*. According with the consumption of phenological stages showed by coccinellid species, it is likely that both *C. bipustulatus* larvae and adults use the black scale as a frequent food resource and *S. mediterraneus*, *S. subvillosus* and *S. interruptus* use the black scale as an occasional food resource. Therefore, the occurrence of different coccinellid species may result in a more effective and complementary suppression of the pest through the consumption of different nymphal stages (Santos, 2007). The importance of coccinellids as natural enemies of *S. oleae* was also explored by several authors in different Mediterranean countries (Argyriou and Katsoyannos, 1977; Velimirovic, 1994; Ba M'Hamed and Chemseddine, 2001; Ba M'Hamed and Chemseddine, 2002; Santos, 2007).

Within Neuroptera, it must be referred that relatively high numbers of the coniopterygid *Conwentzia psociformis* (Curtis) were found in a survey carried out during the period of 2004 to 2006, in several groves of the region, in coincidence with the first instars of *S. oleae* (Torres *et al.*, 2007b), indicating that in future works it would be interesting to critically assess the role of this predator on the olive agro-ecosystem.

As previously said, predators can have an important role in the mortality of the olive fruit fly when larvae leave the fruit. The study of the complex of predators during the overwintering period of the olive fruit fly showed that generalist edaphic organisms namely, ants, carabids, staphylinids and spiders can be important predators of the olive fruit fly (Gonçalves *et al*, unpublished data). In fruits, in a study conducted in 2006, only 3% of the immature stages examined showed signs of predation (Gonçalves *et al.*, 2007a). Sometimes, ants and wasps were seen removing larval specimens from the fruit and accordingly Kapatos & Fletcher (1986) some birds can have the same behavior.

This phase also comprises habitat management practices relating to conservation biological control that were tested in the olive grove in two different approaches a) the effect of ground cover manipulation with non-crop vegetation and b) the use of artificial food sprays on abundance of natural enemies and olive pests, namely the olive moth.

3.2.1.3. Ground Cover Manipulation

The general function of ground cover with non-crop vegetation is to provide beneficial arthropods with structures offering additional food (e.g. nectar and pollen), shelter and alternative hosts (Landis *et al.*, 2000, Fiedler *et al.*, 2008). These structures can increase prey availability for polyphagous predators, influencing its fitness and to facilitate reinvasion of areas where population reductions have occurred (Pfiffner & Luka, 2000, Griffiths *et al.*, 2008). In the olive grove, the effect of ground cover manipulation on abundance of natural enemies was tested in two different approaches: creation of flowering vegetation strips, consisting of buckwheat plants *Fagopyrum esculentum* Moench, and preservation of strips with natural vegetation. The use of sticky traps showed that, in general, natural vegetation strips were more attractive for hymenopterous parasitoids and chrysopids and *F. esculentum* plots were more attractive for coccinellids (Jorge *et al.*, 2003). However, ground cover manipulation also favoured the olive moth population with a subsequent higher fruit infestation resulting in no net gain in biological control. In part, these results can be related with the landscape heterogeneity and complexity of Trás-os-Montes region, where olive groves are included in a typically small-scale diversified landscape mosaic (Fleskens, 2007) with annual flowering plants, shrubs and tree fulfilling hedgerows. According with Thies & Tscharntke (1999), increasing non-crop habitats in simple landscapes can have an impact on biological control but in structurally diverse landscapes that already have a high proportion of non-crop areas, increasing of the proportion of non-crop areas further has little effect. It appears, therefore, that the key to successful natural control is in increasing functional diversity of different guilds (e.g. soil and canopy predators, spring and summer breeders, day and night active species, prey preferences) of the natural enemies of the pests (Schmidt *et al.*, 2003, Weibull *et al.*, 2003, Griffiths *et al.*, 2008) and further research is required to explore the long-term effect of these measures for sustainable pest control.

3.2.1.4. Artificial Food Sprays

Artificial food sprays can act as an attractive food supplement that aims to increase the abundance and effectiveness of natural enemies. They typically consist of carbohydrate and/or protein-rich ingredients in liquid formulation and are sprayed to the foliage on one or more occasions (Wade *et al.*, 2008). Many predators and parasitoids of pests may include supplementary foods such as pollen, nectar or honeydew in their immature and/or adult diet (Coll & Guershon, 2002). The effect of artificial food sprays on olive foliage on the activity (predation and parasitism) of natural enemies and infestation levels of the anthophagous and carpophagous generations of the olive moth was explored in field experiments, using different foods or combinations of foods: a) *protein hydrolysate + acid-hydrolyzed L-tryptophan*, b) *protein hydrolysate*, c) *protein hydrolysate + honey*, d) *sugar + yeast autolysate + acid-hydrolyzed L-tryptophan* e) *yeast autolysate*, f) *acid-hydrolyzed L-tryptophan + sugar*, g) *sugar + yeast autolysate*, h) *honey*, i) *acid-hydrolyzed L-tryptophan*, j) *sugar*. Results suggested that the combination '*sugar + yeast autolysate + acid-hydrolyzed L-tryptophan*' was especially attractive for Hymenoptera parasitoids (presenting a higher rate of parasitism), Coccinellidae, Chrysopidae, Hemiptera, Diptera and Lepidoptera. The combinations '*protein hydrolysate + honey*' and '*honey*' also showed good results for Hymenoptera parasitoids and Coccinellidae. It was also observed that predation on olive moth eggs enhanced in all food types (17% to 21%) but no effect on flower and fruit infestations was observed (Bento *et al.*, 2004). Thus, the food sources tested attracted key natural enemies, however the impact on their activity remains unclear since no increase in mortality or reduction of the infestation of the pest was observed. In parallel with field experiments, laboratory bioassays were developed to determine whether these artificial food sources might eventually also act as food sources for olive moth adults, leading to increased crop damage and females and males fed with honey presented a significantly higher longevity (Bento *et al.*, 2007).

Besides their high price, artificial food sprays may have several undesirable side effects such as the promotion of sooty moulds on foliage and the influx of insects from surrounding unsprayed plots, resulting in reduced predation and parasitism and therefore increased crop damage in the latter (Jervis *et al.*, 1992).

3.3. Third-Phase Strategies

This phase involve releasing commercially available live agents into organic fields to control pests. In the olive grove, different experiments were carried out to test the effect of three commonly used biocontrol agents against the olive moth: the insecticidal bacterium *Bacillus thuringiensis* Berliner, and the entomophagous insects *C. carnea* and *Trichogramma cacaeciae* Marchal.

The entomopathogen *B. thuringiensis* is the most wide and successful commercial biological control agent of lepidopterous pests (Frederici, 1999). Till present, this is the only direct or suppressive strategy of protection allowed against the olive moth in organic olive production. The efficacy of sprays of *B. thuringiensis* ssp. *kurstaki* against the flower generation of *P. oleae* and their side effects on natural enemies was investigated in organic groves in northeastern Portugal. Bento *et al.* (1999b) found a mortality rate for the antophagous generation between 60% and 80% on *B. thuringiensis* plot. In another experiment, the number of males captured on delta pheromones traps was significantly lower on *B. thuringiensis* plot as also as the percentage of attacked fruits (14% on *B. thuringiensis* plot to 32% in the control plot) (Cabanas *et al.*, 2003). Considering beneficials, it was observed that coccinellids were present in lower numbers on *B. thuringiensis* plot, but other natural enemies like ants, spiders, chrysopids and parasitoids were not affected (Cabanas *et al.*, 2003).

The entomophagous insect *C. carnea* was field released in the beginning of the egg-laying period of the carpophagous generation of the olive moth (Bento *et al.*, 1999a). Different plots with increasing densities of *C. carnea* larvae were tested (0-control, 90, 180, 360, 720 and 900 larvae/tree) and results showed that the percentage of *P. oleae* eggs predation significantly increased from 32% on the control plot to 50.1% on the 180 larvae/tree plot and 68.8% on the 900 larvae/tree plot. In addition, fruit infestation significantly decreased from 36.6% on the control plot to 28.4% on the 180 larvae/tree plot and 17.2% on the 900 larvae/tree plot.

Mass-produced *Trichogramma* were also released against both the antophagous and the carpophagous generations of *P. oleae*. After promising preliminary results obtained with *T. cacaeciae* Marchal releases (Bento *et al.*, 1998a; 2001), research was enlarged within EU-funded project "TRIPHELIO: Sustainable control of lepidopterous pests in olive groves – integration of egg parasitoids and pheromones". During the years of 2002-2004 a survey was carried out in olive groves from the countries participating in the project (Portugal, Greece, Egypt and Tunisia), mainly by exposure of sentinel eggs (*Sitotroga cerealella* Olivier or *Ephestia kuehniella* Zeller) (Sakr *et al.*, 2000), in order to identify native species associated with the olive agro-ecosystem, as they are assumed to have higher potential than those from other areas and crop systems (Glenn *et al.* 1997). The results showed that the *Trichogramma* parasitoids fauna of the olive groves studied was species-poor (Herz *et al.*, 2006). In Portugal, only two species were identified: *T. cacaeciae* Marchal, representing 92% of the individuals, and T. *nerudai* Pintureau & Gerding (8%) (Herz *et al.*, 2006; Gonçalves *et al.*, 2006b). Field releases of mass-reared *Trichogramma* were carried out during the period of 2002-2004, with a density of 900,000 wasps per ha and release. Four releases of *T. cacaeciae* (a commercial strain collected in tomato) were done 2002, using as releasing devices folded cardboard cards ('TrichoKarte' AMW-Nützlinge, Pfungstadt, Germany). A high ant predation on parasitoids inside the releasing devices (Pereira *et al.*, 2004), hampered the trials, stressing the need for better protection. In each of the following two years, six releases were done with modified releasing devices. In 2003, a local Portuguese strain of *T.cacaeciae* collected from eggs of the olive moth in 2002, was tested, while in 2004, three strains were used (*T. cacoeciae* – German strain, used also in 2002; *T. cordubensis* – Portuguese strain, collected in 1992 in tomato, and *T. nerudai* – Portuguese strain, collected in olive groves in 2002 & 2003). Even under best conditions (no egg predation, high quality of the material) observed levels of egg parasitism were low and no control of the target pest was achieved. It must be noted that these trials were carried out under environmental conditions highly unfavorable for *Trichogramma* survival and reproduction, mainly during the carpophagous generation of *P.oleae,* when results were poorer. So, the experimental grove covered an area of 6 ha and was surrounded by olive groves, all them with the soil vegetation removed by mechanical methods. Then adult wasps had scarce possibility for finding carbohydrate sources, which seems to be important for their longevity and fecundity (Hegazi *et al.*, 2000; Romeis *et al.*, 2005; Fuchsberg *et al.*, 2007). Moreover, minimum relative humidity reached values as low as 15%, for a temperature up to 34.6°C, conditions worsened as the grove was not irrigated and the tree density was low (124 trees/ha), not allowing the canopies to touch. Heat and low humidity have been a problem in use of *Trichogramma* elsewhere (Lopez & Morrison, 1980; Ramesh & Baskaran, 1996; Maisonhaute *et al.*, 1999; Hoffmann & Hewa-Kapuge, 2000). For *T. cacaeciae* strain used, Herz & Hassan (2006) have found a reduction in mean longevity of 56% and fecundity of 77%, when kept under the temperature of 35°C and the low humidity of <40% RH for 6 h per day ('arid stress') in comparison with the individuals maintained under 25°C and 70% RH used as control. Even for the native *T. nerudai*, which seemed to be better adapted to the semi-arid conditions of the olive growing region, a reduction of 37% in mean longevity and 53% in mean fecundity was observed, in the tested conditions (Herz & Hassan, 2006).

3.4. Fourth-Phase Strategies

In this phase, pheromones for mating disruption, pheromones and/or food attractants for attract and kill and mass trapping strategies and the application of insecticides of biological and mineral origin were tested in the olive grove.

3.4.1. Mating Disruption with Sex Pheromones

Mating disruption dispensers spreads a large amount of sex pheromone in the environment that confounds males and limits their ability to locate calling females. This strategy is considered an environmentally friendly method because it simply confuses male insects from accurately locating a mating partner and has the benefit of only affecting a limited number of closely related insects while leaving other species unaffected (Amaro, 2003). The use of synthetic pheromones is allowed in organic agriculture because they are contained in dispensers and therefore do not come into contact with crops (Zehnder *et al.*, 2007).

Mating disruption trials were carried out to control the olive moth. Polypropylene dispensers with the pheromone component, (Z)-7-tetradecenal, formulated into a ß-cyclodextrin-complex were applied at the end of the flight period of the flower generation, at a rate of 40 g (a.i.) per hectare. The success of mating disruption was evaluated by comparing pheromone trap catches of males and fruit infestation on the treated plot with the control plot. After three years of field studies, excellent rates of male disorientation were obtained (> 90%), but fruit damage reached 54.7% and remained above the recommended economic threshold of 20% to 40% (Gomes & Cavaco, 2003, Bento *et al.*, 2005). The highly variable climatic conditions can represent a severe constrain for the success of mating disruption. Therefore, based on available mating disruption technique, the application of the sex pheromone needs to be optimized for use in a wider range of climatic conditions. In addition, a strong reduction of the target pest densities before pheromone application is essential for an effectiveness increase of the mating disruption technique. This can be achieved, for instance, through a prior application of *B. thuringiensis* against the flower generation. These results stress the need for long-term and large-scale application of this method to restrain the pest population under the economic threshold, since the application in a small area runs the risk of invasion by mated females from neighboring groves, diminishing the pheromone effect.

3.4.2. Mass-Trapping

Mass-trapping refers to the use of devices designed to attract and kill adults of both sexes. Normally, these devices are filled with different ammonia based products that attracts both males and females of the olive fruit fly. Mass-trapping is the result of an ongoing effort to extend the efficacy of the attractants used in baits and eliminate the adverse side effects of their application directly on the foliage (Broumas *et al.*, 2002).

Olipe traps are a low cost product that can be easily acquired by farmers. The effects of different trap-attractants (biammonium phosphate 4%, ammonium phosphate 5%, endomosyl 5% and urea 5%) that can be used in Olipe traps were studied on olive fruit fly and on natural enemies (Pereira *et al.*, 2007, Pavão *et al.*, 2007). After three year of studies, promising results were obtained for the olive fly mass trapping: a) yellow sticky traps, with pheromone, captured a significantly higher number of flies in the control plot than in the mass-trapping plots, b) the efficacy of Olipe traps remained during two months after their installation and c) fruit infestation was slightly lower in the Olipe traps with *biammonium phosphate* than in the others trap-attractants tested, which in turns were lower than in the control plot (Pavão *et al.*, 2007). Considering natural enemies, ants were mainly attracted by *biammonium phosphate*. On the other hand, *urea* attracted significantly more spiders and hymenopterous parasitoids and was especially harmful for chrysopids. In general, *ammonium phosphate* attracted a smaller percentage of natural enemies (0.9-1.4%) when compared with the other trap-attractants used (16.5-41.6% in *biammonium phosphate*, 47.7-76.7% in *endomosyl* and 5.4-9.7% in Urea) (Pereira *et al.*, 2007). Olipe traps are simple devices that have been developed and used in organic farming to control olive fruit fly infestations and can be seen as an alternative method to chemical insecticides.

Other devices use a food attractant and a pheromone to attract and an insecticide to kill the pest. Two commercially available traps, Eco-Trap (Bento *et al.*, 1998b) and Magnet OLI (AgriSense, UK) (Gonçalves & Torres, submitted-a) were tested in Portuguese regions as well as in other countries (Baldacchino & Simeone, n/d; Broumas *et al.*, 2002; Mazomenos *et al.*, 2002). ECO-Trap is an insecticide (deltamethrine)-treated paper bag of green colour baited with a food attractant (ammonium bicarbonate) and a sex pheromone that attracts and kills the olive fruit fly. Magnet OLI consists on a square brown-coloured carton (19x19) which is impregnated on both surfaces with deltramethrin and carries an ammonium bicarbonate dispenser. One in each three devices has a sex pheromone dispenser. Bento *et al.* (1998b) observed a significant reduction in the level of infestation when one ECO-Trap per tree was used. In Greece, Ragoussis (2002, 2003) also obtained good results after the application of the ECO-Traps that provided an efficient control of the insect, keeping the fly population in low levels. Furthermore, the ingredients of the ECO-Trap were maintained active during the period of the olive fruit fly. Advantages of the ECO-Trap have been referred as the safe and easy application for the farmer, the efficient control of the olive fly and the protection of natural enemies. The efficiency of Magnet OLI (AgriSense, UK) for the control of the olive fly was evaluated during 2002 and 2003. In 2002, neither total nor active infestation differed statistically between the treatments. Active infestation reached at harvest, 17.1% of infested drupes in the experimental plot and 18.4% in the control. However, in 2003, both total and active infestation levels were significantly higher in the control than in the experimental plot, showing a good efficiency of the tested method. In this year total infestation at harvest was of 5.6% in the experimental plot, while in the control reached 18.2%. On the other hand active infestation was respectively of 3.2% and 11.5% (Gonçalves

& Torres, submitted-a). Recently, it was demonstrated in two Italian regions, Toscana and Liguria, that the results of mass-trapping can be maximized if it is applied in a large area (Petacchi *et al.*, 2003). Moreover, the best results were obtained in trials where traps were timely placed by all farmers, which required appropriate organization and competent technical advisors (Broumas *et al.*, 2002, Petacchi *et al.*, 2003). The success of the application was also highly dependent on the density of the pest (Petacchi *et al.*, 2003).

3.4.3. Application of Insecticides of Mineral Origin

In organic olive groves, when all methods used in preceding phases were inefficient to control the olive fruit fly, cooper formulations are used by farmers as the last option for pest control. Thus, the effect of a spray with copper oxychloride on the olive infestation by the olive fly was evaluated in 2003 and 2004 (Gonçalves & Torres, submitted-b), with interesting results, which is in accordance with those obtained in several studies conducted in Italy (Belcari & Bobbio, 1999; Belcari *et al.*, 2003; Caleca & Rizzo, 2006). Copper was reported as having an ovipositional deterrent action against *B. oleae* (Prophetou *et al.*, 1991) as well as an important role as symbionticide, destroying or inactivating bacteria associated with the phytophagous (Belcari & Bobbio, 1999; Belcari *et al.*, 2003; Caleca & Rizzo, 2006). Elevated concentrations of this element are reported to be toxic and when found in soils may result in undesirable effects, including reduced biological activity and subsequent loss of fertility (Dumestre *et al.*, 1999). However adequate levels of copper content in plant tissue have been linked to greater resistance to diseases or physiological disorders in several crops. In northeastern Portugal, the incidence of chestnut blight, caused by *Cryphonectria parasitica*, on chestnut orchards was found to be correlated with very low levels of cooper either on the soil or on the tree phloem (Portela & Louzada, 2005). In olive groves, copper-based products are likely to be the most effective options for main fungal diseases control, such as the olive anthracnose, *Gloeosporium* spp., the olive leaf spot, *Cycloconium oleaginum* and the cercosporiosis, *Pseudocercospora cladosporioides* (Gomes & Cavaco, 2003). Moreover they may reduce infection of the tree by *Pseudomonas syringae* pv. *savastanoi*, as well as the development of the sooty mould fungi associated with the olive black scale, *Saissetia oleae* (Trapero & Blanco, 2004 cf.). It must also be noted that, in a recent study carried out by Iannotta *et al.* (2007), copper oxychloride have shown only few negative effects on non target arthropods when used in olive groves with a grass cover, which is supposed to reduce the effects of the active agent by furnishing a shelter against the direct contact with it.

4. FUTURE PROSPECTS

Future prospects should be focus on a wider utilization of recent advances and the refinement of promising results. Thus, the exploitation of subjects like the use of composts previously produced in long term-field experiments, the enhancement of soil quality and the implementation of functional biodiversity are key aspects for future studies. In this scope, it is important to understand how the management of soil quality will help to prevent pest outbreaks. Namely, how do physical, chemical or biological properties of the soil influence plant resistance or tolerance to olive pests since, for other crops (Altieri & Nicholls, 2003, Phelan *et al.*, 1995, Phelan *et al.*, 1996, Blumberg *et al.*, 1997, Alyokhin *et al.*, 2005, Altieri

& Nicholls, 2006) it was demonstrated that the optimization of these parameters negatively affected the performance of phytophagous insects. In the same way, it is necessary to improve knowledge about the impact of natural enemies in the olive agroecosystem. As advocated by Jervis *et al.* (1992) priority should be given to natural enemies known to effect significant control of pests under unmanipulated conditions. Thereafter, relevance should be given to the component members of the complexes: "*P. oleae – C. carnea – C. eleaphilus – A. fuscicollis – E. flabellatus*" and "*S. oleae – M. lounsburyi – C. lycimnia – S. nigra – C. bipustulatus – S. interruptus – S. subvillosus – S. mediterraneus*".

For the olive moth, the priority is to test the effect of mating disruption and *Trichogramma* releases against the carpophagous generation. In addition, since *Trichogramma* sp. longevity is very short, limiting its impact on the target and requiring accurate monitoring so field releases coincide with the peak in host egg densities, parasitoid's longevity and fecundity should be improved by providing it access to nectar-producing plants. Therefore, another approach is the identification of native plant species that can provide essential food sources (e.g. nectar, pollen and honeydew) to natural enemies, thus enhancing their survival, fecundity, longevity and behavior, to increase their effectiveness. The habitat manipulation is desired and aims to attain optimal natural enemy diversity and abundance through the management of plant biodiversity and ecological infrastructures.

The use of insecticides of biological and mineral origin is allowed in organic farming, but it will be desirable to integrate their use with other strategies, such as timing applications to minimize harmful effects on beneficial organisms as well as to validate models of population dynamics adjusted to local and seasonal conditions aiming to increase the crop protection level.

A promising alternative to conventional chemical pesticides is the increasing interest in use of entomopathogenic fungi against pests. In course, is investigation to identify indigenous fungal species isolated from field collected dead olive fruit fly larvae and pupae (Gonçalves *et al.*, 2007). In future research, we expect to evaluate the virulence of the isolated fungi against olive fruit fly thought contact and ingestion bioassays. Moreover, in that fungi considered important as biological control agent, the effects of environmental conditions and agricultural practices must also be analyzed, in order to improve their efficiency either by conservation or augmentation. Also for the olive fruit fly, it is necessary to optimize the correct time of harvest for minimizing damaging effects of the pest in the pulp.

5. Conclusion

Organic farming involves science-based and ecologically sound measures but there are still important knowledge gaps in composting of organic matter and in insect pest management. The success of an integrated arthropod pest management results from the combination in practice of methods from various phases described in this chapter. However, potential and spread of organic farming will depend on policies, attitude changes on the part of researchers and policy makers, existence of markets for organic produce, and also the organization of farmer and consumer movements that demand a more healthy and sustainable agriculture.

REFERENCES

Alburquerque, J.A., Gonzálvez, J., García, D. & Cegarra, J. (2007). Effects of a compost made from the solid by-product (''alperujo'') of the two-phase centrifugation system for olive oil extraction and cotton gin waste on growth and nutrient content of ryegrass (*Lolium perenne* L.). *Bioresource Technology, 98,* 940–945.

Altieri, M.A. & Nicholls, C. (2003). Soil fertility and insect pests: harmonizing soil and plant health in agroecosystems. *Soil Tillage Research, 72*, 203–211.

Altieri, M.A. & Nicholls, C. (2006). Optimizando el manejo agroecológico de plagas a través de la salud del suelo. *Agroecologia, 1*, 29-36. available from: http://www.um.es/ojs/index.php/agroecologia/article/view/18/6.

Altieri, M.A., Ponti, L. & Nicholls, C. (2005). Enhanced pest management through soil health: toward a belowground habitat management strategy. *Biodynamics, (Summer),* 33-40.

Alyokhin, A., Porter, G., Groden, E., Drummond, F. (2005). Colorado potato beetle response to soil amendments: a case in support of the mineral balance hypothesis? *Agriculture Ecosystems and Environment, 109*, 234–244.

Amaro, P. (2003). A protecção integrada. Lisboa: ISA Press.

Anonymous (2007). *Olivicultura: Diagnóstico Sectorial.* Lisboa: Gabinete de Planeamento e Políticas do Ministério da Agricultura, do Desenvolvimento Rural e das Pescas.

Arambourg, Y. (Ed.) (1986). *Traite d'entomologie oleicole.* Madrid: International Olive Oil Council.

Argyriou, L.C. & Katsoyannos, P. (1977). Coccinellidae species found in the olive-groves of Greece. *Annals de L'Institut Phytopathologique Benaki, 11*, 331-345.

Ba M'Hamed, T. & Chemseddine, M. (2001). Assessment of temperature effects on the development and fecundity of *Pullus mediterraneus* (Col., Coccinellidae) and consumption of *Saissetia oleae* eggs (Hom, Coccoida). *Journal of Applied Entomology, 125*, 527-531.

Ba M'Hamed, T. & Chemseddine, M. (2002). Selective toxicity of some pesticides to *Pullus mediterraneus* Fabr. (Coleoptera: Coccinellidae), a predator of *Saissetia oleae* Bern. (Homoptera: Coccoidea). *Agricultural and Forest Entomology, 4*, 173-178.

Baldacchino, F. & Simeone, V. n/d. Controllo della mosca delle olive in olivicoltura biologica in Puglia: esperienze preliminari, *http://adatti.sssup.it/olivobio/relazioni/baldacchino.htm.*

Belcari, A. & Bobbio, E. (1999). L'impiego del rame nel controlo della mosca delle olive, *Bactrocera oleae*. *Informatore fitopatológico, 49*, 52-55.

Belcari, A., Sacchetti, P., Marchi, G. & Surico, G. (2003). La mosca delle olive e la simbiosi batterica. *Informatore Fitopatologico, 53*, 55-59.

Bengtsson, J., Ahnström, J. & Weibull, A.C. (2005). The effects of organic agriculture on biodiversity and abundance: a meta-analysis *Journal of Applied Ecology, 42*, 261–269.

Bento, A. (1999). *Contribuição para o estabelecimento de um programa de protecção integrada contra a traça-da-oliveira,* Prays oleae *(Bern.) em Trás-os-Montes.* PhD Thesis. Universidade de Trás-os-Montes e Alto Douro, Vila Real. Portugal

Bento, A., Torres, L. & Lopes, J. (1998a). Results from *Trichogramma cacoeciae* March. Releases to control the carpophagous generation of *Prays oleae* (Bern.) in Trás-os-Montes. *Revista das Ciências Agrárias, 21*, 207-211.

Bento, A., Torres, L., Lopes, J. Pereira, J & Rocha, M. (1998b). Ensaio de captura em massa contra a mosca da azeitona *Bactrocera oleae* (Gmel.). *Revista das Ciências Agrárias, 21*, 231-235.

Bento, A., Lopes, J., Passos-Carvalho, P. & Torres, L. (1999a) Biological control of *Prays oleae* (Bern.) by Chrysopids in Trás-os-Montes region (Northeastern Portugal). Proceedings of the 3rd International ISHS Symposium on Olive Growing. *Acta Horticulturae, 474*, 535–539.

Bento, A., Torres, L., Pereira, J.A. & Lopes, J. (1999b). Avaliação da eficácia de *Bacillus thuringiensis* (Berliner) contra a geração antófaga de *Prays oleae* (Bern.) em condições de campo. *Proceedings of the V Encontro Nacional de Protecção Integrada*, Bragança, pp. 239-244.

Bento, A., Pereira, J.A., Torres, L. M. & Torres, R. (2001). Trials to control the olive moth, *Prays oleae* (Bern.), with two *Trichogramma* species. *Egg Parasitoid News, 13*, 20-21.

Bento, A., Pereira, J. Cabanas, J., Pinto, A. & Torres, L. (2003). Sensibilidade de diferentes cultivares de oliveira aos ataques da mosca da azeitona, *Bactrocera oleae* (Gmel.) e da traça da oliveira *Prays oleae* (Bern.). *III Simpósio Nacional de Olivicultura,* Castelo Branco.

Bento, A., Cabanas, J.E., Pereira, J.A., Torres, L., Herz, A. & Hassan, S.A. (2004). Effects of different attractive sources on the abundance of olive predatory arthropods and possible enhancement of their activity as predators on eggs of *Prays oleae* Bern. *5th International Symposium on Olive Growing*. Izmir, p. 73.

Bento, A.A., Pereira, J.A., Cabanas, J.E., Konstantopoulou, M., Torres L. & Mazomenos, B. (2005). Mating disruption trials for the olive moth, *Prays oleae* (Bern.), (Lep.:Yponomeutidae) in Trás-os-Montes olive groves (Northeast of Portugal). *V Congresso Ibérico de Ciências Hortícolas, IV Congresso Iberoamericano de Ciências Hortícolas*. Porto.

Bento, A., Pereira, S., Cabanas, J.E. & Pereira J.A. (2007). Effects of different sugars and pollens on the olive moth, *Prays oleae* Bern. and their parasitoids *Elasmus flabellatus* Westw. *3rd European Meeting of the IOBC/WPRS Working Group* p. 119.

Blumberg, A.Y., Hendrix, P.F., Crossley, J.D.A. (1997). Effects of nitrogen source on arthropod biomass in no-tillage and conventional tillage grain sorghum agroecosystems. *Environmental Entomology*, 26, 31–37.

Böller, E.F., Häni, F. & Hans-Michael, P. (eds) (2004). *Ecological infrastructures: Ideabook on functional biodiversity at farm level. Temperature zones of Europe*. Swiss Centre for Agricultural Extension and Rural Development. Switzerland.

Broumas, T., Haniotakis, G., Liaropoulos, C., Tomazou, T. & Ragousis, N. (2002). The efficacy of an improved form of mass-trapping method, for the control of the olive fruit fly, *Bactrocera oleae* (Gmelin) (Diptera: Tephritidae): pilot-scale feasibility studies. *Journal of Applied Entomology, 126,* 217-223.

Cabanas, J.E., Pereira, J.A., Bento, A.A., Pereira, S. & Torres, L. (2003). Efficacy of *Bacillus thuringiensis* to control the olive moth, *Prays oleae* (Bern.) and side effects on the olive tree arthropodofauna. *1st European Meeting of the IOBC/WPRS Study Group-Integrated Control in Olives. Chania*, 70.

Caleca, V. & Rizo, R. (2006). Effectiveness of clay and cooper products in the control of *Bactrocera oleae* (Gmelin). Olivebioteq 2006, Mazara del Vallo – Marsala, II, 19-25.

Cardoso, C., Bento, A. & Torres, L. (2006). Infestation of the olive fly, *Bactrocera oleae* (Gmelin), in the cultivars "Cobrançosa", "Madural" and "Verdeal Transmontana". *Melhoramento, 41*, 124-130.

Civantos, M. (1998). *Olive pest and disease management.* Madrid: International Olive Council.

Coll, M. & Guershon, M. (2002). Omnivory in terrestrial arthropods: mixing plant and prey diets. *Annual Review of Entomology, 47*, 267–297 (online Supplemental table, 7pp).

Cunha-Queda, A.C. (1999). Nitrogen dynamics during putrescible biomass composting. Ph.D Thesis. Instituto Superior de Agronomia, Universidade Técnica de Lisboa, Lisbon, Portugal

Cunha-Queda, A.C., Silva, C., Cordeiro, A., Teodoro, A., Morais, M.C., Ferreira, J., Nunes, M., Coelho, D., Wiersma, H.L., Correia de Oliveira, J. & Brandão de Menezes, J. (2006). Production of high quality composts from food industry residues and sub products for fertilisation of organic farming grown vegetables and olive trees. In: *Proceedings Book of 5th International Conference ORBIT 2006 – Biological Waste Management – From Local to Global*, Part 2, pp. 457-463.

Daane, K.M. & Caltagirone, L.E. (1989). Biological control of black scales in olives. *California Agriculture, January-February*, 9-11.

DG Env. A.2. (2001). Biological treatment of biowaste, 2nd Draft of biological treatment – Working document. Available from: http://www.compost.it/www/pubblicazioni on_line/biod. pdf.

Dumestre, A., Sauvé, S., McBride, M., Baveye, P., Berthelin, J. (1999). Copper speciation and microbial activity in long-term contaminated soils. *Archives of Environmental Contamination and Toxicology, 36*, 124-131.

Fabbri, A. & Ganino, T. (2002). Organic olive growing in Italy. *Advances in Horticultural Sciences, 16*, 204-217.

Ferreira, J.C., Strecht, A., Ribeiro, J.R., Soeiro, A. & Cotrim, G. (2002). *Manual de Agricultura Biológica – Fertilização e protecção das plantas para uma agricultura sustentável.* Lisboa: Agrobio.

Fiedler, A.K., Landis, D.A. & Wratten, S.D. (2008). Maximizing ecosystems services from conservation biological control: The role of habitat management. *Biological Control, 45*, 254-271.

Fleskens, L. (2007). Conservation scenarios for olive farming on sloping land in the Mediterranean. PhD Thesis. Wageningen University. Wageningen. The Netherlands.

Frederici, B.A. (1999). *Bacillus thuringiensis* in biological control. In T.S. Bellows & T.W. Fisher (Eds.), *Handbook of Biological Control* (pp. 575-593). San Diego, California: Academic Press.

Fuchsberg, J.R., Yong, T.H., Losey, J.E., Carter, M.E., Hoffmann, M.P. (2007). Evaluation of corn leaf aphid (*Rhopalosiphum maidis*; Homoptera: Aphididae) honeydew as a food source for the egg parasitoid *Trichogramma ostriniae* (Hymenoptera: Trichogrammatidae). *Biological Control, 40*, 230–236.

Gea, T., Artola, A., Sort, X. & Sánchez, A. (2005). Composting of residuals produced in the Catalan wine industry. *Compost Science and Utilization, 13*, 168-174.

Glenn, D.C., Hercus, M.J. & Hoffmann, A.A. (1997). Characterizing *Trichogramma* (Hymenoptera: Trichogrammatidae) species for biocontrol of the Light Brown Apple Moth (Lepidoptera: Tortricidae) in grapevines in Australia. *Annals of the Entomological Society of America, 90*, 128-137.

Gomes, H.B. & Cavaco, M. (2003). *Protecção Integrada da Oliveira – Lista dos Produtos Fitofarmacêuticos e Niveis Económicos de Ataque*. Oeiras: Ministério da Agricultura, Desenvolvimento Rural e Pescas – Direcção Geral de Protecção da Culturas.

Gonçalves, M.F.M. (2002). Estudo sobre a traça-da-oliveira, *Prays oleae* (Bern.), em Trás-os-Montes: Bioecologia, estimativa do risco e luta por confusão sexual. Relatório Final de Estágio em Engenharia Agrícola. UTAD, Vila Real. Portugal.

Gonçalves, F., Oliveira, A., Herz, A. & Torres, L. (2006a). Contribuição para o conhecimento da bioecologia da traça-da-oliveira, *Prays oleae* (Bernard) em Trás-os-Montes. IV Simpósio Nacional de Olivicultura, Elvas, 22 a 24 de Novembro, *Melhoramento, 41*, 146.

Gonçalves, F., Herz, A. & Torres, L. (2006b). Survey of native *Trichogramma*-species (Hymenoptera, Trichogrammatidae) associated with the olive ecosystem in the northeast of Portugal. *Melhoramento. 41*, 139-145.

Gonçalves, F., Oliveira, A. & Torres, L. (2007a). Contribuição para o conhecimento da bioecologia da mosca-da-azeitona *Bactrocera (Daculus) oleae* Gmelin na Terra Quente Transmontana. In: Cunha-Queda, C. & Rodrigues, M.F. *II Colóquio Nacional de Horticultura Biológica. Actas Portuguesas de Horticultura,* Lisboa, 10, 312-319.

Gonçalves, F., Rodrigues, C., Torres, L. & Pereira, J.A. (2007b). Survey of entomopathogenic fungi associated with the olive fruit fly, *Bactrocera oleae* (Gmelin) in the northeast of Portugal. *3rd European Meeting of the IOBC/WPRS Working Group-Integrated Protection of Olives Crops*. Bragança, 87.

Gonçalves, F. & Torres, L. (submitted-a) Efficiency of Attract & Kill traps (AgriSense, UK) to control *Bactrocera oleae*, at Trás-os-Montes region (Northeastern Portugal). *I Congresso Nacional de Produção Integrada / VIII Encontro Nacional de Protecção Integrada*. Ponte de Lima.

Gonçalves, F. & Torres, L. (submitted-b) Effect of copper oxychloride on the olive infestation by *Bactrocera oleae* in Northeastern Portugal. *Acta Horticulturae*.

González, R. & Campos, M. (1990). Evaluation of natural enemies of the *Phloeotribus scarabaeoides* (Bern.) (Col: Scolytidae) in Granada groves. *Acta Horticulturae*. 286, 355-358.

Gouveia, J.M.B & Soeiro, A. (1997). Protected designations of origin for olive oils in Portugal. *Olivae, 66*, 25-27.

Griffiths, G.J.K., Holland, J.M., Bailey, A. & Thomas, M.B. (2008). Efficacy and economics of shelter habitats for conservation biological control. *Biological Control, 45*, 200-209.

Guardia, A., Tremier, A. & Martinez, J. (2004). Indicators for determination of stability of composts and recycled organic Wastes In: P. Lens, B. Hamelers, H. Hoitink & W. Bidlingmaier (Eds.), *Resource Recovery and Reuse in Organic Solid Waste Management* (pp. 338-376). London: IWA Publishing.

Hegazi, E.M., Khafagi, W.E. & Hassan, S.A. (2000). Studies on three species of *Trichogramma*. I. Foraging behaviour for food or hosts. *Journal of Applied Entomology, 124*, 145–149.

Herrick, J.E. (2000). Soil quality: an indicator of sustainable land management. *Applied Soil Ecology, 15*, 75-83.

Herz, A. & Hassan, S.A. (2006). Are indigenous strains of *Trichogramma* sp. (Hym., Trichogrammatidae) better candidates for biological control of lepidopterous pests of the olive tree? *Biocontrol Science and Technology, 16,* 841-857.

Herz, A., Hassan, S.A., Hegazi, E., Khafagi, W.E., Nasr, F.N., Youssef, A.I., Agamy, E., Blibech, I., Ksentini, I., Ksantini, M., Jardak, T., Bento, A., Pereira, J.A., Torres, L., Souliotis, C., Moschos, T. & Milonas, P. (2006). Egg parasitoids of the genus *Trichogramma* (Hymenoptera, Trichogrammatidae) in olive groves of the Mediterranean region. *Biological Control, 40*, 48-56.

Hoelmer, K.A., Kirk, A.A. & Pickett, C.H. (2007). Ecology of the olive fruit fly and its parasitoids in wild olives in Southern Africa. In Hopper, K.R., Kester, K.M., Hoelmer K.A. XIV International Entomophagous Insects Workshop. 25pp. *Journal of Insect Science.* 7, 16. available from: insectscience.org/7.16.

Hoffmann, A.A. & Hewa-Kapuge, S. (2000). Acclimation for heat resistance in *Trichogramma* nr. *brassicae*: can it occur without costs? *Functional Ecology, 14*, 55-60.

Hole, D.G., Perkins, A.J., Wilson, J.D., Alexander, I.H., Grice, P.V. and Evans, A.D. (2005). Does organic farming benefit biodiversity? *Biological Conservation, 122*, 113-130.

Holland, J.M. (2004). The environmental consequences of adopting conservation tillage in Europe: reviewing the evidence *Agriculture, Ecosystems and Environment, 103*, 1–25.

Iannotta,N., Belfiore, T., Noce, M.E., Scalercio, S. & Vizzarri, V. (2007). The impact of some compounds utilized in organic olive groves on the non-target arthropod fauna: canopy and soil levels. *Ecoliva 2007, VI Jornadas Internacionales de Olivar Ecologico*, Puente de Génave (Jaén), España. Available from: http://www.ecoliva.info/index.php?option=com_content&task=blogcategory&id=17&Itemid=38.

INE (2007b). Estatísticas Agrícolas (2006). Instituto Nacional de Estatística Portugal. Lisbon. Available from *http://www.ine.pt*

Jarraya, A. (1986). Scolytidae. In Y. Arambourg, (Ed.), *Traité d'entomologie oleicole.* (pp 21-36). Madrid. Conseil Oleicole International.

Jervis, M.A., Kidd, N.A.C., McEwen, P., Campos, M. & Lozano, M. (1992). Biological control strategies in olive pest management. In: Haskell, P.T. (Ed). Research Collaboration in European IPM Systems. BCPC Monograph nº 52. *Proceedings of Brighton Crop Protection Conference – Pests and Diseases*, 31-39.

Jonsson, M., Wratten, SD., Landis, D.A. & Gurr, G.M. (2008). Recent advances in conservation biological control of arthropods by arthropods. *Biological Control, 45*, 172-175.

Jorge S., Bento, A. & Torres, L. (2003). Preliminary results on the effect of the creation of vegetation islands with flowering plants on beneficial insects associated with the olive agroecosystem. *1st European Meeting of the IOBC/WPRS Study Group-Integrated Control in Olives*. Chania, 44.

Kapatos, E.T. & Fletcher, B.S. (1986). Mortality factors and life-budgets for immature stages of the olive fly, *Dacus oleae* (Gmel.) (Diptera, Tephritidae), in Corfu. *Journal of Applied Entomology, 102*, 326-342.

Kladivko E. (2001). Tillage systems and soil ecology. *Soil Tillage Research, 61*, 61–76.

Laga-Merkblatt M 10 (1995). Qualitätskriterien und Anwendungsempfehlungen für Kompost. Müll-Handbuch Lfg. 5/95, Kennziffer 6856, Berlin: Verlag

Landis, D.A., Wratten, S.D. & Gurr, G.M. (2000). Habitat management to conserve natural enemies of arthropod pests in agriculture. *Annual Review of Entomology, 45*, 175-201.

Lobry de Bruyn, L.A. (1999). Ants as bioindicators of soil function in rural environments. *Agriculture Ecosystems and Environment, 74*, 425-441.

Lopez JD Jr. & Morrison RK. (1980). Effects of high temperatures on *Trichogramma pretiosum* programmed for field release. *Journal of Economic Entomology, 73*, 667-670.

Loumou, A. & Giourga, C. (2003). Olive groves: "The life and identity of the Mediterranean". *Agriculture and Human Values, 20*, 87–95.

Lousão, C.A.R. (2006). Efeito da não mobilização do solo do olival na diversidade de artrópodes em comparação com a mobilização tradicional. MSc. Thesis. Universidade dos Açores. Horta.

Lozano, C., Morris, T., Campos, M., Pereira, J.A. & Bento, A. (2002). Detection by ELISA of predators of *Prays oleae* (Lepidoptera: Plutellidae) in a Portuguese olive orchard. *Acta Horticulturae, 586*, 831-834.

Maisonhaute, C., Chirhrane, J. & Lauge, G. (1999). Induction of thermotolerance in *Trichogramma brassicae* (Hymenoptera: Trichogrammatidae). *Environmental Entomology, 28*, 116-122.

Malavolta C., Delrio, G. & Böller, E.F., (2002). IOBC Technical Guidelines III. Guidelines for the Integrated Production of Olives. *IOBC/WPRS Bulletin. 25*, 1-8.

Mazomenos, B.E., Pantazi-Mazomenos, A. & Stefanou, D. (2002). Attract and kill of the olive fruit fly *Bactrocera oleae* (Gmel.) in Greece as a part of an integrated control system. *In* Witzgall, P., Mazomenos, B. & Konstantopoulou, M. (Eds). *Pheromones and other semiochemicals in integrated pest control. IOBC/wprs Bulletin, 25*, 137-147.

Morris, T.I., Symondson, W.O.C., Kidd, N.A.C., Jervis, M.A. & Campos, M. (1998). Are ants significant predators of olive moth, *Prays oleae*? *Crop Protection, 17*, 365-366.

Morris, T.I., Campos, M., Kidd, N.A.C. & Symondson, W.O.C. (1999a). What is consuming *Prays oleae* (Bernard) (Lep.: Yponomeutidae) and when: a serological solution? *Crop Protection, 18*, 17-22.

Morris, T.I., Campos, M., Kidd, N.A.C., Jervis, M.A. & Symondson, W.O.C. (1999b). Dynamics of the predatory arthropod community in Spanish olive groves. *Agricultural and Forest Entomology, 1*, 219–228.

Morris, T.I., Symondson, W.O.C., Kidd, N.A.C. & Campos, M. (1999c). Heterópteros depredadores presentes en el olival y su incidencia sobre *Prays oleae* (Lepidoptera, Plutellidae). *Nutri-Fitos, 99*, 32-41.

Neuenshwander, S., Michelakis, S. & Kapatus, E. (1986). Tephritidae. *Dacus oleae* (Gmel.). In Y. Arambourg, (Ed.), *Traité d'entomologie oleicole* (pp 115-159). Madrid: Conseil Oleicole International.

Niggli, U., Earley, J. & Ogorzalek K. (2007). Issues paper: Organic agriculture and environmental stability of the food supply. 21 pp. Available from: http://www.orgprints.org/10752.

Panis, A. (1981). Action des fourmis sur la biocénose parasitaire de la cochonilhe noire des agrumes en France (Homoptera, Coccidea, Coccidae). *Fruits, 36*, 47-48.

Passos-Carvalho, J., Torres, L.M., Pereira, J.A. & Bento, A.A. (2003). *A cochonilha-negra Saissetia oleae (Olivier, 1791) (Homoptera: Coccidae)*. Lisboa: INIA/UTAD/ESAB.

Pavão, F., Pereira, J.A. & Bento, A. (2007). Mass-trapping of the olive fly with Olipe traps in Trás-os-Montes region (Northeast of Portugal). *3rd European Meeting of the IOBC/WPRS Working Group-Integrated Protection of Olives Crops*. Bragança, 97.

Peck, S.L., McQuaid, B. & Campbell, C.L. (1998). Using ant species (Hymenoptera: Formicidae) as a biological indicator of agroecosystem condition. *Environmental Entomology, 27*, 1102-1110.

Pera, A., Vallini, G., Frassinetti, S. & Cecchi, F. (1991). Co-composting for managing effluent from thermophilic anaerobic digestion of municipal solid waste. *Environmental Technology, 12*, 1137-1145.

Pereira, J.A.C. (2004). Bioecologia da cochonilha-negra, *Saissetia oleae* (Olivier), na oliveira, em Trás-os-Montes. PhD Thesis. Universidade de Trás-os-Montes e Alto Douro. Vila Real.

Pereira, J.A., Bento, A., Sousa, D., Campos, M. & Torres, L. (2002). Preliminary study on the ants (Hymenoptera: *Formicidae*) associated to the olive grove of the Terra Quente Transmontana (Northeast Portugal). *Bol. San. Veg. Plagas, 28*, 357-365.

Pereira, J.A., Bento, A., Cabanas, J.E, Torres, L.M., Herz, A. & Hassan, S.A. (2004). Ants as predators of the egg parasitoid Trichogramma cacoeciae (Hymenoptera: Trichogrammatidae) applied for biological control of the olive moth, Prays oleae (Lepidoptera: Plutellidae) in Portugal. *Biocontrol Science and Technology, 14*, 653-664.

Pereira, J.A., Pavão, F. & Bento, A. (2007). Effects of different attractants used in Olipe traps for olive fly mass-trapping on beneficial arthropods. *3rd European Meeting of the IOBC/WPRS Working Group-Integrated Protection of Olives Crops*. Bragança, 93.

Petacchi, R., Rizzi, I. & Guidotti, D. (2003). The "lure and kill" technique in *Bactrocera oleae* (Gmel.) control: effectiveness indices and suitability of the technique in area-wide experimental trials. *International Journal of Pest Management, 49*, 305-311

Pfiffner, L. & Luka, H. (2000). Overwintering of arthropods in soils of arable fields and adjacent semi-natural habitats. *Agriculture, Ecosystems and Environment, 78*, 215-222.

Pfiffner, L., Merkelbach, L. & Luka, H. (2003). Do sown wildflower strips enhance the parasitism of lepidopteran pests in cabbage crops? *IOBC/WPRS Bulletin, 26*, 111-116.

Phelan, P.L., Mason, J.F. & Stinner, B.R. (1995). Soil-fertility management and host preference by European corn borer on *Zea mays*: a comparison of organic and conventional chemical farming. *Agriculture, Ecosystems and Environment, 56*, 1–8.

Phelan, P.L., Norris, K.H. & Mason, J.F. (1996). Soil-management history and host preference by *Ostrinia nubilalis*: evidence for plant mineral balance mediating insect-plant interactions. *Environmental Entomology, 25*, 1329–1336.

Portela, E. & Louzada, J. (2005). Nutrient status of chestnut orchards. I. Relationship with incidence of blight. In, Abreu, C.G., Rosa, E. & Monteiro, A.A. (eds). Proceedings of the IIIrd International Chestnut Congress. *Acta Horticulturae, 693*, 557-561.

Prophetou-Athanasiadou, D.A., Tsanakakis, M.E., Myroyannis, D. & Sakas, G. (1991). Deterrence of oviposition in *Dacus oleae* by copper hydroxide. *Entomologia Experimentalis et Applicata, 61*, 1-5.

Ragoussis, N. (2002) Eco-Trap: efficient tool for the control of the olive fruit fly *Bactrocera oleae* in the Mediterranean area. *IOBC/WRRP Bulletin, 25*, 195-201

Ragoussis, N. (2003). Contribution to the biological olive agriculture. Efficient control of the olive fruit fly by the ECO-Trap. *1st European Meeting of the IOBC/WPRS Study Group-Integrated Control in Olives*. Chania, 11.

Ramesh, B. & Baskaran, P. (1996). Developmental response of four species of *Trichogramma* (Hym.: Trichogrammatidae) to heat shocks. *Entomophaga, 41*, 267-277.

Ramos, P., Campos, M. & Ramos, J. (1983). Present status of research on biological control of theolive moth in Spain. Proceedings of the E.C.- Experts Meeting "Entomophages and biological methods in integrated control in olive orchards". Chania, 127-144.

Raviv, M., Oka, Y. Katan, J., Hadar, Y., Yogev, A., Medina, S., Krasnovsky, A. & Ziadna, H. (2005). High-nitrogen compost as a medium for organic container-grown crops. *Bioresource Technology, 96*, 419–427.

Redolfi, I., Tinaut, A., Pascual, F. & Campos, M. (1999). Qualitative aspects of myrmecocenosis (Hym., Formicidae) in olive orchards with different agricultural management in Spain. *Journal of Applied Entomology, 123*, 621-627.

Roig, A., Cayuela, M.L. & Sánchez-Monedero, M.A. (2006). An overview on olive mill wastes and their valorisation methods. *Waste Management, 26*, 960–969.

Romeis, J., Babendreier, D., Waeckers, F.L. & Shanower, T.G. (2005). Habitat and plant specifity of Trichogramma egg parasitoids—underlying mechanisms and implications. *Basic and Applied Ecology, 6*, 215–236.

Sánchez-Monedero, M.A., Roig, A., Paredes, C. & Bernal, M.P. (2001). Nitrogen transformation during organic waste composting by the Rutgers system and its effects on pH, EC and maturity of the composting mixtures. *Bioresource Technology, 78*, 301-308.

Santos, S.A.P. (2007). Action of predators against the black-scale, *Saissetia oleae* (Oliv.) in Trás-os-Montes olive groves. PhD Thesis. Universidade de Aveiro. Aveiro.

Santos, S.A.P., Pereira, J.A., Torres, L.M. & Nogueira A.J.A. (2007a). Evaluation of the effects, on canopy arthropods, of two agricultural management systems to control pests in olive groves from north-east of Portugal. *Chemosphere, 67,* 131-139.

Santos, S.A.P., Cabanas, J.E. & Pereira, J.A. (2007b). Abundance and diversity of soil arthropods in olive grove ecosystem (Portugal): the pitfall trap type effect. *European Journal of Soil Biology, 43*, 77-83.

Sakr, H.E.A., Hassan, S.A. & Zebitz, C.P.W. (2000). A new device to capture and monitor the activity of *Trichogramma* in the field. *Mitt. Biol. Bundesanstalt, 376*, 578.

Schmidt, M.H., Lauer, A., Purtauf, T., Thies, C., Schaefer, M. & Tscharntke, T. (2003). Relative importance of predators and parasitoids for cereal aphid control. *Proceedings of the Royal Society of London Series B-Biological Sciences, 270*, 1905-1909.

Stelzl, M. & Devetak, D. (1999). Neuroptera in agricultural ecosystems. *Agriculture, Ecosystems and Environment, 74*, 305–321

Straub, C.S., Finke, D.L. & Snyder, W.E. (2008). Are the conservation of natural enemies biodiversity and biological control compatible goals? *Biological Control, 45*, 225-237.

Thies, C. & Tscharntke, T. (1999). Landscape structure and biological control in agroecosystems. *Science, 285*, 893-895.

Torres, L., Gonçalves, F. & Oliveira, A. (2007). On the occurrence of *Conwentzia psociformis* (Curtis) (Neuroptera: Coniopterygidae) on olive groves from Trás-os-Montes region. In: Cunha-Queda, C. & Rodrigues, M.F. *II Colóquio Nacional de Horticultura Biológica, Actas Portuguesas de Horticultura*. Lisboa, 195-201.

Trapero, A. & Blanco, M.A. (2004). Enfermedades. In D. Barranco, R Fernández Escobar & L. Rallo (Eds.), *El Cultivo del Olivo*. (pp 517-664). Madrid: Junta de Andalucía Y Mundi-Prensa.

Velimirovic, Y. (1994). Black scale *Saissetia oleae* Oliviere, significant olive pest in the area of Yugoslav seaside. *Acta Horticulturae, 356*, 407-410.

Wade, M.R., Zalucki, M.P., Wratten, S.D. & Robinson, K.A. (2008). Conservation biological control of arthropods using artificial food sprays: Current status and future challenges. *Biological Control, 45*, 185-199.

Weibull, A.C., Ostman, O. & Granqvist, A. (2003). Species richness in agroecosystems: the effect of landscape, habitat and farm management. *Biodiversity and Conservation, 12*, 1335-1355.

Wyss, E., Luka, H., Pfiffner, L., Schlatter, C., Uehlinger, G & Daniel, C. (2005). Approaches to pest management in organic agriculture: a case study in European apple orchards. *Organic-Research, (May)*, 33-36.

Zehnder, G., Gurr, G.M., Kühne, S., Wade, M.R., Wratten, S.D. & Wyss, E. (2007). Arthropod pest management in organic crops. *Annual Review of Entomology, 52*, 57-80.

In: Organic Farming: Methods, Economics and Structure ISBN 978-1-60692-864-6
Editors: Michael Nelson and Ivan Artamova

Chapter 2

THE PARADIGMATIC NATURE OF ORGANIC FARMING: EVIDENCE FROM AGRICULTURAL PROFESSIONALS' ATTITUDES TOWARDS ORGANIC FARMING AND GENETIC ENGINEERING

Sarah Wheeler

Centre for Regulation and Market Analysis, School of Commerce
University of South Australia, Adelaide, SA 5000

ABSTRACT

This paper discusses the concept of scientific paradigms and draws parallels with paradigms in agriculture, and in particular focuses upon organic farming and genetic engineering. There is anecdotal evidence that they are both potential new agricultural paradigms. Different writers have referred to both as new 'paradigms' in agriculture, though the qualitative evidence is stronger for organic farming than genetic engineering. The attitudes of agricultural professionals are important as they are responsible for creating, developing and diffusing innovations as well as informing and educating farmers and the public. This paper summarizes past research on the attitudes of Australian agricultural professionals (n=185) towards organic farming (Wheeler 2008) and genetic engineering (Wheeler pending). These two sets of results are compared to assess if professional attitudes to the two innovations are driven by similar influences or are quantitatively significantly different. In addition, analysis using bivariate ordered probit regression is conducted in this study on the nature of professionals' answers to the net benefits of the two innovations themselves. Do professionals tend to be negative or positive about both innovations, or positive about one and negative about the other? From a comparison of the results, it is shown there is some quantitative evidence that organic farming represents a potential new sub-paradigm in agriculture, but only very weak evidence that genetic engineering is a potential new paradigm.

Introduction

Most scientists believe that science operates via the rationalistic 'scientific method', which postulates that science is pure, and that propositions and methods are arrived at only by efficient, logico-empirical operations. Personal, political and power influences play no role. However, there is considerable evidence to suggest that science does not always operate in a manner consistent with this methodology.[1] The question is, why? Part of the reason includes professionals' individual psychological factors and belief systems. In particular, the existence of paradigms may help explain why the rationalistic scientific method does not always prevail in science.

In 1962, Thomas Kuhn wrote what is claimed to be his classic work: *The Structure of Scientific Revolutions,* in which he introduced the term 'paradigm'. Kuhn used the term to explain the nature of scientific debates. A paradigm was a dominant theory in science, or 'normal science'. The term is still used today to convey the implicit assumptions underlying and guiding scientific inquiry, as well as describing the fundamental currents of social change. In Kuhn's view, science evolves in the following way:

Pre-science (pre-paradigm) → normal science (paradigm) → crisis → revolution → new normal science → new crisis (and so on)…..

In the pre-science state, a number of schools of thought compete against each other for dominance. The scientific field becomes more structured and directed when one school of thought overshadows the others, and becomes the dominant paradigm or 'normal science'. The framework helps define the nature of the world and a way of looking at things: it is a set of shared assumptions, beliefs, dogmas, conventions and theories. Working within a paradigm implies rules and boundaries, which limits what questions can be asked, how research is conducted, and the type of answers that are found. Kuhn argues that scientists are not bold adventurers discovering new truths; rather they are puzzle-solvers working within an established view of the world. The normal state of science is said to be dominated by one paradigm, although it was later agreed by Kuhn that a number of sub-paradigms can also exist at one point in time. Lakatos (1970) agrees that a number of research programmes can exist at the same time, and his research programmes closely resemble Kuhn's paradigms. Some writers prefer Lakatos' description of research programmes to Kuhn's paradigms as being more descriptive of scientific history, especially in the field of economics (i.e. Blaug 1975).

Paradigms can be a useful framework in science. They can eliminate the need to continually question basic assumptions or issues and therefore reduce wasted time. However, the existence of dominant theories and power structures in science may result in the suppression of undesirable opinions, unoriginality in research and emphasis on trivial topics, a denigration of non-specialists scientific knowledge, and an increase in fraud and favouritism in hiring and promotion (Martin 1999, Manwell & Baker 1988, Broad & Wade 1982, Martin 1979, de Grazia 1966). Another potential consequence of paradigms in science is path-

1 Bauer (1984), Burr (2002) and Krimsky (2000) provide histories of a number of innovative scientists whose ideas were rejected outright by mainstream science. Other deviations include: suspicions attached to theories which originate in particular countries; less rigorous checks on scientists of long standing and demonstrated ability; the withholding of theories until the release impact can be maximised; profit-orientated research goals; and reluctance to let go of particular theories no matter what the evidence suggests.

dependence, where one research area becomes superior to another simply because of the research funds spent and its future use and development. However, had the other new theory had similar levels of funds, use and development, it would have been socially superior. The implications are that socially beneficial research and innovations may be lost. The path-dependence argument has been strongly criticised (i.e., Liebowitz & Margolis 1994 and 1990), yet it remains as a potential negative consequence of the existence of paradigms.

A new paradigm often attracts controversy and criticisms when the intellectual attraction of the research network is exhausted, or when external conditions lead to questioning the application of the current paradigm. A 'paradigm shift' is a change in the dominant theory of science. One of the questions Kuhn seeks to answer is why individual scientists change their allegiance from one paradigm to the other (or why they work in a particular field). Kuhn suggests that such a change is not purely rational, that other factors come into play, as scientists have different values, standards, goals and principles. In one way, opponents and proponents of different paradigms speak different languages. Each group cannot *compel* the other to accept their view, but they can attempt to *persuade* each other to change views. There will be an increasing shift in professional allegiance as more and more scientists are converted to a particular view. In Kuhn terms, the "revolution" occurs when the majority of scientists have shifted to a particular stance. This may happen quite quickly, but at other times it could take centuries for the shift to occur.

There is no one absolute test to show that a new theory is becoming the new paradigm. A variety of factors will influence how scientists view new paradigms and either maintain or switch allegiance. This paper assembles a number of economic, sociological, philosophical and psychological theories on how scientists react to new ideas and then compares quantitative analysis on two agricultural innovations to indicate the presence of a new paradigm. Previous research in this area has primarily been current or historical case studies and generally qualitative in nature (for example, Barnes *et al.* 1996, Feyerabend 1988, Latour and Woolgar 1986, Latour 1987, Merton 1973).

Organic Farming and Genetic Engineering

Genetic engineering and organic farming represent two key innovations in agriculture that are both lauded and deplored. Different parties represent the two innovations as radical alternative ways to address current sustainability crises in agriculture (i.e., National Academy of Sciences 2000, Pretty 1995). Like Wynen (1996), we draw an analogy of a paradigm shift in science with a paradigm shift in agriculture, although they are strictly not the same thing.[2] But can they be described as potential new paradigms in agriculture? That is, will the current system of conventional agriculture give way to a new system that looks more like organic farming, or a new system of biotechnologically-engineered research and products? Or are both systems simply comparable to the current system?

2 Agriculture is not strictly a scientific field; it draws on knowledge and practices from many different scientific fields. It is better described as a management system, with conventional agriculture representing the dominant management system today.

Organic Farming

Organic farming is being increasingly adopted around the world and it is becoming more mainstream. The goal of organic farming is to give priority to long-term ecological health, such as biodiversity and soil quality, rather than short-term productivity gains. A number of studies have provided scientific evidence on the environmental, economic and health benefits of organic farming (Macilwain 2004, Shepherd *et al.* 2003, Nelson *et al.* 2004, Mäder *et al.* 2002, Stolze *et al.* 2000, O'Riordan & Cobb 2001). However, many agricultural professionals still question its financial viability, environmental credentials and overall efficiency and productivity, and reject it as an alternative farming system (e.g., Trewavas 2001, Avery 1995). Organic farming in Australia receives extremely little government research or extension support (Wynen 2003), far less than its current adoption by farmers.

Organics represents a way of farming – but it can also represent a way of scientific research.[3] Wynen (1996) argued there was strong qualitative evidence of a paradigm shift in agriculture, from the dominant paradigm of conventional agriculture towards organic farming. This qualitative evidence included highlighting problems within the dominant paradigm (such as the crises within conventional farming, along with attempts to 'patch-up' these problems); increasing argument between agricultural professionals and farmers that is unable to be resolved (there is growing polarisation of opinions between conventional and organic proponents which indicates the incommensurability of their views); and the increasing popularisation of the potential new paradigm (popularisation occurred through increasing organic journal and book publications, increased tertiary education, development of extension channels and farmer financial support, and increasing farmer adoption of organic farming). Such indications led Wynen (1996) to suggest that organic farming does represent a paradigm shift in agriculture.

Bues and Dunlap (1990) also sought to find whether conventional agriculture and organic farming represent two different paradigms (but they did not use the exact same definition as Kuhn). They concluded (p. 597): "in short, all indications are that the contemporary debate in agriculture seems to have the makings of a true paradigmatic conflict". Other writers are increasingly referring to organic farming as a potential "new paradigm" in agriculture (Dimara *et al.* 2003, Abaidoo & Dickinson 2002, Allen 1996, Dahlberg 1986, Merrill 1983). There is enough qualitative evidence to suggest that organic farming may represent a potential new paradigm – both as a system of farming and as a new form of agricultural research. Later in this chapter we test how much quantitative evidence exists for organic farming to be a paradigm shift in agriculture.

---

3 Conducting organic agricultural scientific research is difficult for scientists due to the need to consider the whole farm and its ecological system, not just parts of the system. There are five main ways organic research differs from conventional: (1) it applies consistently multidisciplinary approaches to solutions; (2) it integrates stakeholders into research; (3) it rescales continually the focus of research work in the same project; (4) it considers and decides on long term impacts; and (5) it considers individual, social, ethical and philosophical frameworks of agriculture and food in addition to natural science (Niggli 2002).

Genetic Engineering

Genetic engineering (GE) is the modern innovative form of biotechnology, known also as recombinant-DNA technology. Some of the major benefits of genetic engineering that are cited include agronomic (increased yields, reduced costs), health (increased vitamins) and environment (reduced pesiticides, less land needed for production) (King 2004 and 2003, National Academy of Sciences 2000, Beyer *et al.* 2002). Critics argue that its benefits are uncertain and often exaggerated, while potential problems include health, environment, trade and legal issues (Independent Science Panel 2003, Quist & Chapela 2001, Ewens & Pusztai 1999). Australia invests strongly in genetic engineering, with public sector related R&D biotechnology investment in agricultural, vet and environment representing 35% of total R&D expenditure in these fields in 2002-03 (Hopper & Thorburn 2005).

Some scientific writers have described genetic engineering as a new paradigm that is going to change the face of science (Leisenger 2000, ISSC 1993, Miller 1993), but we do not know of any writers in sociology who have applied the term. On the one hand, scientists argue that biotechnology basically involves the same modifications that have been taking place for thousands of years, with the modern form of genetic engineering adding additional information (Borlaug 2000, Prakash 2001). Critics of genetic engineering believe that rDNA techniques are fundamentally different from traditional technology and should be evaluated differently (ISP 2003, Gepts 2002), and that it will not solve the current problems inherent in conventional agriculture today. Research on genetic engineering does not seem to have been restricted or held back, as in the beginning there were no scientific restrictions imposed on molecular biological scientists, and although concerns began to be raised by other scientists in the late 1960s, no direct regulation or public involvement resulted. In addition, funding and research support were never limited. Government involvement and scientific criticism increased in the 1980s but it was not until the 1990s that the public (and other scientists) really became involved in the debate and restrictions started being imposed.

There is some evidence for and against genetic engineering, causing a paradigm shift in agricultural scientific research. Evidence for genetic engineering includes public and some scientist resistance, the ground-breaking developments in scientific research, and the framework around which genetic engineering is considered. Evidence against genetic engineering includes the history of the technology, the ready availability of research funds and the tacit acceptance of the techniques by the majority of scientists. The overall balance of evidence may imply that genetic engineering is an important new innovation/technology only, but not a paradigm shift, at least not in Kuhn's sense of a 'revolution'. Some writers argue that there are indications that it has been absorbed into the current dominant practice of agricultural scientific research and is fast becoming part of that paradigm (Egri 1994). Whether this is the case or not is explored in this paper.

Predicting a Paradigm Change

The current conflict in agriculture between organic farming and conventional agriculture; between genetic engineering and conventional agriculture; and between genetic engineering and organic farming exists at both societal and scientific levels of a debate. At a broad level, the debate is fundamentally philosophical and value-laden, which is hard to resolve with empirical evidence. Indeed, it is impossible to separate out agricultural practices from the beliefs and values that underlie them. The development and acceptance of these innovations has not been a purely rational process, but one that has involved conflicting beliefs, values, perceptions and social interaction.

The following is an attempt to assemble, discuss and state various hypotheses on how scientists react to new ideas, using ideas from philosophers, historians of science and sociologists about the reception of radical changes in scientific beliefs. It is discussed why some scientists may be encouraged to accept the current paradigm and why others consider alternative theories. This paper develops a number of hypotheses for agriculture which are then compared to data for organic farming and genetic engineering. The hypotheses are not by any means considered to be definitive or final; they are open for consideration with the aim to generate further debate, discussion and research.

Status Effects

One of the most cited influences on scientists' acceptance of new paradigms relates to status effects (i.e. Dolby 1979, Merton 1973, Mulkay 1972, Kuhn 1970, McClelland 1969). Philosophers of science suggest that science control is maintained through socialisation; through the exercise of authority; and by means of social exchange. Innovations are said to come predominantly from those who are new to a particular area or new to research altogether and who have not yet acquired a strong commitment to the current paradigm. Nonconformity will therefore generally occur within academic status classes, namely the young and those of high status. Members of the middle-status category are least likely to deviate from group opinion or group norms, while persons of low (either because they are young or new to the area at hand) or high (either those who have been around for a while, are a leader or professor or journal editor, etc.) status are much more likely to be nonconformist. Those of high status obtain decreasing returns from undertaking further research within the current paradigm and are less likely to feel stigma from their colleagues about deviating from current thinking. However, the effect is not certain because those with high status still have a lot to lose in deviating from different research patterns. The young gain little from producing highly predictable results, while they potentially gain a lot from innovating. Kuhn (1970; 151) describes what is known as the "Planck Effect":

> "Max Planck ..... sadly remarked that 'a new scientific truth does not triumph by convincing its opponents and making them see the light, but rather because its opponents eventually die, and a new generation grows up that is familiar with it'".

Evidence such as the age of some famous scientists (Darwin was 29 when he first wrote about natural selection and Einstein was 26 when he published his theories) supports the theory that being younger is associated with revolution. However, Kuhn believed this generalisation badly needed systematic investigation. Popper (1970) suggests from his personal experience that it is young science students that are more likely to be non-revolutionary and accept the ruling dogma of the day – hence age is not a good criterion. Indeed, professionals who are about to retire may be more likely to want to consider alternatives as they no longer feel the pressure associated with self-interest. Merton (1973) found that the evidence on age differentials in receptivity to new ideas in science was thin and uncertain. One of the conflicting corollaries is that age is also associated with other measures of status and specialisation, hence once these effects are removed, being older may be associated with championing new paradigms. Although it is recognised that the effect of age on acceptance of new paradigms is far from certain, because most of the literature has supported the generalised link between being young and accepting alternative paradigms, this is the hypothesis that is tested here.

Finally, many scientists swap fields during their career to work in another scientific area (one that they have not been trained in). Personal reflections of such scientists suggest that they believe they are able to see a broad picture within that field, without being hampered by 'knowing what everyone else knew'. Newcomers have often introduced radical new ideas into their adopted scientific field (Merton 1973). Thus:

H1: Younger professionals are more likely to be positive towards an alternative paradigm than older professionals.

H2: Professionals who have worked longer in their area are less likely to be positive towards an alternative paradigm than professionals who have worked less in their area.

H3: Professionals who are of 'high-status' (classified as a leader in their job) are more likely to be positive towards an alternative paradigm than non-leaders.

Self-Interest

Economic theory suggests that professionals will be influenced by their own self-interest. Self-interest can lie in the salary obtained under the current agricultural system, or the considerable costs associated with tertiary education, or in the funding available to current research (i.e., Barnes *et al.* 1996, Barnes and MacKenzie 1979, Martin 1979, Mulkay 1979). Generally there is pressure on scientists to do certain types of research, which is predisposed towards the types of topics that can be researched and the economic and non-monetary rewards that are available from pursuing certain paths. Kuhn suggests that scientists who have more to lose from abandoning the current paradigm will not support alternative paradigms. It is only rational that professionals can be expected to look after their own best interests, and the more self-interest (via, say, salary or work relevance) professionals have in the research area of the current paradigm, the less likely they will support an alternative.

Scientists have undergone prolonged training to accept and repeat the values, standards and assumptions of science as presented by teachers and the education system. Therefore, there may be a vested interest associated with increased tertiary education and acceptance of new paradigms, with scientists who have incurred large education costs from the current system more likely to reject alternative paradigms. Francis Bacon was the first to suggest that increased specialised education may introduce bias and preconceptions (Riggs 1992). On the other hand, advanced education may be a variable that allows for a more questioning mind. Research on people's consciences and obedience to scientific authority indicates that individuals with advanced education may be more likely to hold on to the perception that they are a legitimate decision-making authority and will therefore question decisions made by people who are meant to be authorities (Milgram 1983). Also, education is a sunk cost, hence should not affect scientists' current or future behavior. Having increased education may mean a professional is more positive towards an alternative paradigm; nevertheless, it is the previous proposition that is tested due to the predominance of writers suggesting this link.

Scientists who are advocates or tacit supporters of a new paradigm believe that they are 'right' because they have higher knowledge about the area. Those who reject an alternative paradigm outright may feel that they do not need to know about it because its underlying principles show that it is clearly wrong. Therefore, increased knowledge of a paradigm increases the likelihood that professionals think more favorably about it. At the same time, critical skeptics of alternative paradigms may argue that scientific knowledge has not been correctly interpreted by the advocates or tacit supporters, and therefore critics rate their knowledge higher and proponents' knowledge deficient, which may weaken the above hypothesis.

H4: Professionals who have a higher salary under the existing scientific structure are more likely to be negative towards alternative paradigms.
H5: Professionals who have a higher percentage of research relevance tied to an alternative paradigm are more likely to be positive towards the alternative.
H6: Professionals who have more tertiary education are more likely to be negative towards an alternative paradigm than professionals with less tertiary education.
H7: Professionals who have a higher knowledge of an alternative paradigm are more likely to be positive towards that paradigm than professionals with less knowledge.

Specialisation

The professionalisation process for individuals has been described by sociologists and others as a process of acculturation that occurs in often subtle ways, resulting in a state of mind that is characteristic of a member of a particular profession or occupational group. Becker and Carper (1956) identified four elements that individuals use to weave occupational identity, which are still used today as a basis for defining the professionalisation process. These elements are (a) occupational title, and associated ideology; (b) commitment to task; (c) commitment to particular organisations or institutional positions; and (d) significance for one's position in the larger society. One's self image or the self image of one's profession will therefore affect one's conduct, attitudes and perspective. Occupational title and organization employment are tested here. Merton (1973) also discusses the importance of the roles played within science. Different professions (such as scientists, researchers, academics and extension officers) within science play different roles, and are employed in organizations such as government, industrial scientific organisations and universities. Popper (1970) distinguishes between 'applied' and 'pure' scientists -- applied scientists are those who are more content to live within 'normal science' and 'pure' scientists are not. Popper argued that pure scientists are comparatively rare within science, hence research scientists are on average expected to be more applied than pure. It is expected that given the nature of their job, extension officers are more applied scientists than pure scientists. Conversely, it is hypothesized that professionals employed within universities are more likely to be pure scientists.

Finally, specialisation within a profession may reinforce the structure and worldview inherent in the selection and organisation of scientific knowledge because of the reduction of contact with wider political and social issues. This may increase the potential loss in personal research from a change in paradigm (Martin 1979, Popper 1970), leading to an immense restriction on the scientist's vision and considerable resistance to paradigm change. Therefore, professionals that work across many areas of agriculture (in, say, fields such as natural resource management) are more likely to be positive towards alternatives. Thus:

H8: Scientists are more likely to be negative towards an alternative paradigm (such a hypothesis assumes that generally there are more 'applied' than 'pure' scientists).
H9: Extension officers are more likely to be negative towards an alternative paradigm.
H10: The existence of any occupational effects for organisation of employment indicates professionals are more likely to be negative towards an alternative paradigm.
H11: Professionals whose job is less specialised (i.e.: they work in the field of natural resource management) are more likely to be positive towards an alternative paradigm.

Gender

Feminist epistemology and the feminist critique of science and technology literature suggest that women and men are fundamentally different when it comes to holding views about research, science and technology. Such literature (reviewed in Buttel and Goldberger 2002; 27) argues that females tend to be "more critical, more questioning and more public-regarding", and hence would be less likely to be captured by the current paradigm. On the other hand, there is an argument in the literature that females are more financially dependent upon their jobs and hence would be less likely to support alternative paradigms where the payoffs are risky. Nevertheless, it is the first proposition that is tested.

H12: Female professionals are more likely to be positive towards an alternative paradigm.

Principles

Agricultural professionals are taught to be useful, generally because they are scientists who are responsive to the demands of specific users. Things that are considered to be 'uneconomic' are often considered not worth pursuing and are not the prime goal of research (Hadwiger 1982; 56). A scientist who is preoccupied with trying to be useful may not notice wider social implications. It is suggested that most get immediate satisfaction out of the belief that their research has made feeding the world's population possible. Krimsky (2000) suggests that a scientist who is traditional, positivist, private and sponsored by government or industry will tend not to become involved in advocating or supporting environmental health hypotheses.

Empirical evidence seems to support Krimsky's above argument. A survey of 70 Australian environmental scientists found that most scientists provided policy advice in the form of apolitical advisory roles rather than active participation. Over half of them (54%) believed that their career prospects or research funding would be jeopardised by speaking out on environmental issues (29% were unsure), while 11% had been directly disadvantaged because of their environmental views (Wilson & Barnes 1995). It is suggested that agricultural environmental views would not be voiced as strongly as agricultural production views.

H13: Professionals are more likely to be positive towards an alternative paradigm if it is seen to be associated with profitability. Their views on such issues will be more important than any other potential benefits.

H14: Professionals who believe conventional agriculture is unsustainable are more likely to be positive towards an alternative paradigm. Such views will not be as strong in influencing overall attitudes as their views on production or profitability.

Kuhn suggests scientists who support alternative paradigms are more involved in crises (problems surrounding the current paradigm) and are more likely to be vocal in advocating new positions. Krimsky (2000) suggests controversies in science arise when facts are uncertain, values are in dispute, stakes are high and decisions are urgent. There are no clear scientific ethics of engagement pertaining to these sorts of controversies. Ethical problems include the release of uninterpreted data, conjecture sensitivity, prepublication release of research, and conjectural risk. Responses of scientists to controversial hypotheses can be classified into four types: (a) advocacy scientists who feel it is their responsibility to support and advocate controversial hypotheses, even if the evidence has not been absolute. They call for regulation of and action against questionable products; (b) tacit supporters who believe that the controversial hypothesis is likely to be true, and therefore rethink their own research; (c) critical skeptics whose arguments lead to the strengthening, weakening or falsification of a controversial hypothesis. Critics support acting only when the evidence is undeniable and the science is 'sound'; and (d) silent skeptics who will not believe that there is enough evidence supporting the controversial hypothesis, and refuse to consider its potential effects (Krimsky 2000). Scott, Richards and Martin (1990) show that when scientists become involved with non-mainstream theories they often become participants in the "fierce fight to construct reality", and their position generally changes from that of neutral, symmetrical analysis to active intervention in the dispute. An outspoken critic of pesticides, the entomologist Van den Bosch (1978; 48-49), states that he understands why many scientists are not willing to participate in public debate on technological issues and that the lack of fair debate over the issue forced him to also abandon reasoned debate.

> ...I have become increasingly involved in the roaring pesticide controversy – a vicious, nerve-wracking imbroglio that has turned my entomological niche into a veritable hornet's nest. What is most saddening is that, as I have become increasingly enmeshed in the pesticide hassle, I have turned into a ruthless gut-fighter in a slugfest without rules or a semblance of fair play. This metamorphosis has been forced upon me because it is the only way that I can hold my own in the shoot-out, or for that matter even survive it. It is difficult for a scientist to play this way. We are largely a tribe of preoccupied people who just want to be let alone to do our thing. Since we lack the appetite for slashing combat, or the skills to survive connivance, we usually avoid confrontation.

Therefore, those scientists who are advocating stronger involvement from government are more likely to be favourable towards alternative paradigms, because scientists within mainstream science are unlikely to actively intervene.

There is currently strong debate over the appropriateness of the increasing use of intellectual property rights and patenting in agriculture and the influence of private funds on research outcomes (Huffman & Tegene 2002, David 2000, Mazzoleni & Nelson 1998, Drahos 1996). Tisdell (1991) suggests that patented inventions and intellectual property rights favour knowledge associated with conventional agriculture, hence private research is distorted. He believes a follow-on effect also often occurs with publicly funded or conducted agricultural research. Huffman (2001) argues that major problems are arising with overlapping patents, excessive breadth, lockouts and issues with freedom to operate. Universities cannot be 'open' if their funding or research agenda has been captured by a number of large companies. Professionals who are critical of certain research influences in the current agricultural paradigm may be more likely to be favourable to alternative paradigms where such issues may not exist.

H15: Professionals who advocate the need for a more pro-active government policy role for the alternative paradigm are more likely to be positive towards it.
H16: Professionals who view the increasing role of intellectual property rights and patenting in agricultural research positively are more likely to be negative towards an alternative.
H17: Professionals who think that private funding in agricultural research does not influence research outcomes are more likely to be negative towards an alternative paradigm.

Information Choices

Due to the quantity and quality of information choices available, professionals' code (i.e., rate or weight) messages received. Agricultural information choices include those from scientific sources (such as conferences, journals, peers, scientific books, university) and other sources (media, farmers, industry, books). Different information channels have differing weights (Allen 1969). Firms and organisations control the intake of information to avoid being overwhelmed by setting up a 'system boundary' to provide barriers between information on the 'inside and on the outside' (Price 1969; 151).

Another form of setting system boundaries is how professionals view the involvement of the public, media and other lay groups in decision making, as compared to recognised scientific sources. One of the main tenants of the rationalistic scientific method is that scientists are the best interpreters of scientific data, and results should only be communicated to the public after it has gone through a rigorous peer review system. However, if scientists do not believe that their concerns have been given a fair consideration within science (for reasons such as those provided in footnote 1), then those scientists often turn to the public for a hearing of their beliefs (Burr 2002, Krimsky 2000, Scott, Richards & Martin 1990). The involvement of the public and other non-scientific sources can be viewed positively or negatively by scientists. If they view the involvement of lay groups as a main barrier against the diffusion or employment of scientific methods, then they are likely to be more favorable to the current paradigm. Thus:

H18: Professionals that gain information primarily from 'inside sources' (primarily scientific sources) are more likely to be negative towards an alternative paradigm.

H19: Professionals who name public attitudes, lobby groups and media influences as the main barrier against the diffusion of innovations are more likely to be negative towards an alternative paradigm.

Overall Changes in Views

Finally, for a paradigm shift to occur in agriculture, professional views towards the innovations must be changing, with more professionals 'converting' to the view. Kuhn suggests that a paradigm change is decided by the number of scientists who advocate the reigning paradigm as against those who advocate the alternative. It is important to note that this hypothesis is not a test of how scientists react to new paradigms per se, but more of an indication of the possibility of a paradigm shift occurring.

H20: More professionals must have positively changed their view towards the alternative paradigm than those that negatively changed their views over the past few years.

There are a number of sociological and philosophical hypotheses that could not be tested within this paper. Some of the more important remaining hypotheses that could be explored in future research include examining the 'gatekeeper effect'. If professionals view themselves as 'gatekeepers of science', (i.e., they may be peer reviewers for a journal, members of an editorial board or members of a scientific group) then the more they may support the current paradigm. Also, the influence of professional goals could not be assessed; that is, how does their stated job goal (i.e., feed the world, scientific curiosity, business development, etc.) affect their views of alternatives? Other factors that have often been mentioned include religious, insecurity issues and political factors. These and other hypothesis tests are left to future research.

Survey Methodology and Model Specification

Acceptance of an agricultural innovation can differ for a number of levels: research, extension, farmer, and community. Agricultural professionals (scientists, extension officers, researchers and academics) play a very important role in creating, diffusing and informing the public about agricultural innovations.[4] Previous papers (Wheeler (2008) and (pending)) have explored the influences on agricultural professionals' beliefs towards organic farming and genetic engineering, respectively. These studies used data from a telephone survey conducted in mid 2004. The sample frame for the survey consisted of two groups: a general and a targeted group. The general group of agricultural professionals was limited to State and Commonwealth bodies based in South Australia (SA). Respondents were randomly selected from each group and telephoned, with 119 surveyed.

4 An agricultural professional was defined as either providing agricultural advice to farmers; conducting agricultural research; or teaching agriculture at university. The sample was limited to public bodies.

To supplement the general sample, professionals who had actual experience with organic farming across Australia were specifically sought for additional telephone interviews. The purpose of this supplementation was to obtain enough records of professionals with organic research/extension experience in order to be able to compare their attitudes and knowledge with those who do not have such experience (as only a small percentage of professionals do have such experience).[5] 66 interviews were conducted with professionals around Australia, giving a total sample of 185 respondents, and a total response rate of 96%.[6]

The dependent variables used by Wheeler (2008 and pending) were the respondents' answers to the questions on whether the benefits of each innovation outweighed its costs or risks. Professionals either responded "no," "unsure," or "yes". The answer "no" indicated a belief in no net social benefits associated with either innovation, "unsure" indicated uncertainty due to unknown factors, and "yes" indicated a belief in net benefits. Therefore as it stands, the dependent variable represents a proxy benefit cost ratio. It attempts to assess all the aspects of each innovation and determine whether overall, the innovation is considered worthwhile. Two ordered probit models were used to evaluate factors that influenced overall views towards each innovation. Appendix 1 provides the variable coding and questions asked in the surveys.

EMPIRICAL RESULTS

With the exception of hypotheses 15, 19 and 20 (which could not be tested in the regressions due to their inappropriateness as independent variables), all the remaining hypotheses can be tested using the ordered probit results. Table 1 depicts and compares the seven broad groups of independent variables that influence views towards organic farming and genetic engineering. Factors significant in the organic farming model included working age, Commonwealth Scientific and Industrial Research Organization (CSIRO) employment, being part of the targeted sample, organic knowledge, information sources, beliefs in the innovation's environment sustainability, food quality, profitability and innovativeness. Factors significant in the genetic engineering model included age, ethnicity, farm background, scientist job title, extension officer, genetic engineering research relevance, views on the sustainability of conventional agriculture and intellectual property rights and patenting in agriculture, and views on genetic engineering's products profitability, environmental friendliness and long term research needs. Further discussion on the regression results is provided later.

The psuedo R-square value describes how well the data fit the model. A significant amount of variation was explained. This prediction result compares favourably with other ordered probits.

5 A targeted database of over 100 academics, scientists, extension officers and policy makers from 16 Australian public organisations was created. Respondents were randomly selected and phoned if permission from the head of the organization had been received for ethics clearance purposes. If not, individuals were passed over if selected. The sample sizes for the random and targeted groups were both statistically significant at a 7% level of precision and a 95% confidence level.

6 The response rates of each population were not statistically different from each other, though the targeted population was slightly higher than the random population (97% vs 96%).

Table 1. Determinants of Views towards Organic Farming & Genetic Engineering (ordered probit regression)

	Organic Farming			*Genetic Engineering*		
	Coef.	z	P>|z|	Coef.	z	P>|z|
Age	-0.13	-1.53	0.13	-0.14	-1.84	0.07
Age^2	0.00	1.66	0.10	0.00	1.97	0.05
Working Age	-0.03	-1.69	0.09	-0.01	-0.55	0.58
Ethnicity	0.63	1.19	0.24	1.55	2.18	0.03
Male	-0.19	-0.67	0.50	0.22	0.99	0.32
Having a Farm Background	-0.02	-0.11	0.91	-0.51	-2.65	0.01
Years of Tertiary Education	0.13	2.12	0.03	-0.06	-0.76	0.45
Salary	0.00	-0.10	0.92	0.00	0.72	0.47
Job title is Scientist	0.00	0.00	1.00	0.63	2.63	0.01
Classified as a Leader in their Position	0.18	0.46	0.65	-0.04	-0.12	0.91
Working predominantly as Extension Officer	-0.06	-0.23	0.82	0.40	1.70	0.09
Employed by a state organization	0.02	0.07	0.94	0.16	0.50	0.62
Working in Natural Resource Management	0.35	1.13	0.26	-0.07	-0.26	0.80
Working in Broadacre and/or Grazing	0.04	0.14	0.89	0.12	0.49	0.62
Employed by CSIRO	-0.89	-1.77	0.08	0.13	0.25	0.80
Part of the Targeted Sample	0.65	1.93	0.05	-0.14	-0.59	0.56
Organic/Genetic Engineering Knowledge	0.24	1.76	0.08	-0.05	-0.41	0.68
Naming Media as first main source of Info.	-0.34	-1.27	0.21	0.29	1.15	0.25
Naming Scientific sources as first main source of Info.	-1.16	-3.55	0.00	-0.07	-0.25	0.80
% of Research Time Relevant to Organic Farming/Genetic Engineering	0.01	1.16	0.25	0.01	2.12	0.03
Views on Conventional Agric. Sustainability	0.12	1.10	0.27	-0.24	-2.63	0.01
Views on agric. intellectual property rights & patents	0.08	0.79	0.43	0.23	2.37	0.02
Views on private funding of agric. research	0.02	0.14	0.89	-0.16	-1.53	0.13
Views on Organic Farming's*/Genetic Engineering's Profitability	0.25	1.65	0.10	0.39	2.60	0.01
Views on Organic Farming's Innovativeness/ Beliefs on uncertainty of GE and need for long-term research	-0.37	-2.70	0.01	-0.65	-4.56	0.00
Views on Organic Farming's/Genetic Engineering Food Quality	0.27	2.33	0.02	0.11	0.74	0.46
Views on Organic Farming's/Genetic Engineering Environmental Sustainability	0.70	5.16	0.00	0.51	3.57	0.00
/cut1	-0.64	1.95		-3.39	2.00	
/cut2	0.65	1.96		-2.06	2.01	
Number obs.	185.00			185.00		
Log Likelihood	-115.44			-137.29		
Wald Chi2	147.91			129.83		
Prob>chi2	0.00			0.00		
Pseduo R2	0.40			0.32		

Sources: Adapted from Wheeler (2008, pending) and further analysis

Notes: All models calculated with Huber/White heteroskedasticity-consistent standard errors and covariance. Cuts report standard errors.

* In order to make similar comparisons between the two, the variable sign for organic farming's profitability was reversed.

In the models in Table 1 above, similar factors are thought to influence both genetic engineering and organic agricultural views. A question may be raised over the degree to which innovation views are inter-related: do professionals tend to be negative or positive about both innovations, or positive about one and negative about the other? This paper employed bivariate ordered probit regression which fits maximum-likelihood two-equation ordered probit models of our two dependent variables (views towards genetic engineering and views towards organic farming) on the independent variables for each model. Results are shown in Appendix 1. It was found that the error terms of the two models were not correlated (by the Wald test of independent equations of if rho is significantly different from zero); hence, believing that genetic engineering has net social benefits has no influence over believing that organic farming has net social benefits. Hence, it is possible to model the two beliefs separately and the results shown in Table 1 can be discussed further with confidence.

Individual Tests of Remaining Hypotheses

Hypothesis 15 suggests professionals who advocate the need for an increased government policy role are more likely to be favourable towards a new paradigm. In the survey, professionals were asked whether they agreed or disagreed with two questions: "Government should take a more pro-active government stance for organic farming (*or genetic engineering*)". Table 2 provides two comparisons of professionals' views on the need for a more pro-active government policy stance: the general sample and the total sample.

Table 2. Professionals' Views on a More Pro-active Government Policy Stance by Type of Innovation

Views	Organic Farming (General)	Genetic Engineering (General)	Organic Farming (Total Sample)	Genetic Engineering (Total Sample)
Positive	3.76	3.48	4.03	3.41
Sample size (n)	*44*	*54*	*94*	*69*
Unknown	3.26	3.57	3.27	2.66
Sample size (n)	*35*	*29*	*46*	*58*
Negative	2.6	2.14	2.59	1.91
Sample size (n)	*40*	*29*	*47*	*58*

Note: Responses were based on a 5 point Likert scale, from Strongly Disagree (1) to Strongly Agree (5).

As hypothesised, the more positive a respondent was towards an innovation, the more they advocated a stronger governmental stance. For example, proponents of organic farming and genetic engineering were significantly more likely to agree with the question (mean 4.03 and 3.41 respectively) than opponents (mean 2.59 and 1.91 respectively) (t tests: $p<0.000$ for both).[7]

7 All t tests were conducted with a student two-tailed distribution.

The second hypothesis (H19) to be examined is the idea that professionals who are favorable to an alternative paradigm are more likely to name 'inside scientific' sources as innovation barriers than 'outside public' sources than professionals negative to an alternative paradigm. Professionals were asked: "What do you think are the major barriers facing the diffusion of genetic engineering (*or organic farming*) in Australia?" Responses were categorised as (1) inside scientific barriers (i.e. lack of scientific knowledge, uncertainty about future benefits/costs, lack of independent verification, scientists' attitudes/perceptions; misinformation by scientists, no long-term scientific studies conducted, information needs, lack of scientific research and development, and lack of extension/scientific support); (2) outside public barriers (i.e., media portrayal, public attitudes/perception/knowledge, lobby group influence, and political regulations); (3) practical innovation issues (market, on-farm issues); and (4) other. The results for (1) and (2) for the total sample are shown in Table 3.

Table 3. Percentage of Inside & Outside Barriers out of Total Barriers Named by Innovation for the Total Sample (n = 185) %

Type of Barrier Named		Positive	Unsure	Negative	Total
Inside Scientific	Organic Farming	27.4	24.1	10.7	22.7
	Genetic Engineering	11.9	22.0	35.3	23.1
Outside Public	Organic Farming	3.3	2.3	1.0	2.6
	Genetic Engineering	61.5	39.4	32.4	44.4

Professionals who were favorable to organic farming were much more likely to name inside sources as barriers to further diffusion than those that felt negatively (27.4% vs 10.7% of the total barriers named, respectively) but were marginally more likely to name outside public barriers than negative respondents (3.3% vs 1.0% respectively). However, those that felt favourably towards genetic engineering were much less likely to name inside barriers than those that felt negatively (11.9% vs 35.3% respectively), and were more likely to name outside barriers than negative respondents (61.5% vs 32.4% respectively).

Overall, professionals were more likely to name inside scientific sources (22.7%) as organic barriers than outside public barriers (2.6%). Conversely, professionals were much more likely to name outside public sources (44.4%) than inside scientific sources (23.1%) as genetic engineering barriers. The quantitative evidence for H19 is therefore much stronger for organic farming.

The final individual hypothesis test (H20) is to consider how professional views on organic farming and genetic engineering are changing. For a paradigm shift in agriculture to occur, then a prerequisite must mean that agricultural professionals have to be "converted" to the new paradigm (Kuhn 1970). Such a hypothesis is extremely hard to test without comparable historical data for Australia. To assess how these professional attitudes have changed recently, two questions were asked: "Over the last 5 years, have your views on organic farming *(or genetic engineering)* become (a) more favorable; (b) less favorable; or (c) stayed the same?" Table 4 shows the results for both samples, which is derived from Wheeler (2007). Professionals were statistically more likely to have positively changed their opinion of organic farming and genetic engineering than negatively (35% vs 13% and 25% vs 16%

respectively). Both innovations therefore provide statistical support for H20, though overall there has been more net favorable change in organic than genetic engineering views.

Table 4. Changes in Agricultural Professionals Views by Innovation

Questions on Agricultural Innovations	Genetic Engineering Answers			Organic Farming Answers		
	General	Target	Total	General	Target	Total
Views becoming more favourable over the past 5 yrs	25%	20%	23%	35%	46%	38%
Views becoming less favourable over the past 5 yrs	16%	32%	22%	13%	6%	10%
Plans to learn more about the innovation in the future	83%	80%	82%	76%	89%	81%

DISCUSSION

Table 5 displays a summary of the hypotheses tests. A tick in Table 5 indicates the hypothesis had the right sign for a new paradigm test and was statistically significant. A cross indicates the hypothesis had the wrong sign and was statistically significant, while a blank indicates either the incorrect sign or statistical insignificance.

Table 5. Summary of Hypothesis Tests for a Paradigm Shift in Agriculture

Paradigm Hypothesis	*Hypothesized Response*	*Organic Farming*		*Genetic Engineering*	
		Actual Sign	*Result*	*Actual Sign*	*Result*
Status					
H1: Younger professionals	+	(-)[1]		-	✗
H2: Professionals with more working experience in area	-	-	✓	(-)	
H3: High- status professionals.	+	(+)		(-)	
Self-Interest					
H4: Professionals who have a higher salary	-	(-)		(-)	
H5: Professionals with higher alternative research experience	+	(+)		+	✓
H6: Professionals who have increased years of tertiary educat.	-	+	✗	(-)	
H7: Professionals with higher alternative paradigm knowledge	+	+	✓	(-)	

Table 5. (Continued)

Paradigm Hypothesis	*Hypothesized Response*	*Organic Farming*		*Genetic Engineering*	
		Actual Sign	*Result*	*Actual Sign*	*Result*
Specialisation					
H8: Professionals who are scientists	-	(+)		+	✗
H9: Professionals who are extension officers	-	(-)		+	✗
H10: The existence of any occupational effects for organisation	-	-	✓	(+)	
H11: Professionals whose job is less specialised	+	(+)		(-)	
Gender					
H12: Female professionals	+	(+)		(-)	
Information Choices					
H18: Professionals gain info. primarily from scientific sources	-	-	✓	(-)	
H19: Professionals name inside more than outside sources	+	+	✓	-	✗
Changes in Views					
H20: Conversion of professionals to the new paradigm	+	+	✓	+	✓
Perceptions					
H13: Professionals views on profitability	+	+	✓	+	✓
H14: Professional views on the sustainability of current agric.	+	(+)		-	✗
Principles					
H15: Professionals who advocate the need for an increased government policy	+	+	✓	+	✓
H16: Professional views on intellectual property rights in agricultural research	-	(+)		+	✗
H17: Professionals views on private funding issues in research	-	(+)		-	✓
Simple Count		✓	8	✓	5
		✗	1	✗	6

Note: 1. Signs in brackets indicate the result was not statistically significant at 10% (or below) level.

Now, the question remains as to how much evidence is required before there is support for an agricultural paradigm shift. Given that a paradigm shift occurs very rarely in science, it is to be expected that applying these tests to potential candidates would mean most would fail. How can we distinguish between an innovation and a paradigm? In other words, what results could we expect from a new theory or method in agriculture that is a new innovation but not a new paradigm? We argue that if a new theory is an innovation only, then applying these tests should result mainly in a lack of statistical significance. There should be no clear pattern among the way that professionals react to new ideas. If there is statistical significance, then it is to be expected that such significance should be randomly distributed and hence there should be an equal chance of satisfying the individual hypothesis. A systematic pattern of reactions should only emerge if professionals are consistently reacting the way the literature suggests. Having said this, it is another thing entirely to suggest how many hypotheses need to be satisfied. Probably at least half the hypotheses need to be met, but the question remains open as to if this is enough. For a new theory to be a new paradigm, it may have to meet most of the hypotheses, while weaker evidence may suggest a sub-paradigm, or even less evidence may suggest a new research program.

Table 5 shows there is statistical support for eight of the organic farming hypotheses, with six others having the expected sign for a paradigm test but were not statistically significant. Only one hypothesis (education effect) in the organic model had the wrong sign for a paradigm test but was statistically significant (see discussion later). These results may indicate evidence to suggest that organic farming fits the mould of a potential new paradigm in agriculture.

There is far less evidence that genetic engineering represents a new paradigm in agriculture, with statistical support for only five hypotheses, while four had the expected sign but were statistically insignificant. Six hypotheses had the wrong sign for a paradigm test and were statistically significant. These results seem to suggest a certain element of randomness in the statistical significance of the genetic engineering hypotheses.

The case for organic farming becomes stronger when we consider the education hypothesis. The hypothesis suggested that increased education would have a negative effect on views towards a new paradigm. However, education could indeed have the reverse effect on views if the Milgram effect dominated and professionals viewed education as a sunk cost. Education is found to be a positive influence in the organic farming model and a negative influence in the genetic engineering model. Given the other results from the models, it is likely that the Milgram effect does indeed dominate, which means that the expected sign on the education paradigm hypothesis is most likely positive, not negative. Increased tertiary education may play a role in encouraging professionals to critically assess current scientific thinking, as well as encouraging a skeptical mind. If this supposition is correct, then there is even stronger statistical evidence for organic farming to represent a new paradigm.

The age result is also interesting. Age was a positive influence on innovation acceptance in both models (though only significant for genetic engineering). These results do not support the generalised hypothesis that younger scientists are more likely to accept radical new ideas, while they support the idea that other factors (that may be correlated with age) play a more important role. Working age seems to be the more important variable, especially for organic farming.

Implications of the Presence of a Dominant Paradigm in Agriculture

There seems to be some statistical support that organic farming may represent a new agricultural paradigm, with far less support for genetic engineering. What can we learn from such a result? Does this indicate a strong conclusion that organic farming may become a dominant paradigm in agriculture in the future? A weaker conclusion may be that there are a number of sub-paradigms (or a number of research programmes) in agriculture, and organic farming is on the way to establishing itself as one such alternative research programme. It can be concluded that agricultural genetic engineering is already well-established as a research programme in agriculture, one that is different and innovative but still shares many common fundamental or 'hardcore' assumptions as the current dominant paradigm (conventional agriculture). Organic farming as a research programme also shares many similar skills and assumptions, but differs on fundamental assumptions. It is not completely mutually exclusive to the current paradigm of conventional agriculture but it is different. Therefore, organic farming has more characteristics to qualify as a sub-paradigm in agriculture than genetic engineering does.

The presence of paradigms may result in a number of problems (such as indirect scientific suppression, the delay of new ideas and the emergence of path dependence). Many writers have highlighted examples of agricultural path dependence (Wilson & Tisdell 2001, Wolffe & Recke 2000, Ison & Russell 2000, Cowan & Gunby 1996, MacIntyre 1987). It may also have serious implications for the adoption of organic farming. If organic farming indeed represents a potential new paradigm in agriculture, then this may have implications for agricultural research and infrastructure. Such research may need to change from its current conventional focus to a system of research that is more holistic, long-term, multi-disciplinary, externality encompassing, on-farm and region specific (such as described in MacRae *et al.* 1989 and 1990). At the very least, there is support for the argument that organic farming should receive more research allocation. Also, government policies, education and extension may need to change.

CONCLUSION

The current conflict in agriculture exists at both levels of paradigmatic debate, societal and scientific. Organic farming and genetic engineering represent two innovations in agriculture that had some *prime a facie* suggestion of being new paradigms. This paper uses Kuhn's analogy of a paradigm shift in science for a paradigm shift in agriculture, though they are not strictly the same thing. Qualitative evidence from a range of different writers provides support for the concept of a paradigm shift in agriculture towards organic farming, but less support for genetic engineering which has more characteristics of an innovation than a new paradigm.

This paper put forward a number of hypotheses to test whether agricultural professionals' reactions to organic farming and genetic engineering fit a predetermined response to a new paradigm. Such hypotheses are not considered to be conclusive in any way, and further work will be needed to establish tests for detecting paradigms. Using empirical results from previous studies, this paper has shown there is relatively strong quantitative evidence to support organic farming as a new paradigm, but only very weak quantitative evidence for genetic engineering. Further bivariate ordered probit regression modeling confirmed that professionals are reacting differently to the two innovations, which provides further evidence of the stability of the results discussed here. The evidence may actually support the presence of research programmes in agriculture rather than them being paradigms (in the strict sense). Therefore, the detection of organic farming as a new research programme or a new sub-paradigm does not necessarily mean organic farming will ever become the 'dominant paradigm' in agriculture, it may just remain a relatively small research programme that is an alternative to more dominant sub-paradigms. At the very least, the results in this study support a change in agricultural research, funding, education and dissemination. Research may need to move from its current conventional focus to a system that is more holistic, long-term, multi-disciplinary, externality encompassing, on-farm and region specific.

REFERENCES

Abaidoo, S. & Dickinson, H. (2002). Alternative and conventional agricultural paradigms: Evidence from farming in southwest Saskatchewan, *Rural Sociology*, Vol. 67, 114-131.

Allen, T. (1969). The Differential Performance of Information Channels in the Transfer of Technology in Gruber W. & Marquis D. (ed) *Factors in the Transfer of Technology*, MIT Press, Massachusetts.

Allen, P. (1996). "Social imperatives in organic agriculture: a reflection", in Østergaard T. (ed) (1996) *Fundamentals of Organic Agriculture*, 11th IFOAM International Scientific Conference August 11-15, 1996, Copenhagen.

Avery, D. (1995). *Saving the Planet with Pesticides and Plastic: The Environmental Triumph of High-Yield Farming*, Hudson Institute, Indianapolis, Indiana.

Barnes ,B., Bloor, D., & Henry, J. (1996). *Scientific Knowledge: a Sociological Analysis*, University of Chicago Press, Chicago.

Barnes, B. & MacKenzie, D. (1979). On the Role of Interests in Scientific Change, in Wallis R. (ed) *On the Margins of Science: The Social Construction of Rejected Knowledge*, University of Keele, Keele.

Becker, H. & Carper, J. (1956). The elements of identification with an occupation, *American Sociological Review*, Vol. 21, 341-348.

Beus, C. & Dunlap, R. (1990). Conventional versus Alternative Agriculture: The Paradigmatic Roots of the Debate, *Rural Sociology*, Vol. 55, 590-616.

Beyer ,P., Al-Babili, S., Ye, X., Lucca, P., Schaub, P., Welsch, R. & Potrykus, I. (2002). Golden Rice: Introducing the Beta-Carotene Biosynthesis Pathway into Rice Endosperm by Genetic Engineering to Defeat Vitamin A Deficiency, *Journal of Nutrition*, Vol. 132, 506-510.

Blaug, M. (1975). Kuhn versus Lakatos, or paradigms versus research programmes in the history of economics, *History of Political Economy*, Vol. 7, 399-433.

Borlaug, N. (2000). Ending world hunger. The promise of biotechnology and the threat of antiscience zealotry, *Plant Physiology*, Vol. 124, 487–490.

Broad ,W. & Wade, N. (1982) .*Betrayers of the Truth: Fraud and deceit in the Halls of Science*, Simon and Schuster, New York.

Burr, C. (2002). *The Emperor of Scent: A Story of Perfume, Obsession, and the Last Mystery of the Senses*, Random House, New York.

Busch, L. & Lacy, W. (1983). *Science, Agriculture and the Politics of Research*, Westview Press Inc, Boulder.

Buttel, F. & Goldberger, J. (2002). Gender and Agricultural Science: Evidence from Two Surveys of Land-Grant Scientists, *Rural Sociology*, Vol. 67, 24-45.

Cowan, R. & Gunby, P. (1996). Sprayed to Death: Path dependence, Lock-in and Pest Control Strategies, *The Economic Journal*, Vol. 106, 521-542.

Dahlberg, K. (ed) (1986). *New Directions for Agriculture and Agricultural Research: Neglected Dimensions and Emerging Alternatives*, Rowman & Allanheld Publishers, New Jersey.

David, P. (2000). The Political economy of public science, in Smith H. (ed) *The Regulation of Science and Technology*, Macmillan, London.

de Grazia, A. (ed) (1966). *The Velikovsky Affair*, Sidgwick & Jackson, London.

Dimara, E., Petrou, A. & Skuras, D. (2003). The socio-economics of niche market creation: A social ecology paradigm for the adoption of organic cultivation in Greece, *International Journal of Social Economics*, Vol. 30, 219-235.

Dolby, R. (1979). "Reflections on Deviant Science", in Wallis R. (ed) *On the Margins of Science: The Social Construction of Rejected Knowledge*, University of Keele, Keele.

Drahos, P. (1996). *A Philosophy of Intellectual Property*, Dartmouth Publishing Company Ltd, Aldershot.

Egri, C. (1994). *Power, Politics and the Innovation Process: Analysis of an Organisational Field in Agriculture*, PhD thesis, Department of Commerce, University of British Columbia.

Ewens, S. & Pusztai, A. (1999). Effect of Diets Containing Genetically Modified Potatoes Expressing Galanthus Nivalis Lectin on Rat Small Intestine, *Lancet*, Vol. 354, 1353-1354.

Feyerabend, F. (1988). *Against Method*, Verso, London.

Gepts, P. (2002). A Comparison between Crop Domestication, Classical Plant Breeding, and Genetic Engineering, *Crop Science* Vol. 42, 1780-1790.

Hadwiger, D. (1982). *The Politics of Agricultural Research*, University of Nebraska Press, London.

Harp, A. & Sachs, C. (1992). Public agricultural researchers: Reactions to organic, low input and sustainable agriculture, *Agriculture and Human Values*, Vol. 9, 58-63.

Hopper, K. & Thorburn, L. (2005). *2005 Bio-Industry Review: Australia & New Zealand*, Innovation Dynamics.

Huffman, W. (2001). Public-private research and development relationships: discussion, *American Journal of Agricultural Economics*, Vol. 83, 754-757.

Huffman, W. & Tegene, A. (2002). Public Acceptance of and Benefits from Agricultural Biotechnology: A Key Role for Verifiable Information, in Santaniello V., Evenson R. and Zilberman D. (eds) *Market Development for Genetically Modified Foods*, CABI Publishing, Wallingford.

Independent Science Panel (2003). *The Case for a GM-Free Sustainable World*, published by Institute of Science in Society, available at http://www.indsp.org/ISPreportSummary.php

Ison, R. & Russell, D. (ed) (2000). *Agricultural Extension and Rural Development: Breaking out of Traditions*, Cambridge University Press, Cambridge.

ISSC (1993). Food Biotechnology: A New Paradigm for Food, the Farm, and the Public, *Bio/Technology*, Vol. 11, 1584-1589.

King, D. (2004). *GM Science Review: Second Report*, GM Science Review Panel for the UK Govt, January.

Krimsky S. (2000). *Hormonal Chaos: The Scientific and Social Origins of the Environmental Endocrine Hypothesis*, The Johns Hopkins University Press, Baltimore.

Kuhn, T. (1970). *The Structure of Scientific Revolutions*, The University of Chicago Press, Chicago.

Latour, B. & Woolgar, S. (1986). *Laboratory Life*, Princeton University Press, New Jersey.

Latour, B. (1987). *Science in Action: How to Follow Scientists and Engineers through Society*, Open University Press, Milton Keynes.

Liebowitz, S. & Margolis, S. (1990). "The Fable of the Keys", *Journal of Law and Economics*, Vol. XXXIII, 1-25.

Liebowitz, S. J. & Margolis, S. E. (1994). 'Network Externality: An Uncommon Tragedy', *Journal of Economic Perspectives*, Vol. 8, 133-150.

Macilwain, C. (2004). "Organic: Is it the future of farming?" *Nature*, Vol. 428, 792-793.

MacIntyre, A. (1987). "Why Pesticides Received Extensive Use in America: A Political Economy of Agricultural Pest Management to 1970", *Natural Resources Journal*, Vol. 27, 533-578.

MacRae, R, Hill, S, Henning, J & Mehuys, G (1989). Agricultural Science and Sustainable Agriculture: a Review of the Existing Scientific Barriers to Sustainable Food Production and Potential Solutions, *Biological Agriculture and Horticulture*, 6, 173-219.

MacRae, R, Hill, S, Henning, J & Bentley, A (1990). Policies, programs, and regulations to support the transition to sustainable agriculture in Canada, *American Journal of Alternative Agriculture*, 5, 76-92.

Mäder, P., Fliebach, A., Dubois, D., Gunst, L., Fried, P. & Niggli, U. (2002). "Soil fertility and biodiversity in organic farming", *Science,* Vol. 296, 1694-1697.

Manwell, C. & Baker, C. (1988). "Information Flow in Agriculture: The Major Interest Groups and their Interactions", *Philosophy and Social Action*, Vol. 14, 51-64.

Martin, B. (1979). *The Bias of Science*, Southwood Press Pty Ltd, Marrickville, Australia.

Mazzoleni, R. & Nelson, R. (1998). "The Benefits and Costs of Strong Patent Protection: A Contribution to the Current Debate", *Research Policy*, Vol. 27, 273-284.

Merrill, M. (1983). "Eco-Agriculture: A Review of its History and Philosophy", *Biological Agriculture and Horticulture*, Vol. 1, 181-210.

Merton , R. (1973). *The Sociology of Science: Theoretical and Empirical Investigations*, The University of Chicago Press, Chicago and London.

Milgram, S. (1983). *Obedience to Authority: An Experimental View*. New York: Harper/Collins.

Miller, H. (1993). "Perception of Biotechnology Risks: The Emotional Dimension", *Bio/Technology*, Vol. 11, 1075-1076.

Mulkay, M. (1972). *The Social Process of Innovation: A Study in the Sociology of Science*, Macmillian, London.

Mulkay, M. (1979). *Science and the Sociology of Knowledge*, Allen & Unwin, London.

National Academy of Sciences (2000). *Transgenic Plants and World Agriculture,* National Academies Press, Washington.

O'Riordan, T. & Cobb, D. (2001). "Assessing the Consequences of Converting to Organic Agriculture", *Journal of Agricultural Economics*, Vol. 52, 22-35.

Popper, K. (1970). "Normal Science and its Dangers", in Lakatos I. & Musgrave A. (eds) (1970) *Criticism and the Growth of Knowledge*, Cambridge University Press, London.

Prakash, C. (2001). "The genetically modified crop debate in the context of agricultural evolution", *Plant Physiology*, Vol. 126, 8–15.

Pretty, J. (1995). "Participatory Learning for Sustainable Agriculture", *World Development*, Vol. 23, 1247-1263.

Price, D. (1969). "The Structures of Publication in Science and Technology" in Gruber W.H. and Marquis D.G. (ed) *Factors in th e Transfer of Technology*, The MIT Press, Massachusetts.

Quist,, D. and Chapela, I. (2001). "Transgenic DNA introgressed into traditional maize landraces in Oaxaca, Mexico", *Nature* Vol. 41, 541–543.

Riggs, P. (1992). *Whys and Ways of Science: Introducing Philosophical and Sociological Theories of Science*, Melbourne University Press, Melbourne.

Shepherd, M., Pearce B., Cormack B., Philipps L., Cuttle S., Bhogal A., Costigan P. and Unwin R. (2003). *An Assessment of the Environmental Impacts of Organic Farming*. DEFRA-funded project OF0405, DEFRA, London.

Stolze, M., Piorr, A., Haring, A. & Dabbert, S. (2000). "The environmental impacts of organic farming in Europe", Organic Farming in Europe, Economics and Policy, 6, University of Hohenheim: Stuttgart-Hohenheim.

Tisdell, C. (1991). *Economics of Environmental Conservation: Economics for Environmental and Ecological Management*, Elsevier, Amsterdam.

Trewavas A. (2001). "Urban myths of organic farming", *Nature*, Vol. 410, 409-410.

Turner, L. (2004). "Biotechnology as religion", *Nature Biotechnology*, Vol. 22, 659-660.

van den Bosch, R. (1978). *The Pesticide Conspiracy: An alarming look at pest control and the people who keep us "hooked" on deadly chemicals*, Doubleday & Company Inc, Garden City, New York.

Wheeler, S. (2007). "Contrasting the beliefs of Australian agricultural professionals about the benefits and costs of genetic engineering and organic agriculture", *Australian Journal of Experimental Agriculture*, Vol. 47, 1389–1396.

Wheeler, S. (2008). "What Influences Agricultural Professionals' Views towards Organic Agriculture?" *Ecological Economics*, Vol. 65, 145-154.

Wheeler, S. (pending). "Exploring the Influences on Australian Agricultural Professionals' Genetic Engineering Beliefs: An Empirical Analysis", *Journal of Technology Transfer*, (DOI: 10.1007/s10961-008-9094-y).

Wilson, S. &Barnes, I. (1995). "Scientists' Participation in Environmental Policy", *Search*, Vol. 26, 270-273.

Wilson, C. & Tisdell, C. (2001) "Why farmers continue to use pesticides despite environmental, health and sustainability costs", *Ecological Economics*, Vol. 39, 449-462.

Wolffe, H. & Recke, G. (2000). "Path Dependence and Implementation Strategies for Integrated Pest Management", *Quarterly Journal of International Agriculture*, Vol. 39, 149-171.

Wynen, E. (1996). "Research Implications of a Paradigm Shift in Agriculture: the case of Organic Farming" *Resource and Environmental Studies*, 12, Centre for Resource and Environmental Studies, ANU.

Wynen, E. (2003). *Organic Agriculture in Australia – Levies and Expenditure*, report prepared for RIRDC, No. 03/2002, March

APPENDIX ONE

Explanatory Variable Table

Variable	*Explanation*
Age Working Age Ethnicity Gender Farm Background Tertiary Education Years Salary	Actual age in years squared Years spent working in current agricultural area Dummy variable, 1 = non-European, 0 = European Dummy variable, 1 = male, 0 = female Dummy variable, 1 = farm background, 0 = otherwise Years of post-secondary education Salary of professionals (midpoint of salary band of job classification level)
Scientist Leader Extension role Natural resource focus CSIRO	Dummy variable, 1 = scientist, 0 = otherwise Dummy variable, 1 = leader, 0 = otherwise Dummy variable, 1 = extension role, 0 = otherwise Dummy variable, 1 = natural resource management focus, 0 = otherwise Dummy variable, 1 = employed at CSIRO, 0 = otherwise
Target OA/GE knowledge OA/GE research relevance Media main source info for GE Scientific info main source for OA/GE OA farms main source info	Dummy variable, 1 = targeted sample, 0 = otherwise Scalar variable of OA (or GE) knowledge, where 1= no knowledge and 5 = detailed and practical knowledge % indicating the research relevance of OA (or GE) in the past 3 years Dummy variable, 1 = media main source for GE info, 0 = otherwise Dummy variable, 1 = scientific information (journals, peers, conferences etc) as first main source for OA (or GE) information, 0 = otherwise Dummy variable, 1 = organic farms first main source for info, 0 = otherwise
Views on CA's sustainability	Scalar variable of response to "Conv. agric. farming in Aus. is generally not environment. sustainable", where 1= strongly disagree to 5 = strongly agree
OA/GE benefit-cost ratio	Ordered dependent variable of response to "do you think the benefits of OA/ GE are greater than their costs/risks?", where 0 = no, 1 = unsure, 2 = yes
Views on OA profitability Views on OA environmental sustainability Views on OA product quality Views on OA innovativeness Views on OA yields	Scalar variable of response to the statement that "CA is more financially profitable than OA", where 1= strongly disagree and 5 = strongly agree Scalar variable of response to the statement that "OA is more environmentally friendly than CA", where 1= strongly disagree and 5 = strongly agree Scalar variable of response to the statement that "OA produces better quality of food than CA", where 1= strongly disagree and 5 = strongly agree Scalar variable of response to the statement that "OA farming is a return to pre-1950s agriculture", where 1= strongly disagree and 5 = strongly agree Dummy variable where 1 = belief that conversion from CA to CA results in falls of yield over 30
Views on agric. IPR and Patents Views on agric. private funding effects	Scalar variable of response to "the increased use of intellectual property rights & patenting activity is a good thing for agric. research", of 1 to 5 Scalar variable of response to the statement that "private funding of agricultural research does not influence research outcomes", of 1 to 5
Views on GE profitability Views on GE environmental sustainability Views on GE product quality Views on GE yields Views on GE long-term research needs	Scalar variable of response to the statement that "GE is more financially profitable than CA", where 1= strongly disagree and 5 = strongly agree Scalar variable of response to the statement that "GE is more environmentally friendly than CA", where 1= strongly disagree and 5 = strongly agree Scalar variable of response to the statement that "GE produces better quality of food than CA", where 1= strongly disagree and 5 = strongly agree Dummy variable where 1 = belief that conversion from CA to GE results in increases of yields over 30% Scalar variable of response to "there needs to be more research on LT consequences of GE products before further general release", of 1 to 5

Bivariate Ordered Probit Regression Results

	Coef.	*z*	*P>|z|*
GENETIC ENGINEERING			
Age	-0.14	-1.84	0.07
Age^2	0.00	1.97	0.05
Working Age	-0.01	-0.55	0.59
Ethnicity	1.52	2.13	0.03
Female	0.22	0.98	0.33
Having a Farm Background	-0.51	-2.66	0.01
Years of Tertiary Education	-0.05	-0.74	0.46
Salary	0.00	0.75	0.45
Job title is Scientist	0.62	2.55	0.01
Classified as a Leader in their Position	-0.05	-0.15	0.88
Working predominantly as an Extension Officer	0.39	1.67	0.09
Employed by a state organization	0.15	0.47	0.64
Part of the Targeted Sample	-0.13	-0.55	0.58
Working in the area of Natural Resource Management	-0.08	-0.31	0.76
Working in the area of Broadacre and/or Grazing	0.13	0.50	0.62
Employed by CSIRO	0.14	0.28	0.78
Genetic Engineering Knowledge	-0.06	-0.46	0.64
% of Research Time Relevant to Genetic Engineering	0.01	2.10	0.04
Views on Conventional Agriculture's Sustainability	-0.24	-2.62	0.01
Views on agric. intellectual property rights & patents	0.23	2.37	0.02
Views on private funding of agricultural research	-0.15	-1.51	0.13
Views on Genetic Engineering's Profitability	0.39	2.69	0.01
Beliefs on uncertainty of GE and need for long-term research	-0.65	-4.59	0.00
Views on Genetic Engineering Food Quality	0.11	0.80	0.43
Views on Genetic Engineering Environmental Sustainability	0.50	3.48	0.00
Naming the Media as the first main source of Info.	0.28	1.11	0.27
Naming Scientific sources as first main source of Info.	-0.08	-0.28	0.78
ORGANIC FARMING			
Age	-0.13	-1.50	0.13
Age^2	0.00	1.63	0.10
Working Age	-0.03	-1.68	0.09
Ethnicity	0.60	1.14	0.25
Female	-0.18	-0.65	0.52
Having a Farm Background	-0.03	-0.13	0.90
Years of Tertiary Education	0.13	2.10	0.04
Salary	0.00	-0.11	0.91
Job title is Scientist	0.00	0.01	0.99
Classified as a Leader in their Position	0.19	0.48	0.63
Working predominantly as an Extension Officer	-0.06	-0.24	0.81
Employed by a state organization	0.02	0.06	0.95
Part of the Targeted Sample	0.64	1.90	0.06
Working in the area of Natural Resource Management	0.35	1.13	0.26
Working in the area of Broadacre and/or Grazing	0.04	0.13	0.89
Employed by CSIRO	-0.88	-1.77	0.08
Organic Agriculture Knowledge	0.24	1.76	0.08
% of Research Time Relevant to Organic Farming	0.01	1.14	0.25
Views on Conventional Agriculture's Sustainability	0.12	1.10	0.27
Views on agric. intellectual property rights & patents	0.09	0.82	0.41
Views on private funding of agricultural research	0.02	0.15	0.88
Views on Organic Farming's Profitability	0.69	5.07	0.00

Bivariate Ordered Probit Regression Results (Continued)

	Coef.	*z*	*P>|z|*
Views on Organic Farming's Innovativeness	0.28	2.36	0.02
Views on Organic Farming's Food Quality	-0.36	-2.68	0.01
Views on Organic Farming's Environmental Sustainability	-0.25	-1.57	0.12
Naming the Media as the first main source of Info.	-1.15	-3.55	0.00
Naming Scientific sources as first main source of Info.	-0.35	-1.32	0.19
/cut11	-3.38	1.99	
/cut12	-2.05	2.00	
/cut21	-0.55	1.94	
/cut22	0.75	1.95	
Number obs.	185.00		
Log Likelihood	-252.59		
Wald Chi2	128.85		
Prob>chi2	0.00		
rho	0.08	0.14	
Wald test of dep. Equns chi2(1)	0.30		
Prob > chi2	0.58		

In: Organic Farming: Methods, Economics and Structure
Editors: Michael Nelson and Ivan Artamova
ISBN 978-1-60692-864-6

Chapter 3

PILL MILLIPEDES – AN OVERVIEW

K.R. Sridhar[1] and B.S. Kadamannaya[1]
Microbiology and Biotechnology, Department of Biosciences,
Mangalore University, Mangalagangotri, Mangalore 574 199, Karnataka
[1]Department of Zoology, Field Marshal K.M. Cariappa College, Madikeri 571 201,
Karnataka, India

ABSTRACT

Global estimate of described millipedes is about 12,000 species out of a total estimate 80,000 species in all the continents except for Antarctica. Millipede collection is available in 268 institutions of 143 countries but understudied in Africa, Central and South America and Asia. The family Sphaerotheriidae (Order: Sphaerotheriida) consists of giant millipedes possessing 13 body segments with 21 pairs of walking legs called 'pill millipedes' as they attain sphere shape on conglobation. The subfamilies and tribes have been separated based on the shape of female vulva and stridulating organ. They predominantly hail from Southern Hemisphere (Indo-Australia, Madagascar and South Africa) and wiped out in the intermediate regions due to unsuitable environment. Limited dispersal ability led to endemism of the Malagasy giant pill millipedes (*Sphaeromimus* and *Zoosphaerium*), which exhibits female gigantism and Batesian mimicry. Pill millipedes are hosts for a variety of microbes and from food to faecal matter the bacterial load increases and fungal load decreases. Presence of ergosterol in the faecal pellets of *Glomeris* indicates intensive digestion of fungi in the gut passage. *Glomeris* consume up to 1.7-10% of annual litter production, which amounts to 10 times the body mass. *Arthrosphaera* prefers mixed litter than monolitter and the mixed litter ingestion and faecal pellet production is higher than other tropical millipedes. Increased nitrogen, phosphorus and potassium, narrow C/N ratio and shift of pH towards neutral in mixed litter compost produced by *Arthrosphaera* reveals its value as alternative to vermicompost. Millipedes are influenced by the forest under storey and act as buffer stock for significant increase of calcium and magnesium and augment mineral recycling. Clear evidence has been emerged on the invasion of *Arthrosphaera* from evergreen and semi-evergreen forests to plantations in Western Ghats in India. Long-term intra-site heterogeneity in organic matter input in mixed plantations might have facilitated such

[1] Corresponding author: E-mail: sirikr@yahoo.com.

invasion. *Arthrosphaera* is known for a narrow range of distribution and exhibits single site endemism in the Peninsular India. They are sensitive to litter chemistry at narrow spatial scales than the earthworms. In spite of conducive ecological conditions, millipedes are highly sensitive and fail to overcome the limitation of even a single edaphic factor (e.g. water deficit, soil texture, litter thickness). Such sensitivity of *Arthrosphaera* qualifies its suitability to serve as indicator of soil quality loss or restoration in managed plantations or forests.

Keywords: pill millipedes, morphology, cytology, diversity, distribution, bioconversion, faecal matter, compost, microbes, edaphic factors, endemism, conservation

INTRODUCTION

Forest ecosystems are the mosaic of sylvogenetic phases and humus profile development is the result of functional connections among vegetation, saprophagous fauna and soil microbes (Oldeman, 1990). Assessment of a few centimetres of topsoil in forest ecosystem reveals such dynamic process (Bernier and Ponge, 1994). Study of humus profiles in natural forests, degraded forests and plantations indicates the human interference on natural ecosystem balance. Decomposition of forest floor litter is dependent on the type of humus. For example, the rate of litter disappearance is slower in moder and mor than mull humus (van der Drift, 1963). Suitable tree plantation is necessary to achieve sufficient amount of litter accumulation to support diverse and abundant litter fauna (Vohland and Schroth, 1999), which helps optimum utilization of natural resources to achieve sustainable benefit in modern agriculture and agroforestry.

Although the global annual production of renewable lignocellulosic residues amounts to about $20\text{-}50\times10^9$ tonnes, only 4×10^9 tonnes are used efficiently (Kelley and Peterson, 1997). The major constraints of digestibility of lignocellulosic residues are transformation of lignin or cellulose. One of the potential means of transforming lignin is composting (to achieve narrow C/N ratio). Composting lignocellulosic materials is possible with combined effort of soil fauna and microorganisms, where millipedes play a vital role. The soil organic matter comprise of carbon reservoir of about 1760 Pg (O'Neill, 1993), which is less than 0.01% of the global carbon pool (more than 99% exists in the ocean and ocean sediments) and amounts to 24% of the terrestrial carbon pool. The high turnover of carbon in soils leads to changes in assimilation and mineralization of global carbon cycle. In the context of climate change, factors influencing the biological immobilization of organic matter by soil fauna are important in controlling the elevation of CO_2 concentrations (Verhoef and Brussard, 1990; O'Neill, 1994; Yeates *et al.*, 1997). Although, several studies on the role of microorganisms in CO_2 immobilization are available (Cheng and Johnson, 1998; Cardon *et al.*, 2001; Cheng, 1999), little attention has been invested to understand the role of meso- and macrofauna in the soil habitat. For instance, the top layer (F layer) of northern temperate forest soil is made up to 90% of the residual organic matter consist of arthropod faecal matter (Bocock, 1963; Nicholson *et al.*, 1966). The precise changes in the biochemical composition of fresh organic matter entering the soil resulting from digestion by arthropods are not clearly known and there is a considerable gap in our knowledge on the initial stages of diagenesis in terrestrial soils (Rawlins *et al.*, 2006).

As global biodiversity is mainly concentrated in tropical rainforests, assessment of impact of tropical rainforest destruction and fragmentation on the biodiversity assumes importance (Gaston, 2000). The loss as well as fragmentation of habitat reduces the complexity of habitat, which affects sustainability of biodiversity (Groombridge and Jenkins, 2000). Such effects on invertebrates are very important as they constitute bulk of the biodiversity and play an important role in ecosystem processes and functioning (Groombridge and Jenkins, 2000; Miller, 1993). In spite of substantial contribution of invertebrates to global biodiversity and ecosystem function, the patterns and causes of saprophagous faunal responses to tropical rainforest destruction and fragmentation remain poorly understood. Among the large animals in soil ecosystems, the earthworms have been most widely investigated on biodegradation of natural residues through the physical alteration of plant tissue and the soil matrix to enhance the microbial activity (Coleman and Crossley, 1996).

The phylum, Arthropoda constitutes earth's major biological diversity (Golovatch *et al.*, 1995; Stock, 1993; Wilson, 1992) and their vital roles in ecosystem services is plant litter decomposition and humus formation (Giller, 1996; Plowman, 1990). Involvement of arthropod assemblages in soil food webs helps to consider them as indicators to assess the human impacts on forest management practices (Kremen *et al.*, 1993). The class, Diplopoda (millipedes) is the largest among animal kingdom and third largest group in terrestrial arthropods after Insecta and Arachnida (Golovatch *et al.*, 1995). The Diplopoda also ranks among the most diverse groups of terrestrial organisms with less understood morphology, phylogeny and diversity compared to other arthropod groups (Sierwald and Bond, 2007). This review aims at providing comprehensive information on pill millipedes (morphology and cytology, diversity and distribution, food and feeding habits, organic matter processing, microbial association, edaphic factors and endemism). In view of importance of pill millipedes in soil qualities, restoration, conservation and human interference have been dealt in general context.

MILLIPEDES

The term 'millipede' is a misnomer because they lack thousand legs, but possess less than 50 pairs or a maximum of 375 pairs (Enghoff, 1990). They have been classified under Diplopoda as they possess two pairs of legs per segment. At present, eight orders of millipedes have been considered as extinct (Amynilyspedida, Archidesmida, Arthropleurida, Cowiedesmida, Eoarthropleurida, Euphorberiida, Microdecemplicida and Pleurojulida) (Sierwald and Bond, 2007). The likely oldest millipedes in mid-Silurian and Lower Devonian were *Cowiedesmus eroticopodus* and *Archidesmus macnicoli* respectively. A wide range of millipedes (including pill millipedes) have been recovered from Carboniferous period (order, Amynilyspedida) (Hannibal and Feldmann, 1981). After Carboniferous period, a decline in fossil records was noted, while in the late Cretaceous, several spirobolids have been described from Mangolia (Kraus, 1974; Dzik, 1975). In the Oligocene, amber preserved millipedes have been collected (Shear, 1981). From Baltic amber, Callipodida, Chordeumatida, Glomerida, Polyxenida and Polyzoniida are known (Sierwald and Bond, 2007).

Depending on the structures, five morphotypes have been distinguished in millipedes by Kime and Golovatch (2000): (1) Bristly millipedes or bark-dwellers (Polyxenidae); (2) Pill millipedes or rollers (Glomeridae and Sphaerotheriidae); (3) Bulldozers or rammers (Julidae); (4) Wedge or litter-splitters (Polydesmidae) and (5) Borers (Platydesmidae). Millipedes are grouped in to five ecomorphological types by Hopkin and Read (1992): (1) Bulldozers or rammers are long cylindrical forms use their broad head as a ram to burrow the soil (Julidae, Spirobolidae and Spirostreptidae); (2) Wedge types are short forms with laterally expanded segments have keels with tapered anterior part helps burrowing (Polydesmidae); (3) Borers have free sternites and pleurites with strongly tapered anterior portion helps forward movement and to widen the crevice (Chordeumatid and Colobognath); (4) Rollers roll into sphere as defence, which reduces water loss by minimizing the exposed surface (Glomeridae and Sphaerotheriidae); (5) Bark dwellers are flattened and small, do not burrow and their flattened nature helps to live beneath the bark.

Millipedes show a wide range of life styles, reproductive strategies and lifespan (between 1 and 11 years) (Dangerfield and Telford, 1991; Carrel, 1990). The extended lifespan has been linked to availability and consumption of poor quality food source (Blower, 1985). The body size ranges from 0.2-30 × 0.1-4 cm (Hopkin and Read, 1992). Increased body size has been noted in individuals inhabiting the high altitudes (Enghoff and Báez, 1993). The major studies on diplopods have been carried out in temperate regions (Cloudsley-Thompson, 1988). Tropical millipedes possess larger body size than temperate counterparts (e.g. Spirostreptidae and Julidae: 40 vs. 2 g) (Dangerfield and Telford, 1989; Lawrence, 1984). The large body size is known to confer several advantages such as tolerance to elevated moisture stress (Baker, 1980) and increase the burrowing ability (Manton, 1977). The body size is also known to affect energy acquisition and utilization, and in turn the rates of decomposition of organic matter (Peterson and Luxton, 1982). The pill millipedes (are also known as rollers) roll into sphere or pill (conglobation) as defence behaviour (e.g. Sphaerotheriidae, Glomeridae).

The global estimates of described millipede nominal species are up to 12,000 (out of 80,000) assigned to 2947 genera and 145 families distributed on all continents except for Antarctica (Bueno-Villegas *et al.*, 2003; Golovatch *et al.*, 1995; Heywood and Watson, 1995; Hoffman 1980; Hoffman *et al.*, 2002; Sierwald and Bond, 2007; Shelly, 2003; WCMC, 1992). Millipede specimen is available from 268 biological collection institutions housed in 143 countries (Sierwald and Reft, 2004). Out of them, nine are major collections possessing over 20,000 specimen and 77 contain over 1000 specimen. Africa, Central and South America and Asia are the underserved with regard to millipedes.

Millipedes contribute significantly to the turnover of organic matter through ingestion and fragmentation of large quantities of plant litter and incorporating into the mineral soil. The mechanical fragmentation of organic matter influences the rates of decomposition through conditioning of plant remains for microbial processing (Swift *et al.*, 1979). Millipedes are known to be highly sensitive even to a single factor (e.g. litter thickness), and thus investigation of their richness and diversity helps to precisely assess the positive or negative impacts of agricultural or forest management practices. Species abundance, dynamics, diversity and biomass of millipedes constitute reliable indices to assess the impact of forest and agricultural perturbations.

Figure 1 gives a general classification scheme of super class, Myriapoda, while Figure 2 the infraclass, Pentazonia.

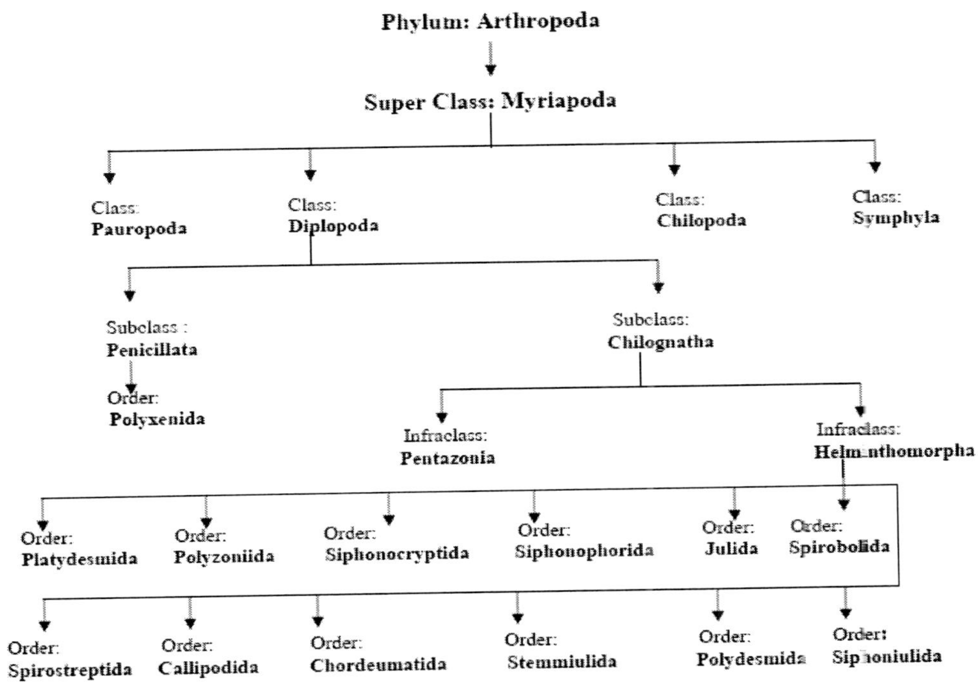

Figure 1. General scheme of classification of super class, Myriapoda.

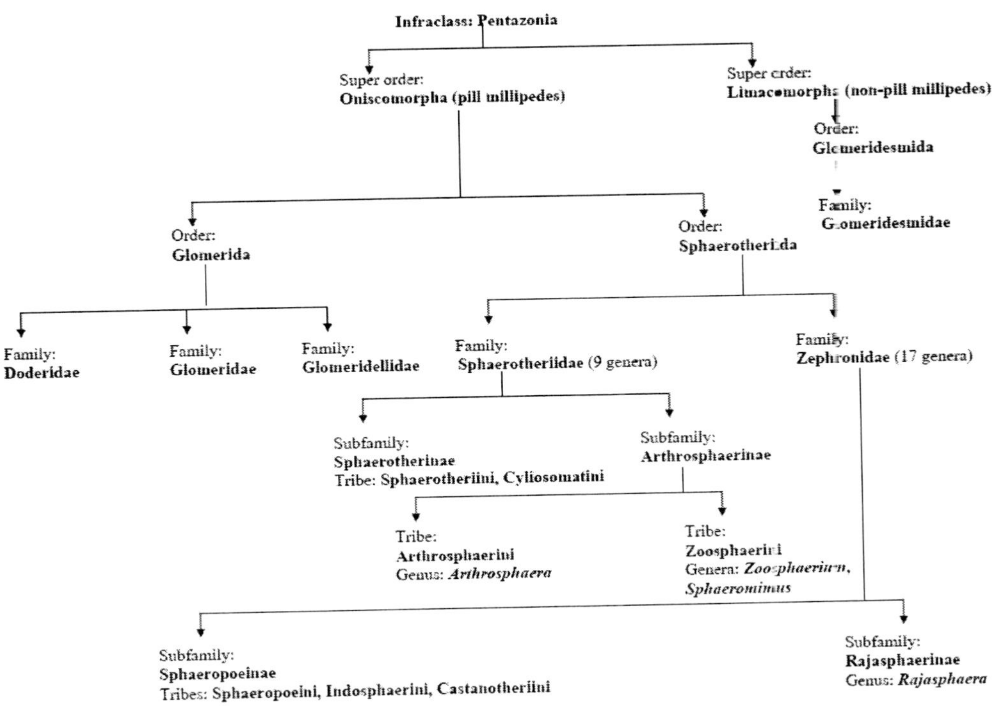

Figure 2. General scheme of classification of infraclass: Pentazonia.

The order Sphaerotheriida consists of two families (Sphaerotheriidae and Zephronidae). Family Sphaerotheriidae has nine genera grouped under two subfamies (Sphaerotherinae and Arthrosphaerinae) and four tribes (Arthrosphaerini, Zoosphaerini, Sphaerotherini and Cyliosomatini). The family Zephronidae is subdivided into two subfamilies (Sphaeropoeinae and Rajasphaerinae) with 17 genera grouped under four tribes (Sphaeropoeini, Indosphaerini, Castanotheriini and Rajasphaerini).

Morphology and Cytology

The body of millipedes is divisible into three parts: a distinct head, trunk and rest of the body. The first trunk segment is legless and called collum. Each of the next three segments possesses one pair of legs, while the remaining segments with two (this is the main diagnostic feature of the Diplopoda). The head consists of a pair of antennae, a lateral patch of kidney-shaped ocelli (found in most millipede groups), which are considered as compound eye derivatives (Hopkin and Read, 1992; Spiess, 1981). At the base of antennae, there is a pair of Tomosvary organ, its function is not clearly known (Hennings, 1906). Mouthparts of millipedes consist of a pair of mandibles and the plate-like gnathochilarium, which has been considered as the fused first maxilla (Kaestner, 1963; Hennig and Mickoleit, 1986). As the mouthparts of millipedes are weak, they tend to confine feeding on highly moist and rotten leaf and woody debris. Structures associated with sperm transfer exist on the ventral side of the body. In all adult males, except for penicillates and pentazonians one or both pairs of limbs on the seventh ring are modified to form species specific gonopods (Hopkin and Read, 1992). The functional morphology of the gonopods is poorly understood (Sierwald and Bond, 2007). In the members of male Pentazonia, the last one or two pairs of legs are modified to form telopods or claspers, which assist in transfer of sperms (Hopkins and Read, 1992). The apertures of the genital ducts are on the coxae of the second legs. The main advantage of millipedes than other terrestrial invertebrates are double leg pairs, which helps in forward movement, calcified head and leverage of labrum drives into rotten wood or soil particles (Enghoff, 1990).

The family Sphaerotheriidae of the order Sphaerotheriida consists of giant pill millipedes, which have 13 body segments and 21 pairs of walking legs. The tergites (segments) are strongly curved and horney. Each tergite (except for pygidium) is overlapped slightly by the preceding tergite. The popular name, 'pill millipede' has been referred to these animals because they are capable of assuming a spherical shape during conglobation and also due to the practice of coating enrolled millipedes with sugar for sale as pills with alleged therapeutic properties (Holloway, 1956). In the process of conglobation, the head and ventral structures are completely covered by the pygidium, which helps in protection of the head and sense organs. Further, special closing ledges (locking carinae) are located on the inner side of the tergites. In Madagascar, representatives of the order Sphaerotheriida exhibit gigantism. Some species reach a length up to 100 mm and comparable to an orange when rolled up (Wesener and Sierwald, 2005a). Stridulation (rubbing the last legs against the sides of last tergites produce sound) is an interesting phenomenon in some pill millipedes. Stridulation was seen in males of *Arthrosphaera* sp. 1 located at the Agumbe, Western Ghats of Karnataka (Ashwini, 2003). In Malagasy (Madagascar) pill millipedes, both sexes can stridulae (Wesener and

Sierwald, 2005a). The male stridulatory organ is the harp, while in female it is wash board which are located at the anterior telopod and subanal plate respectively.

Systematic treatments of millipedes including the order Spharottheriida are scanty (only about 12,000 described species have been assigned to 2947 genera, 145 families with about 68% of all genera having single or two species and 715 generic names are currently considered synonyms) (Shelley, 2003; Sierwald and Bond, 2007). A tangible classification of the order Sphaerotheriida has been given by Jeekel (1974). Subsequently, Hoffman (1976) and Mauries (2001) have modified Jeekel's (1974) classification. The subfamilies and tribes have been separated based on the shape of the female vulva and the stridulating organ. Enghoff's (1984) character matrix of the phylogeny of millipedes has been reexamined by Sierwald *et al.* (2003). The complete listing of millipede families has been given by Shelley (2003) with a classification scheme.

Recently, revision of the genus *Spheromimus* (Sphaerotheriidae) was proposed by Wesener and Sierwald (2005a) in the light of the new species described using the scanning electron microscopic observations. *Sphaeromimus* shares certain characters with members of the Indian genus *Arthrospharea*, which is placed in the tribe Arthrosphaerini. Such characters includes: flat and broad 6th antennomere and four jointed anterior telopods. On the other hand, *Sphaeromimus* shares characters with the genus *Zoosphaerium* (particularly the presence of harp), which belongs to the tribe *Zoosphaeriini*. The members of the order Glomerida are also pill millipedes, which have main distribution at the Northern Hemisphere and Southeast Asia (Golovatch *et al.*, 2006). They are smaller than Sphaerotheriida and also possess 13 body segments, but the second and third tergites are fused to form a very large plate (Hopkin and Read, 1992). They also possess a horseshoe-shaped Tomosvary organ like Sphaerotheriida at the base of the antenna. If ocelli are present, they will be in the form of a row unlike Sphaerotheriida. Sufficient morphological characteristics are lacking in the current classification scheme of pill millipedes and thus demand more characteristics to define monophyletic clades within Sphaerotheriida (Wesener and Sierwald, 2005a). Mouthparts of giant pill millipedes were observed using scanning electron microscopy and illustrated for the first time by Wesener and Sierwald (2005a). Among Arthropoda, based on molecular loci [sequences of three nuclear ribosomal genes, three nuclear protein-coding genes, and two mitochondrial genes (one protein coding, one ribosomal)] with a comprehensive data set of morphological, developmental, ultrastructural and gene-order characters supported the crustacean-hexapod clade and recognizes pycnogonids as sister to other euarthropods indicating monophyly of Myriapoda and Mandibulata (Giribet *et al.* 2001). Recent molecular studies reveal the developmental morphology of *G. marginata* (Janssen and Damen, 2006; Pioro and Stollewerk, 2006; Prpic, 2004). Based on three nuclear protein-encoding genes from 59 myriapod species, Regier *et al.* (2005) tried to throw some light to resolve phylogenetic relationships.

Even though morphological characteristics are of taxonomic significance, chromosome cytology is of immense value in millipede taxonomy (Chowdaiah, 1966a). The early investigations with preliminary techniques on diplopod cytology was established by Oettinger (1909), Sokoloff (1914) and Bessiere (1948), while Indian diplopods by Natarajan (1959), Chowdaiah (1966a, 1966b, 1966c, 1967, 1969), and Chowdaiah and Kanaka (1969). Later, classical cytological techniques have been employed by Chowdaiah and Kanaka (1974, 1979) and those were the basis for significant growth in the cytology of Indian Diplopoda. Achar and Chowdaiah (1979) have successfully introduced air drying technique to study the diplopod chromosomes. This classical technique including Giemsa differential banding (Achar, 1983a, Achar and Chwodaiah, 1980) are the two important developments in the field of diplopod cytogenetics. Subsequently, karyological analysis of some species of Diplopoda including pill millipedes of the Western Ghats has been attempted by Achar (1983b).

The diploid chromosome number ranges between 8 and 30, the highest being represented by the pill millipedes (16-30). It is significant to note that with a few exceptions, all the chromosomal complements are multiples of 4 or 6 (Achar, 1987). The major chromosomal rearrangements in pill millipedes consist of Robertosonian changes, pericentric inversions and polyploidy which appears to have played significant role in the evolution of diplopids (Achar, 1987). The chromosomes of pill millipedes also exhibits interesting phenomenon of heteropycnosis (Achar, 1986) and non-staining gaps in the chromosomal bivalents during pre-metaphase/metaphase stages in *Arthrosphaera disticta* (Achar, 1987). The chromatid bridges in pill millipedes were also reported by Chowdaiah (1969). Table 1 gives chromosome number of some pill millipedes.

Diversity and Distribution

The individuals of the order Sphaerotheriidae are giant pill millipedes (measure up to 10 cm length) found predominantly in Southern Hemisphere and known to be inhabitants of the Indo-Australia, Madagascar and South Africa, subsequently wiped out in the intermediate regions due to unsuitable environmental conditions.

Table 1. Chromosome number of some pill millipedes

Family, genus and species	Chromosome number (2n)	Reference
Glomeridae		
Glomeris annulata	20	Bessiere (1948)
Glomeris connexa	16	Warchalowska-Sliva *et al.* (2004)
Glomeris hexasticha	16	Warchalowska-Sliva *et al.* (2004)
Sphaerotheriidae		
Arthrosphaera bicolor	30	Chowdaiah and Kanaka (1974)
A. craspedata	30	Chowdaiah and Kanaka (1974)
A. dalyi	30	Chowdaiah and Kanaka (1974); Achar (1986)
A. davisoni	26	Achar (1986)
A. disticta	28	Chowdaiah and Kanaka (1974); Achar (1986)
A. gracilis	28	Chowdaiah and Kanaka (1974)
A. hendersoni	30	Chowdaiah and Kanaka (1974)
A. lutescence	26	Achar (1986)
A. magna	30	Achar (1986)
A. nitida	30	Achar (1986)
A. zebraica	26	Chowdaiah (1966a)

In Sphaerotheriida, 330 number of nominal species collection exists, which have distribution in Southern Hemisphere (Southern Africa, Madagascar, India, Scuthern Asia, Australia and New Zealand) (Sierwald and Bond, 2007).

Sphaerotheriidae showed a disjunct distribution in South-Eastern Australia, India, South Africa and Madagascar and New Zealand (Jeekel, 1974). Giant pill millipedes belongs to Sphaerotheriidae are commonly associated with trees in Africa (Haacker and Fuchus, 1972). The first phylogenetic analysis revealed a relationship between Malagasy and Indian taxa (Wesener, 2006). The stridulation organs of Sphaerotheriidae are not homologous, but evolved thrice independently. These organs in the Malagasy-Indian subfamily (Arthrosphaerinae) evolved independently in males as well as females. The significance of the stridulation organs is not well studied, although stridulation behavour plays an important role in the mating. The present distribution of the clades of Sphaerotheriidae coincides perfectly the breaking off of Gondwanaland about 180-90 mya. It means the existing clades now are dating back to the Paleozoic era. Limited dispersal abilities seem to be the main reason for small area of distribution in the Malagasy giant-pill millipedes (*Sphaeromimus* and *Zoosphaerium*). Two more features found in Malagasy giant-pill millipedes include female gigantism and Batesian mimicry.

The genus *Arthrosphaera* (Sphaerotheriidae) is endemic to tropics and abundant in the upper soil horizons of forests in moist regimes. They are confined to widely dispersed lands on tropics of Indo-Australia, South Africa and Madagascar (Pocock, 1892). Peninsular India inhabits about 46 species of *Arthrosphaera* (Ashwini, 2003; Achar, 1980, 1986; Attems, 1936; Chowdaiah, 1969; Janardanan and Ramachandran, 1983; Hosabale, 2000; Pocock, 1892, 1899) (Table 2). They are inhabitants of regions with adequate rainfall in forests of the Western Ghats and Eastern Ghats (Maharashtra, Karnataka, Kerala, Tamil Nadu and Andhra Pradesh) of India and Sri Lanka. Certain species of *Arthrosphaera* are known from high altitudinal range (1430-2130 m) (e.g *A. carinata*, *A. davisoni*, *A. hendersoni*, *A. nitida*, *A. scholastica*). *Arthrosphaera magna* is endemic to forest and coastal locations of Southern India (Maharashtra, Karnataka and Tamil Nadu) (Achar, 1980; Ashwini, 2003; Attems, 1936). The giant pill millipede, *Arthrosphaera brandti* found only in the immediate vicinity of Amani (Tanzania) has never been found elsewhere (Enghoff, 1977). The species is known from Sri Lanka and probably introduced species due to planting a large number of exotic tree species.

A survey of 21 sampling sites of the Western Ghats (500-910 m), foothills of Western Ghats (156-450 m) and west coast (120-150 m) during post-monsoon season revealed the occurrence of a variety of *Arthrosphaera* spp. in the Western Ghats and foothill locations, while the coastal locations were devoid of it (Ashwini and Sridhar, 2008). Adult pill millipedes were usually in upper soil strata along with litter layers, while juveniles restricted to 5-10 cm soil depth. Abundance and biomass was significantly higher in the Western Ghats than in the foothills. Among the *Arthrosphaera* species, *A. magna* showed a wide distribution in foothills followed by *Arthrosphaera* sp. 2 in Western Ghats and foothills. *Arthrosphaera* sp. 3 was found in Western Ghats as well as in foothills, while *Arthrosphaera* sp. 1 confined to a single location of the Western Ghats. None of the samples showed more than one species of *Arthrosphaera* support their habitat specificity and endemism (Ashwini and Sridhar, 2008). Soil organic carbon in the Western Ghats and foothills and biomass of *Arthrosphaera* were significantly correlated (p=0.01; r=0.45). Although the coastal samples lack pill millipedes, other millipedes were abundant indicating the habitat specificity of pill millipedes.

Table 2. Pill millipede, *Arthrosphaera* reported from the Peninsular India

Pill millipede	Distribution	Altitude (m)	Reference
Arthrosphaera atrisparsa Butler	Bombay, Maharashtra, India		Pocock (1899)
Arthrosphaera aurocincta Pocock	East India		Pocock (1899)
Arthrosphaera bicolor Pocock	Salem and Shevaroy Hills, Tamil Nadu, India		Pocock (1899)
	Southern India		Chowdaiah and Kanaka (1974)
Arthrosphaera brandti Humbert	Sri Lanka		Pocock (1899)
	Tamil Nadu, India		Attems (1936)
Arthrosphaera carinata Attems	Karnataka, India	1370-1880	Attems (1936)
Arthrosphaera corrugata Butler	Sri Lanka		Pocock (1899)
Arthrosphaera craspedota Attems	Kerala, India	520-980	Attems (1936)
	Southern India		Chowdaiah and Kanaka (1974)
Arthrosphaera dalyi Pocock	Palani Hills, Tamil Nadu, India	1830	Pocock (1899)
	Southern India		Chowdaiah and Kanaka (1974)
	Kerala, India		Janardanan and Ramachandran (1983)
	Algarkovil Hills, Tamil Nadu, India		Achar (1986); Paulpandian (1966)
Arthrosphaera davisoni Pocock	Annamalai Hills, Tamil Nadu, India	1430	Pocock (1899)
	Khandla Hills, Maharashtra, India		Achar (1986)
Arthrosphaera dentigera Verh.	Sri Lanka		Attems (1936)
Arthrosphaera disticta Pocock	Shevaroy Hills, Tamil Nadu, India		Pocock (1899)
	Southern India		Chowdaiah and Kanaka (1974)
	Kerala, India		Janardanan and Ramachandran (1983)
	Alagarkovil Hills, Tamil Nadu, India		Achar (1986)
Arthrosphaera fumosa Pocock	Coimbatore, Tamil Nadu, India	1430	Pocock (1899)
Arthrosphaera gracilis Attems	Kerala, India		Attems (1936)
	Southern India		Chowdaiah and Kanaka (1974)
Arthrosphaera hendersoni Pocock	Kodaikanal, Tamil Nadu, India	2130	Pocock (1899)
	Southern India		Chowdaiah and Kanaka (1974)
Arthrosphaera heterosticta Newp.	Madras, Tamil Nadu, India		Pocock (1899)
Millipede	Distribution	Altitude (m)	Reference
Arthrosphaera inermis Humbert	Madras, Tamil Nadu, India		Pocock (1899)
	Sri Lanka		Attems (1936)
	Tamil Nadu and Kerala, India		Attems (1936)

Arthrosphaera leopardina Butler	Sri Lanka		Pocock (1899)
Arthrosphaera lutescens Butler	India		Pocock (1899)
	Tamil Nadu and Kerala, India		Attems (1936)
	Thirthahalli, Karnataka, India		Achar (1986)
Arthrosphaera magna Attems	Karkala, Karnataka, India		Achar (1980)
	Karnataka, India		Achar (1986)
	Maharashtra, Tamil Nadu and Andhra Pradesh, India		Attems (1936)
	Karnataka, India	156-200	Ashwini (2003)
	Karnataka, India	91	Ashwini (2003)
Arthrosphaera marginella Silv.	India		Attems (1936)
Arthrosphaera marmorata Butler	India		Pocock (1899)
Arthrosphaera nitida Pocock	Palani Hills, Tamil Nadu, India	2130	Pocock (1899)
	Kerala, India		Janardanan and Ramachandran (1983)
	Wynad, Kerala, India		Achar (1986)
Arthrosphaera noticeps Butler	Sri Lanka		Pocock (1899)
Arthrosphaera rugosa Verh.	Sri Lanka		Attems (1936)
Arthrosphaera pilifera Butler	Sri Lanka		Pocock (1899)
Arthrosphaera pygostolis Attems	Tamil Nadu, India		Attems (1936)
Arthrosphaera rugulosa Butler	Sri Lanka		Pocock (1899)
Arthrosphaera scholastica Attems	Tamil Nadu, India	2130	Attems (1936)
Arthrosphaera severa Attems	Tamil Nadu, India		Attems (1936)
Arthrosphaera thurstoni Pocock	Nilgiri Hills and Shevaroy Hills, Tamil Nadu, India	305	Pocock (1899)
Arthrosphaera transitiva Attems	Tamil Nadu, India		Attems (1936)
Arthrosphaera versicolor White	Sri Lanka		Pocock (1899)
Arthrosphaera wroughtoni Pocock	Maharashtra, India		Pocock (1899)
Arthrosphaera zebraica Butler	Bombay, Maharashtra, India		Pocock (1899)
	Karnataka, India		Chowdaiah (1966a)
Arthrosphaera sp.	India		Chowdaiah (1962)
Arthrosphaera sp.	Karnataka, India		Chowdaiah (1966a)
Arthrosphaera sp.	Karnataka, India		Chowdaiah (1969)
Arthrosphaera sp.	Kerala, India		Janardanan and Ramachandran (1983)
Arthrosphaera sp.	Madikeri, Karnataka, India		Achar (1986)
Arthrosphaera sp.	Karnataka, India		Achar (1986)
Arthrosphaera sp.	Chengancherri, Kerala, India		Achar (1980)
Arthrosphaera sp.	Karnataka, India	635	Hosabaale, 2000
Millipede	Distribution	Altitude (m)	Reference
Arthrosphaera sp. 1	Agumbe, Karnataka, India	863	Ashwini (2003)
Arthrosphaera sp. 2	Balur, Kempuhole, Made and Someshwara, Karnataka, India	167-909	Ashwini (2003)
Arthrosphaera sp. 3	Nittur, Hulikal and Kollur, Karnataka, India	182-870	Ashwini (2003)
Arthrosphaera sp. 4	Tumbri, Karnataka, India	450	Ashwini (2003)

Millipede richness was correlated with the age of forest stands in South Africa (Amer, 1997). Similarly, the richness as well as diversity of millipedes reduced due to decline of woody plant diversity and canopy cover in Zimbabwe (Dangerfield, 1990; Mwabvu, 1997). The mixed evergreen forests of Western Ghats and foothills seem to be hospitable for *Arthrosphaera* against scrub jungle and plantations of the west coast. *Arthrosphaera* spp. in Peninsular India have been recovered at an altitudinal range between 156 and 2130 m (see Table 2). Holloway (1956) also found four species of pill millipedes (belonging to the genus *Procyliosoma*) abundant in decaying leaves and fallen rotten logs of New Zealand forests at an altitude of 700 m.

Over a decade of exclusive organic farming in a mixed plantation of Southern India elevated the richness of *A. magna* (Ashwini and Sridhar, 2005a). Organic farming practices in mixed plantation include: (1) Use of composted coffee husk, coconut husk and poultry manure; (2) Use of farmyard manure; (3) Addition of leaf litter and green manure to the basins twice a year; (4) Intermittent sprinkling during post-monsoon and summer; (5) No tilling, ploughing and weeding. The adult *A. magna* in mixed plantation weighs 7.2 g (n=15; range, 5.1-13.7 g), measures 4.6 cm long (n=15; range, 3.8-6 cm) and 2.2 cm wide (n=15; range, 1.8-2.6 cm) (Ashwini and Sridhar, 2002). Seasonal study on the abundance and activity of *A. magna* in relation to soil edaphic and climatic features in mixed plantation and adjacent semi-evergreen forest reserve has been performed by Ashwini and Sridhar (2005a). The abundance and biomass of *A. magna* were significantly higher in the plantation than in forest. The biomass was significantly differed during post-monsoon, summer and monsoon in the plantation than in the forest. In forest, a maximum of two individuals per sample were recovered, while in plantations it was up to 26. Similarly, the highest biomass of *A. magna* in forest was 9.1 g/sample, while it was up to 66.6 g/sample in plantation. Abundance as well as biomass of *A. magna* in plantation were positively correlated ($p=0.001$) with rainfall, soil moisture and soil temperature, so also the abundance with soil phosphate ($p=0.01$) and biomass with soil calcium ($p=0.01$). Total biomass between plantation and forest was significantly differed ($p<0.001$), so also in post-monsoon, summer and monsoon in plantation ($p<0.001$) indicates the conditions prevailing in the plantation are congenial for survival of *A. magna*. Besides rainfall, other farming practices to maintain the soil moisture and temperature in plantation are also favourable for higher abundance, biomass and extended activities of *A. magna*. Adults were active up to seven months in plantation (late June to end of January), while juveniles emerge in late July, active up to four months and hibernates in November prior to adults. In forest, the adults were active during July through November, while the activity of juveniles was restricted between August and November. The narrow period of activity in the forest location has been predicted due to decline in soil moisture unlike plantation. Dangerfield *et al.* (1992) opined that millipedes with large body size are tolerant to dry conditions and exhibit extended periods of surface activity. It is clear that the pill millipede activity can be extended by organic farming practices and maintaining adequate moisture and organic matter in soil.

Two individuals of *Cyliosomella furciparum* were collected from native woodland (100 and 120 m) and one at a low altitude (87 m) in mixed woodland of Marianne (Seychelles) (Hill *et al.*, 2002). Holloway (1956) revised the New Zealand Sphaerotheriidae and recognized them as *Procyliosoma delacyi delacyi, P. delacyi striolatum, P. leiosomum, P. tuberculatum tuberculatum* and *P. tuberculatum westlandicum*. The *P. tuberculatum* was recorded in Mokoia Island of New Zealand (Meads and Fitzgerland, 2001).

The pill millipede genus, *Sphaeromimus* was known from a single male specimen (*Sphaeromimus musicus* DeSaussure and Zehnter) from Malagasy of Madagascar (DeSaussure and Zehnter, 1897). Subsequently, Wesener and Sierwald (2005a) described two new species (*S. inexpectatus* and *S. splendidus*) from two of the four remaining small patches of littoral forests (Mandena and Sainte Luce) of the southern Madagascar. The *S. inexpectatus* was found in patch of littoral forest of Mandena with 5-30 cm dry leaf litter cover consists of tree fruits and a few winged Blattodea. Juveniles and adults *S. splendidus* were found in 30-80 cm thick wet leaf litter along with a large number of associated fauna (Spirostreptida, Isopoda, winged Blattodea, Diplura and Collembola) at Sainte Luce littoral rainforest. As these species have very restricted distribution, human interference of littoral forest patches and mining projects may leads to endanger these millipedes. All three species of *Sphaeromimus* were collected from forests over 1,000 m elevation and these species seems to be restricted to semi-humid habitats of the forests (Wesener and Sierwald, 2005a). Interestingly, members of the genus, *Zoosphaerium* were found in three collections sites, where *Sphaeromimus* were found. There are no geographic barriers (e.g. rivers, hills) between the Mandena and Sainte Luce littoral rainforests, but annual precipitation differs and probably the extent of precipitation might geographically restrict these pill millipedes. *Sphaeromimus* shares some characteristics with *Arthrosphaera* and *Zoosphaerium* (Wesener and Sierwald, 2005a).

Six species of *Sphaerotherium* have been reported from north-eastern South Africa. *Sphaerotherium giganteum* and *S. punctulatum* were widely distributed in post-mining rehabilitated, spontaneously regenerated and mature sites of dune forests (Redi *et al.*, 2005). *Sphaerotherium* sp. 1, *Sphaerotherium* sp. 2 and *Sphaerotherium* sp. 3 were confined to mature sites, while *Sphaerotherium* sp. to post-mining rehabilitated and mature sites. *Sphaerotherium* species are also known from New Zealand (*S. angulatum*, *S. leiosomum* and *S. novae-zealandiae*) (Holloway, 1956). The sphaerotheriid, *Epicyliosoma excavatum* is endemic to Queensland and New South Wales (Jeekel, 1986).

Zephronia siamensis is known from Kosichang – Chantaboon, Thailand (Engohff, 2005; Hirst, 1907). *Zephronia moderata* and *Z. tivia* are known from Java and Sumatra respectively (Chamberlin, 1945). A new Spherotherida, *Prionobelum joliveti* from Hainan (China) has been described and compared with *Zephronia hainani* recently (Mauriès, 2001). The status and composition of species of the genus, *Zoosphaerium* are unresolved. Revision of 34 nominal species of this genus is with insufficient description. Monophyletic species need to be identified to accommodate the described species as well as estimated 70 to be described species belonging to the *Zoosphaerium* species complex. Till date, 33 species of Malagasy (Madagascar) pill millipede genus, *Zoosphaerium* were known from only museum specimen (Wesener and Sierwald, 2005b). *Zoosphaerium alluaudi* (DeSaussure and Zehntner, 1897) is endimic to the moist littoral forest of Petriky (Madagascar) with 30 cm thick litter along with huge number of eggs (Wesener and Sierwald, 2005b). The vicinity of this millipede is under great threat (e.g. grazing animals, mining) and thus became endangered. *Zoosphaerium villosum* and *Z. arborealis* were described recently from Madagascar (Wesener and Sierwald, 2005b). The former species possess dense hair coverage on the tergites belong to the six largest sphaerotheriid species of the world and known only from a small littoral forest of Toamasina (Madagascar), while the latter has a wide distribution in the montane littoral forests of PN Andringitra (Madagascar) at more than 1200 m altitude. *Zoosphaerium arborealis* was also collected from the Madagascar littoral forests Madena and Sainte Luce with other pill millipede, *Spharomimus splendidus*. Young individuals were often associated with dead, rotten wood (e.g. dead, excavated stems of *Pandanus* sp.). The hollow stems were filled with faecal pellets indicating the soft content constitutes suitable food for these millipedes. During tree climbing, *Z. arborealis* do not conglobates into a pill on mechanical disturbance, while only the head is tucked under the thoracic shield. Such interesting arboreal adaptation or behavior was also reported from the South-African giant pill millipedes of the genus *Sphaerotherium* (Haacker and Fuchs, 1972).

A variety of pill millipedes belonging to Sphaerotheriida with attractive colouration has been reported from different parts of the world: green with prominent orange border (http://kids.mongabay.com/elementary/animals/pill_millipede.html), green with slight black/orange border (http://www.arachnoboards.com/ab/showthread.php?t=27693) and black pill millipedes (http://www.zi.ku.dk/personal/drnash/pics/ Madagascar/ pages/pill_ millipede_2.htm) from Madagascar; maroon *Procyliosoma tuberculata* and *P. striolatum* with black border from New Zealand (http://www.terranature.org/gigantism.htm; http:// soilbugs.massey.ac.nz/diplopoda.php); black with ivory white from the Western Ghats, India (http://www.search.com/reference/Millipede); orange with maroon border (http://www. mantidforum.com/forum/viewtopic.php?p=22323andsid=2cfd94bfa8d500d29b036a0aeb3693 c5) and emerald green pill millipedes (http://www.intenseherp.com/gallery.php? sectionID=6) from Tanzania; orange to maroon with ash coloured from Tanzania (http://www.atshq. org/forum/showthread.php?t=12381); black and brown coloured *Procyliosoma leae* and *P. tasmanicum* from Tasmania (http://www. qvmag.tas. gov.au/zoology/multipedes/tasmilli/ milsph/milsph.html).

Individuals of the order Glomeridae confined to Northern Hemisphere and Southeast Asia. There are 271 number of nominal species collection exists, which are smaller than Sphaerotheriidae (measures 2 cm length). Like Sphaerotheriidae, Glomeridae also consists of a variety of species exhibiting different colouration (http://www.arkive.org/species /ARK/invertebrates_terrestrial_and_freshwater/Glomeris_marginata/more_still_images.html; http://www.wildaboutbritain.co.uk/gallery/browse-29-albums.html; http://window boxwild life.blogspot.com/2007_03_01_archive.html). There are three families of Glomerida with wide distribution: Glomeridae (Europe, North/Central America: South and West Europe; Southeast Asia; Southeast USA: California to Guatemala); Glomeridellidae (Europe: South central Europe); Doderiidae (Europe/Africa: South Europe and North Africa (http://www. Field museum.org/research/zoology/zoo_sites/millipeet/pdfFullarticles/MP10Tablke 3.pdf). *Glomeris marginata* is the most studied species among Glomeridae (Sierwald and Bond, 2007). The genus *Hyleoglomeris* is one of the largest and most widespread in the order Glomerida (Golovatch *et al.*, 2006). This genus consists of about 60 nominal species, distributed from Greece in the West to Japan in the East and the Sunda Archipelago (Sulawesi) in the Southeast. Golovatch *et al.* (2006) reviewed this genus and described six new species confined to the caves of Southern China and Laos. Mikhaljova and Lim (2006) also described six new species (*H. alutacea, H. buana, H. confragosa, H. emarginata H. obscura* and *H. unicolorata*) of the genus *Hyleoglomeris* from Korea. The monophyletic *Glomeris alluaudi* group comprises of six species (*G. alluaudi, G. canariensis, G. gomerana, G. hierroensis, G. speobia* and *G. vicenteae*), which are endemic to the Canary Islands (Golovatch and Enghoff, 2003).

BIOCONVERSION

Millipedes are known to ingest plant detritus and transform into mineral-rich faecal pellets. Ashwini and Sridhar (2005b) studied the pattern of ingestion by *A. magna* offering five kinds of partially decomposed and freshly fallen leaf litter of acacia (*Acacia auriculiformis*), areca (*Areca catechu*), banana (*Musa parasidisiaca*), cashew (*Anacardium occidentale*) and coconut (*Cocos nucifera*) simultaneously for 24 hr. Millipedes preferred partially decomposed over fresh litter. Ten days litter feeding trial with individual and mixed litter diet revealed the highest ingestion rate as well as faecal pellet production on partially decomposed mixed litter diet. Faecal pellets generated on ingesting fresh litter consists of higher calcium than partially decomposed litter, while magnesium was high in banana, cashew and mixed litter diets. Ten days feeding trial with monolitter revealed highest litter ingestion and quantity and number of faecal pellets production between 2 and 4 days and dropped subsequently. On mixed litter diet, the ingestion rate and number and quantity of faecal pellet production reached highest on first day and remain steady until 10 days. The quantity of faecal pellet production in *A. magna* (340-570 mg/animal/day) is equivalent to 5.36-9% of its body mass (Ashwini and Sridhar 2005b). The quantity of mixed litter ingestion by *A. magna* was many folds higher than other tropical millipedes (561.1-598.2 vs. 1.02-157 mg/animal/day), so also the quantity of faecal pellets (550-570 vs. 0.97-126.05 mg/animal/day) (Ashwini and Sridhar 2005b). Thus, *A. magna* preferred spatially and temporally heterogeneic mixed litter. It is likely the nutrients in mixed litter meet the requirement for growth, maturity and reproduction of *A. magna*. High richness and diversity of *Arthrosphaera* spp. only in mixed forests and mixed plantations in Southern India justifies its nutritional requirements.

Partially decomposed acacia, areca, banana, cashew, cocoa and coconut litter mixed separately with sandy loam soil (1:1 w/w), humus and sandy loam soil (without amendments) were offered to *A. magna* for four days (Ashwini and Sridhar, 2006c). Consumption of humus yielded the highest quantity of faecal pellets, while it was least in soil (0.67 vs. 0.03 g/animal/day). Feeding leaf litter along with soil resulted in faecal matter production between 0.35 and 0.47 g/animal/day. Faecal pellet production declined gradually from day one to fourth in all leaf litters with soil except for areca. The organic matter in faecal pellets of millipede ingested acacia, banana, cashew and coconut litter amended with soil was higher (70.1 to 75.8%) than raw soil (9.3%). It is interesting to note that consumption of acacia litter along with soil by *A. magna* resulted in faecal pellet production similar to other plantation litters even though acacia is an exotic plant species. The weight of faecal pellets produced on feeding acacia, banana cashew and coconut litter amended with soil versus humus diet was significantly different ($p=0.03-0.04$) and it was coincided with the high organic matter in faecal pellets indicating the importance of quality of leaf litter in millipede diet. Offering mixed litter along with soil is of great importance to explore the rate of consumption and conversion efficiency of *A. magna* in mineral cycling and compost production in plantations. Polyphenols and tannins are responsible for little attraction of fresh leaf litter to millipedes (Sakwa, 1974). Decrease in phenolics of partially decomposed litter might be one of the reasons for higher palatability than fresh litter by *A. magna*. In addition, fungi and bacteria in partially decomposed litter might play significant role in millipede nutrition as well as food conversion efficiency. Fungal colonized leaf litter consists of rich proteins, lipids (Hodgeman *et al.*, 1959) and enzymes. Fungal lipids, particularly ergosterol may be important in metamorphosis of pill millipedes.

Up to 30% assimilation of litter has been reported in millipedes. Bocock (1963) estimated that consumption in *Glomeris* is up to 1.7-10% of annual litter production. *Glomeris marginata* consumes leaf litter about 10 times of its body mass per annum (van der Drift, 1975) and known to assimilate 18% of beech litter (Bignell, 1989). David and Gillon (2002) found that the mean weight-specific consumption rate in *G. marginata* of holm oak (*Quercus ilex*) was 14 g/g animal/year under Mediterranean conditions. The consumption rate was high in spring and lowest in summer and the assimilation efficiency averaged 6% for the year. Food preference experiment showed that *G. marginata* preferred more decomposed leaves over fresh leaves. Smaller millipedes have high assimilation efficiency than the larger ones (Crawford *et al.*, 1987; Dangerfield and Milner, 1993; Striganova and Prishutova, 1990) indicating the efficient digestive tactics to exploit poor quality food (Sibly, 1981). The structural components of plant litter are degraded in millipede-harboured microbes. Breakdown of cellulose and hemicellulose takes place in millipede gut through microbes (Taylor and Crawford, 1982). However, xylanase, α-amylase and cellulase were detected in the gut walls of *G. marginata* after bacteriostatic treatment indicating such enzymes are the product of millipedes (Urbasek and Tajovský, 1991). In *G. marginata*, optimum temperature for carbon mineralization was 15-23°C. Cellulose loss was independent of temperature (4-23°C) and it was preferentially decomposed by microorganisms in faeces and suggests the occurrence of cellulolytic microorganisms in the gut (Coûteaux *et al.*, 2002). Therefore, microbes colonizing the leaf litter are valuable source of nutrition to millipedes and assimilate with high efficiency. The ^{14}C Labelled microbial (bacteria and fungi) and plant fibre ingestion by *G. marginata* revealed that uptake of microbial tissue (bacteria, 72.2-95.9%; fungi, 69.2-73.5%) by millipede was found to be over three times (18.3%) that of plant fibre assimilation (Bignell, 1989). Anderson *et al.* (1985) demonstrated that feeding activity of *G. marginata* significantly elevate the nitrogen mineralization. Three species of glomerids fed leaf litter (chestnut, beech and holm-oak; C/N ratio, 28:39.1) and generated faecal pellets with substantially low C/N ratio (20.25:27.95) (Marcuzzi, 1970). *Glomeris marginata* is the most important saprophagous macroarthropod of the beech wood sites on a basalt lime stone gradient in Germany (Maraun and Scheu, 1996). The leaf litter feeding activity of *G. marginata* in deciduous woodlands resulted in marked mobilization of calcium and potassium, which elevated the nutrient fluxes (Anderson *et al.*, 1983).

Faecal Material and Compost

It is known that the South African millipede, *Paratophilus* sp. produced maximum faecal mass (126.1 mg/animal/day) followed by an endemic millipede of Seychelles, *Seychelleptus seychellarum* (Sprostreptidae) (111 mg/animal/day) (Lawrence and Samways, 2003). In faecal pellets, the capacity of moisture retention increases with decrease in particle size of organic matter (McBeyer, 1973) followed by increased microbial growth (McGonigle, 1995; Hanlon, 1981; Witkamp and Crossley, 1966). To generate millipede compost by *Arthrosphaera magna* in pilot-scale, partially decomposed plantation leaf litter (acacia, areca, cocoa and cashew), areca nut husk and cocoa pod husk were used (Ashwini and Sridhar,

2006a). Three combinations (w/w) of each equivalent to 5 kg in cement compost tanks [(1) areca leaf litter and areca nut husk (1:1); (2) cocoa leaf litter and cocoa pod husk (1:1); (3) mixed leaf litter (areca, acacia, cocoa and cashew) (1:1:1:1)] were supplied to *A. magna* with adequate moisture up to two months. Adult individuals (10 males + 10 females per tank) acclimatized on organic matter were introduced into compost tanks. Up to two months, the contents of each tank were sprayed with water (500 ml/day). On harvesting, weight of particles less than 5 mm in millipede compost was five times higher than control treatments (without *A. magna*). Organic matter and C/N ratio were considerably decreased in treated residues. Significant difference was seen in total nitrogen (p=0.004), phosphate (p=0.0006) and C/N ratio ($p=1.19\times10^{-6}$) between control and treated residues. Organic matter as well as C/N ratio decreased in treated residues over the control. In compost generated from areca leaf litter and areca nut husk, the moisture and total nitrogen were elevated, while phosphate, calcium and magnesium in mixed litter compost. The nitrogen, phosphorus and potassium were increased and C/N ratio was decreased in mixed leaf litter compost. The significant difference in total nitrogen (p=0.004) and phosphorus (p=0.0006) in compost with/without millipedes indicate the importance of composting plantation lignocellulosic residues through *A. magna*. Elevation of nitrogen, phosphorus and potassium, narrow C/N ratio and pH shift towards neutral in mixed litter compost revels the possibility of production of millipede compost as an alternative to vermicompost.

Five combinations of composts [(1) formyard manure (FYM); (2) areca compost (by *A. magna*); (3) mixed litter compost (by *A. magna*); (4) FYM + areca compost; (5) FYM + mixed litter compost] were tested on dry matter yield of black gram (*Phaseolus mungo*) and finger millet (*Eleusine coracana*) (Ashwini and Sridhar, 2006b). After treatment up to three weeks (finger millet) and six weeks (black gram), shoot length and total dry matter yield in black gram elevated in FYM + areca and FYM + mixed litter composts. In finger millet, shoot and root lengths were highest in plants grown in areca compost, while the dry matter yield in FYM + mixed litter compost. Therefore, pill millipede compost produced from the plantation residues in Southern India is capable to compensate the nutrient deficiency in FYM and may be an ideal alternative to inorganic fertilizers. Although FYM serves as a good source of plant nutrients (Palaniappan and Natarajan, 1993), it meets partial requirements of nitrogen, phosphorus and potassium of plantation crops (Chowdappa *et al.*, 1999). Amending millipede compost in appropriate proportions with FYM helps to overcome the deficiencies of nitrogen, phosphorus and potassium. It is possible to produce millipede compost on-farm within three months duration during monsoon and post-monsoon seasons in South Western India and thus, becomes an added advantage to recycle the plantation residues more effectively. As the ingestion of leaf litter and faecal matter production by *A. magna* is higher than other tropical millipedes (Ashwini and Sridhar, 2005b; Palaniappan and Natarajan, 1993), *A. magna* will serve as a suitable candidate for on-farm composting of agricultural wastes in Southern India. Suitable natural or agroclimatic conditions for sustenance of pill millipedes prevail in the plantations in Western Ghats or foothill of Western Ghats helps to support the activities of pill millipedes and in turn the improvement of soil fertility.

Faecal pellets of *G. marginata* consist of leaf fragments (0.001-0.1 mm diameter), pollen grains, algae, bacteria and fungal mycelia (Nicholson *et al.*, 1966). The surface to volume ratios of faecal pellets in *Glomeris* was 3.2:1 (Nicholson *et al.*, 1966, Webb, 1977). The faeces of *G. marginata* consisted of 30 times more bacteria than litter and humification occurs through bacterial action on the defecated faeces (ven der Drift, 1951). Tajovský *et al.* (1992)

studied the decomposition of faecal pellets of *Glomeris hexasticha* in forest soil. Faecal decomposition showed microbial succession, decrease in total mass, decline of carbon and nitrogen up to 44-50% (C/N ratio: 23.7 vs. 20.9). Heterotrophic bacteria utilized easily degradable organic substrates at first stage, followed by domination of actinomycetes and oligotrophic bacteria, which are known to degrade more complex substrates in the second stage. Activity of nematodes on faecal pellets increased followed by nematophagous fungi. Changes in physical and chemical features and microbial succession on faecal pellets of *G. marginata* were studied under field conditions by Nicholson *et al.* (1966). In calcareous soils, faecal pellets of *G. marginata* were stable due to elevated calcium concentration (Kubikov and Rusek, 1976). The carbon mineralization rates of faecal pellets of *G. marginata* and earthworm (*Octolasion lacteum*) fed with mechanically fragmented litter showed almost equal mean rates of CO_2-C production (Scheu and Wolters, 1991). Bonkowski *et al.* (1998) investigated the pattern of leaf litter decomposition in microcosms with and without earthworms (*O. lacteum*) and millipedes (*G. marginata*). Millipedes along with earthworms significantly increased the rate of litter decomposition. Leaf litter and faecal pellets of millipedes stable on the surface of soil without earthworms, while with earthworms they were mixed with soil. The faecal pellets of *G. marginata* constitute an important nutrient source for earthworms and resulted in increased biomass of *O. lacteum* on beech leaf litter (Bonkowski *et al.*, 1998). Bonkowski *et al.* (1998) have considerd *O. lacteum* as a secondary decomposer as it feeds on organic matter fragmented by other saprophagous macroinvertebrates. Occurrence of *G. marginata* along with *O. lacteum* significantly elevated the breakdown of beech litter and ultimately improved the soil fertility. Earthworms (*Lumbricus castaneus* and *O. lacteum*) incorporated faecal pellets of *G. marginata* into soil and enhanced the carbon mineralization within two weeks duration (Scheu and Wolters, 1991).

Rawlins *et al.* (2007) amended soil with oak leaf litter and faecal material of *G. marginata* to study the degree of stabilization of organic matter and found that the faecal material is highly stabilized and recalcitrant form of organic matter with the changes in biochemical composition brought about by gut passage is a significant factor in stabilization. Rawlins *et al.* (2006) studied biochemical transformation of oak leaf litter by *G. marginata*. Sterols were converted to cholesterol to fulfil biochemical demands, short chain fatty acids were directly converted into phospholipid membranes or catabolised as an energy source, triacylglycerols were catabolised for energy or stored, carbohydrates are reduced in concentration by 85% (likely utilized for energy) and decreased amino acids suggesting selective assimilation.

Microbial Association

Pill millipedes have been associated with a variety of microbes (bacteria, actinobacteria and fungi). Several bacteria have been isolated from the faeces of *Arthrosphaera* spp. of the Western Ghats, India: Actinobacteria (*Gordonia terrae, Microbacterium imperiale, M. trichotecenolyticum, Rhodococcus fascians* and *Streptomyces griseodotifer*), Alphaproteo bacteria (*Methylobacterium populi* and *Novosphingobium subarcticum*), Bacilli (*Bacillus arsenicus, B. licheniformis* and *B. subtilis*), Betaproteobacteria (*Cupriavidus basilensis* and *Pseudomonas putida*), Gammaproteobacteria (*Citrobacter amalonaticus, Klebsiella pneumoniae, Pantoea agglomerans* and *Pseudoxanthomonas koreensis*) (S. Kadamannaya, unpublished observation). Yellow coloured bacterial strain was isolated recently from the faecal pellets of Indian *Arthrosphaera magna* (Kämpfer *et al.*, 2006). Based on analysis of 16S rRNA gene sequences, it has been assigned to new species *Chitinophaga skermanii*, which has extensive intrageneric phylogenetic heterogeneity. A fungus belonging to Laboulbeniales (*Triainomyces hollowayanus*) has been isolated from pill millipede, *Procyliosoma tuberculatum* (Sphaerotheriidae) in New Zealand (Rossi and Weir, 1998). More than 50% of millipedes preferred feeding ascospore-releasing structures of fungi.

Bacteria and fungi of leaf litter became valuable source of nutrition to *G. marginata* (Bignell, 1984). Based on feeding ^{14}C labelled bacterium and fungus along with litter to *G. marginata* and analysis of its faecal pellets and residual gut content, it has been ascertained that microbes colonizing leaf litter were valuable nutritional source (Bignell, 1989). The hindgut of *Glomeris* sp. consists of six bacteria, six actinomycetes and two fungi (Gebhardt *et al.*, 2002). *Pseudomonas putida* and *P. stutzeri* survived passage through the gut of *Pachyiulus flavipes* and increased in fresh excrements (Byzov *et al.*, 1996). Bacterial population has been elevated between 10 and 100 fold in faeces after litter passage through gut of *G. marginata* (Anderson and Bignell, 1980; Anderson and Ineson, 1983). The viable count of bacteria was higher in the gut and faeces than in leaf litter. Thus, it is believed that in millipedes bacterial proliferation takes place in the hindgut and faeces. Feeding activity of millipedes stimulates microbes and leads to substantial breakdown of organic matter (van der Drift, 1951; Anderson and Bignell, 1980). The extent of microbial colonization influences the palatability of litter in *G. marginata* (Anderson and Bignell, 1982). One of the recent studies showed that *Bacillus* sp. confined endosymbiotically to the hind gut of *Glomeris* sp., which produced a strong antifungal metabolite, bacillomycin D (Gebhardt *et al.*, 2002). The gut nocardioforms of the forest litter feeding millipedes represent a taxonomically heterogeneous group (Chu *et al.*, 1987). In the gut as well as fresh faecal matter of *Glomeris hexasticha*, Gram-positive nocardioform actinomycete was found, which seems to obligate as it was not found in the soil and litter fed by the host.

Bacterial load increased from food to faecal material in *G. marginata*, while fungal load decreased (Anderson and Bignell, 1980). Gut passage of beech leaf litter by *G. marginata* considerably decreased the fungal biomass (leaf litter, 445.6 μg/g; faecal pellets 131.2 μg/g) indicating partial digestion in the gut (Maraun and Scheu, 1996; Anderson *et al.*, 1983). Maraun and Scheu (1996) measured ergosterol (lipid specific to filamentous fungi) of faecal pellets of *G. marginata* and opined that fungi were more intensively digested than bacteria in gut passage. Many microorganisms living exclusively in the intestine of millipedes are unable to survive in the soil (Márialigeti *et al.*, 1985) indicates that millipede gut provides congenial environment for certain microbial survival and function.

Edaphic Factors

The richness and diversity of soil macroarthropods severely influenced by the exotic plantations in Southern Africa (van Aarde *et al.*, 1996a) and eucalypt in Zimbabwe (Dangerfield, 1990). In Puerto Rico, the high nitrogen, phosphorus, potassium and calcium concentrations in monoculture *Leucacena* litter layers supported millipedes than monoculture

Casuarina and monoculture *Eucalyptus* (Warren and Zou, 2002). Thus, the characteristics of understorey profoundly influence the millipede population (e.g. texture, moisture, humidity, temperature, calcium, magnesium). For instance, diverse millipede population was confined to calcareous soils (Branquart *et al.*, 1995) and plantations with high soil calcium (Pramanik *et al.*, 2001). Millipedes act as buffer stock for significant increase of calcium and magnesium in their biomass (Reichle *et al.*, 1969; Teuben and Verhoef, 1992) and augment mineral recycling in soils. Soil magnesium and potassium concentrations in coastal sand dune forests of South Africa (Smit and van Aarde, 2001) and nitrogen concentration and C/N ratio of litter layer in Puerto Rico (Warren and Zou, 2002) were mainly responsible for significant increase in biomass of millipedes. Besides soil texture, soil moisture, temperature, humidity, humus and soil minerals (particularly calcium and magnesium) influenced the millipede abundance and diversity. Investigations in Belgium indicated that most diverse millipede population confined to calcareous soils (Branquart *et al.*, 1995). The litter strata of Western Ghats and foothill encompass faecal pellets of *Arthrosphaera* and earthworm castings (Ashwini and Sridhar, 2008). It was predicted that soil enrichment in Western Ghats and foothills takes place due to combined activity of *Arthrosphaera* and earthworms. Interestingly, in spite of higher earthworm abundance than *Arthrosphaera* in Western Ghats, their biomass was low, while in foothills, both abundance and biomass of earthworm was higher than *Arthrosphaera*. This clearly reveals the necessity of future studies to relate the activities and functions of earthworms and pill millipedes in Western Ghats of India.

High population of arthropods (including Diplopoda) was correlated to the soil nutrients (mainly calcium) in *Cassia* plantations by Pramanik *et al.* (2001). Similarly, palatability of leaf litter to some millipedes (*Diploiulus londonensis caeruleocinatus*) was correlated to high calcium (Neuhauser and Hartenstein, 1978). Leaf litter feeding by *Glomeris marginata* in deciduous woodlands resulted in high mobilization of calcium and potassium (Anderson *et al.*, 1983). Calcium assimilation in millipedes takes place as transformation to calcium oxalate through fungal activity. High calcium was found in diplopods with highly calcified exoskeletons (327.33 mg/g ash-free dry mass) (Reichle *et al.*, 1969). Calcium concentration of millipedes (122 mg/g dry mass) due to highly calcareous exoskeletons indicates the importance of millipedes in calcium recycling in soil (Carter and Garg, 1976). Many oxalate decomposers have been identified among macroarthropods (Cromack *et al.*, 1977). Calcium oxalate and oxaloacetic acid will be decomposed into ionic calcium by microbes in the digestive tract of soil invertebrates, which influence the calcium cycle in soil. Millipede faecal pellets are attractive to endogeic earthworms (Bonkowski *et al.*, 1998). Some earthworms (*Lumbricus castaneus* and *Octolasion lacteum*) incorporated faecal pellets of *G. marginata* into soil and facilitate to elevate carbon mineralization (Scheu and Wolters, 1991).

Endemism

High degree of speciation and evolution of a large number of endemic millipedes resulted due to their limited mobility (Hopkin and Read, 1992). Millipedes, unlike other animal groups, have a narrow adaptive zone and often exhibit single-site or narrow range endemism or scattered or disjunct pattern of distribution. For example, East African coastal forests are known for diversity of endemic millipedes (Burgess, 1998; Hoffman, 1993). A large number

(50%) of described millipede fauna (243 species) is known from a single locality or localities separated by less than 10 km (Hamer and Slotow, 2002). The species level endemism of millipedes has been classified into four classes in South Africa: (1) Site endemics (<10 km); (2) Local endemics (<70 km); (3) Regional endemics (<150 km); (4) South African endemics (within the country) (Hamer and Slotow, 2002). This approach seems to be reasonable instead of considering the area of occupancy due to lack of data on presence or absence of a species from an area between two or more locality points. Endemic millipedes have several specific ecological and life-history traits: poor powers of dispersal, confinement to discontinuous habitats, highly seasonal, active only during cooler/wet periods and low levels of fecundity (Harvey, 2002). High density and large body size of *Seychelleptus seychellarum* (Sprostreptidae), a giant millipede endemic to islands of Seychelles showed a major impact on litter consumption and faecal pellet output (Lawrence and Samways, 2003). In the entire Cousine Island, these millipedes ingest about 73.57 kg litter/day and become an important 'keystone species'. Similarly, *Dolichoiulus* (Julidae) is also endemic to Canary Islands out of 46 species, most live in one island (Enghoff and Báez, 1993). Cave millipedes have also been studied extensively in US and France (Shear, 1969, 1984; Geoffroy, 1984) and found that they are highly endemic due to isolated habitats. Recently, Wesener and Sierwald (2005a, 2005b) described the endemic nature of *Sphaeromimus* and *Zoosphaerium* to Madagascar. Attems (1936) opined that the genus *Arthrosphaera* also known for a narrow range of distribution in the Peninsular India and exhibits mono-site endemism or scattered or disjunct distributions.

CONSERVATION

Millipedes are conservative unlike earthworms and woodlice. The desiccation of climate is one of the major threats for millipedes by human interference. They are sensitive to litter chemistry at narrow spatial scales (meters to decameters) than the earthworms (Warren and Zou, 2002). Pill millipedes belong to *Arthrosphaera* also reflect such phenomenon as they are dominant in mixed plantation, while earthworms in forest (Ashwini, and Sridhar 2005a). Long-term intra-site heterogeneity in organic matter input in mixed plantations might have facilitated the invasion, survival and activity of *A. magna* in the foothill of Western Ghats. In spite of conducive ecological conditions, millipedes are highly sensitive and fail to overcome the limitation of a single edaphic factor *viz.,* water deficit, soil texture and litter thickness (Kime and Golovatch, 2000). Such sensitivity of *Arthrosphaera* to narrow ecological factor/s qualifies them to serve as indicators of soil quality restoration in managed plantations or forests. According to Goodman and Benstead (2005), priority of conservation depends on two important criteria: (1) Estimates of species diversity and (2) Levels of endemism. Unfortunately, biological exploration in tropics is still underway and it is at an infant stage although endowed with high biodiversity. Thus, investigation on endemic organisms becomes top priority and restoration and conservation priorities need to be reworked for endemism (e.g. millipedes) as and when sufficient data available.

Significant correlation was found between soil organic carbon and biomass of *Arthrosphaera* in the Western Ghats and foothill of Western Ghats (Ashwini and Sridhar, 2008). An important hotspot of diversity, Western Ghats is under threat of transformation towards commercial forests (*Tectona grandis* and *Hevea brasiliensis*) and plantations: coffee

(*Coffea arabica* and *C. robusta*), cocao (*Theobroma cacao*), citrus (*Citrus lemon* and *C. reticulata*), cashew (*Anacardium occidentale*), cinnamon (*Cinnamomum malabatrum*), cardamom (*Elettaria cardamomum*) and areca *(Areca catechu*). Monocultures of either forest tree or plantation species limit the diversity of litter and addition of agricultural chemicals exerts severe pressure on survival of pill millipedes. Thus, habitat preservation becomes utmost importance to restore or conserve pill millipedes. For example, forest floors of some foothill locations of the Western Ghats (e.g. Mundaje, Gundya, Sampaje) consist of boulders and rocks, which supported *Arthrosphaera* possibly due to accumulation or retention of adequate organic matter in the crevices (Ashwini and Sridhar, 2008). Transformation or disturbance of such natural structures will exert negative pressure on the habitat preference of pill millipedes. Similarly, changes in soil conditions by habitat destruction or fragmentation also seem to have major impact on pill millipedes as they are highly sensitive. In view of restoration of pill millipedes in Western Ghats, forest and plantation practices need to follow specific measures. Many plantation practices: (1) Windrowing (provides shelter by accumulation of rotting logs); (2) Mound ploughing (creates long moist furrows of litter); (3) Pruning (generates large amount of rotting wood); (4) Thinning (adds rotting logs with persistent bark and leaves) facilitates reinvasion of soil invertebrates to plantations from the adjacent native forests and lowers the risk of extinction of invertebrates. These examples clearly reveal the impacts of forest devastations, limits and management practices to be followed to restore the soil fauna in modern silviculture.

FUTURE OUTLOOK

There are several gaps in our knowledge on pill millipedes. Global millipede collection is not adequate because of lack of sufficient collections from different countries (e.g. Africa, Central and South America and Asia). Traditional approaches employed in delineating millipede species is rely solely on genitalic and other morphological characters. This approach may underestimate the diversity of millipedes (Bond and Sierwald, 2002, 2003). To support morphological studies, scanning microscopy and molecular approaches are emerging in Myriapodology to understand the phylogeny and diversity. Three nuclear protein-encoding genes: elongation factor-1α (EF-1α), RNA polymerase II (Pol II) and elongation factor-2 (EF-2) from 59 myriapod species were assessed to resolve phylogenetic relationships by Regier *et al.* (2005) and Pentazonia showed some interclass relationships. Pill millipedes needs special consideration and priority for further investigation as they are highly endemic and sensitive to environmental perturbations. Future studies should be designed to understand the impact of nature of forest understorey in moist and dry regimes on pill millipedes. Enghoff's (1990) observation on colonization of higher elevations accompanied by increased body size of millipedes need to be addressed at the Western Ghats as it has a steep altitudinal gradient of over 2000 m. Ponge (1999a) hypothesized that mull, moder and mor humus formation is the result of multiple interactions of soil fauna, flora and free living/symbiotic microbes. If this hypothesis is acceptable, the events such as human disturbances, addition of toxic chemicals and invasion of foreign organisms to which soil organisms have not adapted destabilize the ecosystem by negative feedback loop, which is usually catastrophic. Humus formation in forests is the result of intricate long-term biological process by the interaction of plants, animals and microbes. In Scots Pine forest, the faecal matter is an important part of the humus for ion exchange and acts as buffering agent for roots. The entire forest ecosystem is a feedback loop formed by a mosaic of vegetation, soil fauna and microbes in and around the humus. Decomposition is one of the important ecological services for the whole ecosystem as 60-90% of terrestrial primary production is decomposed mainly in soils. Physical changes in litter decomposition are governed by chemical composition of litter, range of decomposers and environmental regimes at the site of decomposition. In tropical forests, the complexity of community structure (e.g. bacterial, fungal, faunal), their interaction and involvement in organic matter transformation and nutrient cycling are the essential parts of litter studies. Soil invertebrates enhance the rate of decomposition and microbial activity, increase nitrogenous energy and eliminate bacteriostasis and mycostasis. The role of different trophic levels to nutrient cycling varies in relation to the density, biomass and longevity of soil fauna. For instance, even though the ecological gains occur at a slow pace, they are significant for soil conservation with increased nutrient status, stability and diversity of decomposer community.

Millipedes ingest leaf litter along with soil, increase soil respiration, leaching (e.g. calcium, magnesium, nitrate) and enhance microbial activity in forest soil (Kaneko, 1999). Mean rates of carbon mineralization of faecal pellets of pill millipede (*Glomeris marginata*) and earthworm (*Octolasion lacteum*) fed with mechanically fragmented litter showed almost equal mean rates of CO_2-C production (Scheu and Wolters, 1991). Many microorganisms exist in the intestine of millipedes will not survive in the soil (Márialigeti *et al.*, 1985). Thus, millipede gut provides an appropriate stable environment for microbial function that cannot be achieved elsewhere. Endosymbiotic bacteria (*Bacillus* sp.) isolated from the hindgut of *Glomeris* produced a strong antifungal compound (bacillomycin D) (Gebhardt *et al.*, 2002). Even though saprophagous fauna (e.g. Diplopoda), are ideal to investigate ecological process associated with faunal changes in conserved forests, they are not often employed as bioindicators compared to Lepidoptera and Coleoptera (David *et al.*, 1993). Millipedes are highly influenced by litter chemistry at narrow spatial scales than earthworms (Wallwork, 1976) and very sensitive to even a single edaphic factor (e.g. soil texture/litter thickness) (Kime and Golovatch, 2000), thus, pill millipedes may become authentic indicators of soil quality restoration.

ACKNOWLEDGMENTS

Authors are grateful to Mangalore University for permission to carry out studies on pill millipedes of Western Ghats and west coast of Karnataka. This study constitutes part of Ph.D. thesis of BSK conducted during the tenure of Faculty Improvement Programme (University Grants Commission, New Delhi). BSK is thankful to the Principal, FMKMC College (Mangalore University), Madikeri for support.

References

Achar, K.P. 1980. The use of Air-drying and Giemsa Differential Banding Techniques in the Cytological Studies of some Indian Diplopoda (Myriapoda). Ph.D. Thesis, Bangalore University, Bangalore, India.

Achar, K.P. 1983a. The use of G-banding technique in the chromosome studies of a millipede species – Sprostreptus asthenes. *Current Science* 52, 540-433.

Achar, K.P. 1983b. Karyological studies in nine species of Indian Diplopoda (Myriapoda). Nucleus 26, 191-197.

Achar, K.P. 1986. Analysis of male meiosis in seven species of Indian pill-millipede (Diplopoda: Myriapoda). *Caryologia* 39, 89-101.

Achar, K.P. 1987. Chromosomal evolution in Diplopoda (Myriapoda: Arthropoda). Caryologia 40, 145-155.

Achar, K.P. and Chowdaiah, B.N. 1979. The use of air drying in the study of diplopod chromosomes. In: Myriapod Biology (Ed. Camatini, M.), Academic Press, London, 21-23.

Achar, K.P. and Chowdaiah, B.N. 1980. The use of C-banding technique in the chromosome studies of a millipede species – Carlogonus acifer. *Caryologia* 33, 185-191.

Amer, L.H. 1997. A preliminary assessment of the southern African millipede fauna: Diversity and conservation (Diplopoda). *Entomol. Scand. Suppl.* 51: 209-217.

Anderson, J.M. and Bignell, D.E. 1980. Bacteria in the food, guts and faeces of the millipede Glomeris marginata (Villers). *Soil Biol. Biochem.* 12, 251-255.

Anderson, J.M. and Bignell, D.E. 1982. Assimilation of ^{14}C-labelled leaf fibre by the millipede Glomeris marginata (Diplopoda, Glomeridae). *Pedobiologia* 23, 120-125.

Anderson, J.M. and Ineson, P. 1983. Interactions between soil arthropods and microbial populations in carbon, nitrogen and mineral nutrient fluxes from decomposing leaf litter. In: Nitrogen as an Ecological Factor (Ed. Lee, J. and McNeill, S.), Blackwell Scientific, Oxford, 413-432.

Anderson, J.M., Ineson, P. and Huish, S.A. 1983. Nitrogen and cation mobilization by soil fauna feeding on leaf litter and soil organic matter from deciduous woodlands. *Soil Biol. Biochem.* 15, 463-467.

Anderson, J.M., Leonard, M.A., Ineson, P., Huish, S., 1985. Faunal biomass: A key component for a general model of nitrogen mineralization. *Soil Biol. Biochem.* 17, 735-737.

Ashwini, K.M. 2003. Ecological Studies on Indian Pill Millipede, Arthrosphaera magna. Ph.D. Thesis, Mangalore University, Mangalore, India.

Ashwini, K.M. and Sridhar, K.R. 2002. Towards organic farming with millipede - Arthrosphaera magna. *Current Science* 82, 20-22.

Ashwini, K.M. and Sridhar, K.R. 2005a. Seasonal abundance and activity of pill millipedes (Arthrosphaera magna) in mixed plantation and semi-evergreen forest of southern India. *Acta Oecologica* 29, 27-32.

Ashwini, K.M. and Sridhar, K.R. 2005b. Leaf litter preference and conversion by a saprophagous tropical pill millipede, Arthrosphaera magna Attems. *Pedobiologia* 49, 307-316.

Ashwini, K. M. and Sridhar, K.R. 2006a. Breakdown of plantation residues by pill millipedes (Arthrosphaera magna) and assessment of compost quality. Current Science 90, 954-959.

Ashwini, K.M. and Sridhar, K.R. 2006b. Evaluation of pill millipede (Arthrosphaera magna) compost on plant growth and dry matter yield. *Electronic J. Environ. Agric. Food Chem.* 5, 1323-1329.

Ashwini, K.M. and Sridhar, K.R. 2006c. Feeding response of an endemic pill millipede Arthrosphaera magna Attems to plantation litter. Journal of Plantation Crops 34, 58-62.

Ashwini, K.M. and Sridhar, K.R. 2008. Distribution of pill millipedes (Arthrosphaera) (Family: Sphaerotheriidae) and associated soil fauna in the Western Ghats and west coast of India. *Pedosphere* 18, 749-757.

Attems, C. 1936. Diplopods of India. Memoirs of the Indian Museum 11, 133-167.

Baker, G.H. 1980. The water and temperature relationships Bernier, N. and Ponge, J.-F. 1994. Humus form dynamics during the sylvogenetic cycle in a mountain spruce forest. *Soil Biol. Biochem* 26, 183-220.

Bernier, N. and Ponge, J.-F. 1994. Humus form dynamics during the sylvogenetic cycle in a mountain spruce forest. *Soil Biol. Biochem* 26, 183-220.

Bessiere, CL. 1948. La spermatogenese de quelques Myriapodes. *Arch. Zool. Expl. Gen.* 85, 149-236.

Bignell, D.E. 1984. The arthropod gut as an environment for microorganisms. In: Invertebrate-Microbial Interactions, British Mycological Society Symposium # 6 (Ed. Anderson, J.M., Rayner, A.D.M. and Walton, D.W.H.), Cambridge University Press, 205-227.

Bignell, D.E. 1989. Relative assimilation of ^{14}C-labelled microbial tissue and ^{14}C-plant fibre ingested with leaf litter by the millipede Glomeris marginata under experimental conditions. *Soil. Biol. Biochem.* 21, 819-827.

Blower, J.G. 1985. Millipedes. Linnean Society Synopses of the British Fauna (New Series), Number 35, E.J. Brill/Dr. W. Backhuys, London.

Bocock, K.L. 1963. The digestion of food by Glomeris. In: Soil Organisms (Ed. Doeksen, J. and van der Drift, J.), Elsevier-North Holland, Amsterdam, 85-91.

Bond, J.E. and Sierwald, P. 2002. Cryptic speciation in the Anadenobolus excisus millipede species complex on the island of Jamaica. *Evolution* 56, 1123-1135.

Bond, J.E. and Sierwald, P. 2003. Molecular taxonomy of the Anadenobolus excisus (Diplopoda: Spirobolida, Rhinocricidae) species group on the Caribbean island of Jamaica. *Invertebr.* Syst. 17, 515-528.

Bonkowski, M., Scheu, S. and Schaefer, M. 1998. Interactions of earthworms (Octolasion lacteum), millipedes (Glomeris marginata) and plants (Hordelymus europaeus) in a beech wood on a basalt hill: implications for litter decomposition and soil formation. *Appl. Soil Ecol.* 9, 161-166.

Branquart, É., R. D. Kime, M. Dufrêine, J. Tavernier and G. Wauthy 1995. Macroarthropod-habitat relationships of oak forests of South Belgium. 1. Environments and communities. Pedobiologia 39, 243-263.

Bueno-Villegas, J., Sierwald, P. and Bond, J.E. 2003. Diplopoda. In: Biodiversidad, Taxonomia y Biogeografia de Artrópodos de México (Ed. Bousquets, J.L. and Morrone, J.J.), 569-599.

Burgess, N.D. 1998. Coastal forests of eastern Africa: status, endemism patterns and their potential causes. *Biol. J. Linn. Soc.* 64, 337-367.

Byzov, B.A., Claus, H., Tretyakova, E.B., Zvyagintsev, D.G. and Filip, Z. 1996. Effects of soil invertebrates on the survival of some genetically engineered bacteria in leaf litter and soil. *Biol. Fertl. Soils* 23, 221-228.

Cardon, Z.G., Hungate, B.A., Cambardella, C.A., Chapin, F.S., Field, C.B., Holland, E.A. and Mooney, H.A. 2001. Contrasting effects of elevated CO_2 on old and new soil carbon pools. *Soil Biol. Biochem.* 33, 365-373.

Carrel, J.E. 1990. Chemical defence in the pill millipede Glomeris marginata. In: Proceedings of the 7th International congress of Myriapodology (Ed. Minelli, A.), E.J. Brill, Leiden, 157-164.

Carter, A. and Grag, J.B. 1976. Concentrations and standing crops of calcium, magnesium, potassium and sodium in soil and litter arthropods and their food in an aspen woodland ecosystem in the Rocky Mountains (Canada). *Pedobiologia* 16, 379-388.

Chamberlin, R.V. 1945. On some diplopods from the Indo-Australian archipelago. American Museum Novitates # 1282, 1-43.

Cheng, W.X., 1999. Rhizosphere feedbacks in elevated CO2. Tree Physiol. 19, 313-320.

Cheng, W.X. and Johnson, D.W. 1998. Elevated CO_2, rhizosphere processes, and soil organic matter decomposition. *Plant and Soil* 202, 167-174.

Chowdaiah, B.N. 1962. Cytological Studies of Some Indian Diplopoda (Myriapoda). Ph.D. Thesis, University of Mysore, India.

Chowdaiah, B.N. 1966a. Chromosome studies in two species of pill-millipedes (Diplopoda-Myriapoda). *Caryologia* 19, 135-141.

Chowdaiah, B.N. 1966b. Cytological investigations on some Indian Diplopoda (Myriapoda). *Nature* 210, 847.

Chowdaiah, B.N. 1966c. Cytological studies of some Indian Diplopoda – I (Myriapoda). Cytologia 31, 294-301.

Chowdaiah, B.N. 1967. Chromosome status (Diplopoda (Myriapoda): A note on the abnormal sex chromosome behaviour in the male of Thyropygus sp. *Current Science* 36, 105.

Chowdaiah, B.N. 1969. A note on the occurrence of chromatid bridges in the male of Arthrosphaera sp., a pill-millipede. *Current Science* 7, 169-170.

Chowdaiah, B.N. and Kanaka, R. 1969. Cytological studies of Indian Diplopoda – II (Myriapoda). Bull. Mus. Nat. E'His. *Naturelle* (Supple.) 41, 43-47.

Chowdaiah, B.N. and Kanaka, R. 1974. Cytological studies in six species of pill-millipedes (Diplopoda-Myriapoda). *Caryologia* 27, 55-64.

Chowdaiah, B.N. and Kanaka, R. 1979. Chromosome cytology of seven species of Indian Diplopoda (Myriapoda). In: Myriapod Biology (Ed. Camatini, M.), Academic Press, London, 9-20.

Chowdappa, P., Biddappa, C.C. and Sujatha, S. 1999. Efficient recycling of organic wastes in arecanut (Areca catechu) and cocoa (Theobroma cacao) plantation through vermicomposting. *Indian J. Agric. Sci.* 69, 563-566.

Cloudsley-Thompson, J.L. 1988. Evolution and Adaptation of Terrestrial Arthropods. Springer-Verlag, Berlin.

Coleman, D.C. and Crossley, D.A. 1996. Fundamentals of Soil Ecology. Academic Press, San Diego, CA.

Coûteaux, M.-M., Aloui, A. and Kurz-Besson, C. 2002. Pinus halepensis litter decomposition in laboratory microcosms as influenced by temperature and a millipede, Glomeris marginata. *Appl. Soil Ecol.* 20, 85-96.

Crawford, C.S., Bercovitz, K. and Warburg, M.R. 1987. Regional environments, life history patterns and habitat use of Spirostreptid millipedes in arid regions. *Zool. J. Linn. Soc.* London 89, 63-88.

Cromack, K., Sollins, P., Tood, R.L., Fogel, R., Todd, A.W., Fender, W.M., Crossley, M.E. and Jr. Crossley, D.A. 1977. The role of oxalic acid and bicarbonate in calcium cycling by fungi and bacteria – Some possible implications for soil animals. In: Soil Organisms as components of Ecosystems (Ed. Lohm, U. and Persson. T.), Ecological Bulletin # 25.

Chu, T.L., Szabó, I.M. and Szabó, I. 1987. Nocardioform gut actinomycetes of Glomeris hexasticha Brandt (Diplopoda). *Biol. Fert. Soils* 3, 113-116.

Dangerfield, J.M. 1990. Abundance, biomass and diversity of soil macrofauna in savanna woodland and associated managed habitats. Pedobiologia 34, 141-150.

Dangerfield, J.M. and Milner, A.E. 1993. Ingestion and assimilation of leaf litter in some tropical millipedes. J. Zool. (London) 229, 683-693.

Dangerfield, J.M. and Telford, S.R. 1989. Are millipedes important for soil fertility? Zim. Sci. News 23, 66-68.

Dangerfield, J.M. and Telford, S.R. 1991. Seasonal activity patterns of Julid millipedes in Zimbabwe. *J. Trop. Ecol.* 7, 281-285.

Dangerfiled, J.M., Milner, A.E. and Mathews, R. 1992. Seasonal activity pattern and behaviour of juliform millipedes in south-eastern Botswana. *J. Trop. Ecol.* 8, 451-464.

David, J.-F. and Gillon, D. 2002. Annual feeding rate of the millipede Glomeris marginata on holm oak (Quercus ilex) leaf litter under Mediterranean conditions. *Pedobiologia* 46, 42-52.

David, J.-F., Ponge, J.-F. and Delecour, F.1993. The saprophagous macrofauna of different types of humus in beech forests of the Ardenne (Belgium). Pedobiologia 37, 49-56.

DeSaussure, H. and Zehnter, L. 1897. Atlas de Phistoire naturelle des Myriapodes. Histoire Physique, Naturelle et Politique de Madagascar 27, 1-12.

Dzik, J. 1975. Sprobolid millipedes from the late Cretaceous of the Gobi desert, Mongolia. Palaeontologia Polonica 33, 17-24.

Enghoff, H. 1977. Arthrosphaera brandti (Humbert), a giant pill-millipede found in Tanzania, probably introduced from Sri Lanka (Diplopoda: Sphaerotheriidae). *Rev. Zool. Afr.* 91, 997-999.

Enghoff, H. 1984. Phylogeny of millipedes: A cladistic analysis. *Zool. Z. Syst. Evol. Forsch.* 22, 8-26.

Enghoff, H. 1990. The ground-plan of chilognathan millipedes (external morphology). In: Proceedings of the 7th International Congress of Myriapodology (Ed. Minelli, A.), E.J. Brill, Leiden, 1-24.

Enghoff, H. 2005. The millipedes of Thailand (Diplopoda). *Steenstrupia* 29, 87–103.

Enghoff, H. and Báez, M. 1993. Evolution and habitat patterns in endemic millipedes of the genus Dolichoiulus (Diplopoda: Julidae) on the Canary Islands. With notes on distribution patterns of other Canarian species swarms. *Biological Journal of the Linnean Society* 49, 277-301.

Gaston K. J. 2000. Global patterns in biodiversity. *Nature* 405, 220-227.

Gebhardt, K., Schimana, J., Müller, J., Fiedler, H.-P., Kallenborn, H.G., Holzenkämpfer, M., Krastel, P., Zeeck, A., Vater, J., Höltzel, A., Schmid, D.G., Rheinheimer, J. and Dittner, K. 2002. Screening for biologically active metabolites with endosymbiotic bacilli isolated from arthropods. *FEMS Microbiol. Lett.* 217, 199-205.

Geoffroy, J.J. 1984. Opisthoceiron canayerensis (Diplopoda: Craspedosornida): repartition de l'epèce et variation de la pigmentation. Mémoires de Biopéology 11, 295-302.

Giribet, G., Edgecombe, G.D. and Wheeler, W.C. 2001. Arthropod phylogeny based on eight molecular loci and morphology. *Nature* 413, 157-161.

Giller, P.S. 1996. The diversity of soil communities, the 'poor man's tropical rainforest'. Biodivers. *Conser.* 5, 135-168.

Golovatch, S.I. and Enghoff, H. 2003. Pill-millipedes of the Canary Islands: the Glomeris alluaudi-group (Diplopoda, Glomeridae). *Vieraea* 31, 9-25.

Golovatch, S.I., Hoffman, R.L. and de Morais, J.W. 1995. Identification plate for the millipede orders populating the neotropical region south of Central Mexico (Myriapoda, Diplopoda). *Studies on the Neotropical Fauna and Environment* 30, 159-164.

Golovatch S.I., Geoffroy J.-J. and Mauriès J.-P. 2006. Review of the millipede genus Hyleoglomeris Verhoeff 1910 (Diplopoda, Glomerida, Glomeridae), with descriptions of new species from caves in Southeast Asia. Zoosystema 28, 887-915.

Goodman, S.M. and Benstead, J.P. 2005. Updated estimates of biotic diversity and endemism for Madagascar. *Oryx* 39, 73-77.

Groombridge B. and Jenkins M.D. 2000. Global Biodiversity:Earth's Living Resources in the 21st Century. World Conservation Press, Cambridge.

Haacker, U. and Fuchs, S. 1972. Tree climbing in pill-millipedes. Oecologia, 10, 191-192.

Hamer, M.L. and Slotow, R.H. 2002. Conservation application of existing data for South African millipedes (Diplopoda). *Afr. Entomol.* 10, 29-42.

Hanlon, R.D.G. 1981. Some factors influencing microbial growth on soil animal faeces, II. Bacterial and fungal growth on soil animal faeces. *Pedobiologia* 21, 264-270.

Hannibal, J.T. and Feldmann, R.M. 1981. Systematics and functional morphology of Oniscomorph millipedes (Arthropoda: Diplopoda) from the Carboniferous of North America. *J. Palaeontol.* 55, 730-746.

Harvey, M.S. 2002. Short-range endemism in the Australian fauna: some examples from non-marine environments. *Invertebrate Systematics* 16, 555-570.

Hennings, C. 1906. Das Tomosvaryische organ der Myriopoden II. Zeit. Wissen. Zool. 80, 576-641.

Hennig, W. and Mickoleit, G. 1986. Wierbellose II: Gliedertiere. Jena VEB Gustav Fischer Verlag.

Heywood, V.H. and Watson, R.T. 1995. Global Biodiversity Assessment. United Nations Environment Programme, Cambridge University Press.

Hill, M.J., Matyot, P., Vel, T.M., Parr, S.J. and Shah, N.J. 2002. Marianne. In: National Museum of Natural History Smithsonian Institution, Washington DC, Attoll Research Bulletin # 495, 157-175.

Hirst, A.S. 1907. On four new pill-millipedes from the Malay Peninsula and Siam. – *Ann. Magazine Nat. Hist.* 7th Series 20, 215-219.

Hoffman, R.L. 1993. Biogeography of East African montane forest millipedes. In: Biogeography and Ecology of the Rain Forests of Eastern Africa (Ed. Lovett, J.C. and Wasser. S.K.), Cambridge University Press, Cambridge, 103-115.

Hoffman, R.L. 1976. The systematic status of the millipede genus Rajasphaera Attems 1935. *Entomol. Mitt. Zool. Mus. Hambg.* 5, 117-126.

Hoffman, R.L. 1980. Classification of the Diplopoda. Muséum d'Histoire Naturelle, Genève, Switzerland.

Hoffman, R.L., Golovatch, S.I., Adis, J. and de Morais, J.W. 2002. Diplopoda. In: Arachnida and Myriapoda (Ed. Adis, J.), Pensoft, Sofia-Moscow, 501-533.

Holloway, B.A. 1956. Revision of the New Zealand pill millipedes (Oniscomorpha, Sphaerotheridae). *Trans. Royal Soc of New Zealand* 84, 431-446.

Hopkin, S.P. and Read, H.J. 1992. The Biology of Millipedes. Oxford University Press, Oxford.

Hosabaale, M.H. 2000. Paschima Ghattada Jeeva Vaividhya (Biodiversity of Western Ghats). Dr. Shivarama Karantha Adhyayana Kendra, Vivekananda College, Puttur, Karnataka, India.

Janardanan, K.P. and Ramachandran, P. 1983. Observations on two new species of cephaline gregarines (protozoa: Sporozoa) of the genus Stenoductus Ramachandran, 1976 from sphaerotherid millipedes in Kerala, India. *Archiv für Protistenkunde* 127, 405-412.

Janssen, R. and Damen, W.G.M. 2006. The ten Hox genes of the millipede Glomeris marginata. Dev. Genes Evol. 216, 451-465.

Jeekel, C.A.W., 1974. The taxonomy and geography of the Sphaerotheriida (Diplopoda). *Symposium of the Zool. Soc. London* 32, 41-52.

Jeekel, C.A.W., 1986. Millipedes from Australia, 10: Three interesting new species and a new genus (Diplopoda: Sphaerotheriida, Spirobolida, Polydesmida). *Beaufortia* 36, 35-50.

Kämpfer, P., Young, C.-C., Sridhar, K.R., Arun, A.B., Lai, W.A., Shen, F.T. and Rekha, P.D. 2006. Transfer of (Flexibacter) sancti, (Flexibacter) filiformis, (Flexibacter) japonensis and (Cytophaga) arvensicola to the genus Chitinophaga and description of Chitinophaga skermanii sp. nov. *Int. J. Syste. Evolution. Microbiol.* 56, 2223-2228.

Kaneko, N. 1999. Effect of millipede Parafontaria tonominea Attems (Diplopoda: Xystodesmidae) adults on soil biological activities: A microcosm experiment. *Ecol. Res.* 14, 271-279.

Kelley, J. and Peterson, R. 1997. Crop residues as a resource – The use of fungi to upgrade lignocellulosic wastes. *Biology International* 35, 16-20.

Kime, R.D. and Golovatch, S.I., 2000. Trends in the ecological strategies and evolution of millipedes (Diplopoda). *Biol. J. Linn. Soc.* 69, 333-349.

Kraus, O. 1974. On the morphology of Palaeozoic Diplopids. Symposia of the Zoological Society of London 32, 13-22.

Kremen, C., Colwell, R.K., Erwin, T.L., Murphy, D.D., Noss, R.F. and Sanjayan, M.A. 1993. Terrestrial arthropod assemblages: their use in conservation planning. *Conserv. Biol.* 7, 796-808.

Kubíková, J. and Rusek, J. 1976. Development of xerotermic rendzinas-A study in ecology and soil microstructure. Rozpravy ČSAV, Mat. Přír. 86, 1-78.

Lawrence, J.M. 1984. The Centipedes and Millipedes of Southern Africa: A Guide. A.A. Balkema, Cape Town and Roterdam.

Lawrence, J.M. and Samways, M.J. 2003. Litter breakdown by the Seychelles giant millipede and the conservation of soil process on Cousine Island, Seychelles. *Biological Conservation* 113, 125-132.

Manton, S.M. 1977. The Arthropoda. Oxford University Press, Oxford.

Maraun, M. and Scheu, S. 1996. Changes in microbial biomass respiration and nutrient status of beech (Fagus sylvatica) leaf litter processed by millipedes (Glomeris marginata). Oecologia 107, 131-140.

Marcuzzi, G. 1970. Experimental observations on the role of Glomeris spp. (Myriapoda, Diplopoda) in the process of humification of litter. *Pedobiologia* 10, 401-406.

Márialigeti, K., Conteras, E., Barabás, Gy., Heydrich, M. and Szabó, I.M. 1985. True intestinal actinomycetes of millipedes (Diplopoda). *J. Invert. Pathol.* 45, 120-121.

Mauriès, J.-P. 2001. Sur l'identité de Zephronia hainani Gressitt, 1941, à propos de la description d'un nouveau Prionobelum (Diplopoda, Sphaerotheriida, Sphaeropoeidae) de Haïnan, Chine. *Zoosystema* 23, 131-142.

McBrayer, J.F. 1973. Exploitation of deciduous leaf litter by Apheleria montana (Diplopoda: Eurydesmidae). *Pedobiologia* 13, 90-98.

McGonigle, T.P. 1995. The significance of grazing on fungi in nutrient cycling. *Can J. Bot.* (Suppl. 1), S1370-S1376.

Meads, M.J. and Fitzgerald, B.M. 2001. List of invertebrates on Mokoia Island, Lake Rotorua. Conservation Advisory Science Notes No. 343, Department of Conservation, Wellington.

Mikhaljova E.V. and Lim K.-Y. 2006. New species of the genus Hyleoglomeris from Korea (Diplopoda: Glomerida: Glomeridae). *Zootaxa* 1224, 45-58.

Miller J.C. 1993. Insect natural history, multi–species interactions and biodiversity in ecosystems. *Biodivers. Conser.* 2, 233–241.

Mwabvu, T. 1997. Millipedes in small-scale farming systems in Zimbabwe: Abundance and diversity (Diplopoda, Spirostreptidae). *Entomologica Scandinavica* 51, 287-290.

Natarajan, R. 1959. Cytological studies in Diplopoda (Myriapoda). *J. Zool. Soc. India* 11, 9101.

Neuhauser, E.F. and Hartenstein, R. 1978. Phenolic content and palatability of leaves and wood to soil isopods and diplopods. *Pedobiologia* 18, 99-109.

Nicholson, P.B., Bocock, K.L. and Heath, O.W. 1966. Studies on the decomposition of the faecal pellets of a millipede (Glomeria marginata (Villers)). *J. Ecol.* 54, 755-766.

Oldeman, R.A.A. 1990. Forests: Elements of Silvology. Springer, Berlin.

O'Neill, P., 1993. Environmental Chemistry, second ed. Chapman and Hall, London.

O'Neill, E.G. 1994. Responses of soil biota to elevated atmospheric carbon dioxide. Plant and Soil 165, 55-65.

Oettinger, R. 1909. Zur Kenntnis der spermatogenese bei de Myriapoden. Samenreifung und Samenbilung be Pachyiulus varius Fabre. Arch. F. Zellforschung 3, 563-626.

Palaniappan, S. P., Natarajan, K. 1993. Practical aspects of organic matter maintenance in soil. In: Organics in Soil Health and Crop Production (Ed. Thampan, P.K.), Peekay Tree Crops Development Foundation, Cochin, 23-41.

Paulpandian, A.L. 1966. A note on the respiration of the pill-millipede Arthrosphaera dalyi (Pocock). Current Science 35, 178-179.

Petersen and Luxton, 1982. A comparative analysis of soil fauna populations and their role in decomposition processes. Oikos 39, 287-388.

Pioro, H.L. and Stollewerk, A. 2006. The expression pattern of genes involved in early neurogenesis suggests distinct and conserved functions in the diplopod Glomeris marginata. *Dev. Genes Evol.* 216, 417-430.

Plowman, K.P. 1990. Forest-floor decomposers. In: Australian Tropical Reinforests: Science-Values-Meaning (Ed. Webb, L.J. and Kikkawa, J.), CSIRO Publications, East Melborne, 98-104.

Pocock, R.I. 1892. Report upon two collections of Myriapoda sent from Ceylon by Mr. E.E. Green and from various parts of Southern India by Mr. Edger Thurston of the Government Central Museum, *Madras. Bombay Nat. Hist. Soc.* 17, 131-175.

Pocock, R.I. 1899. A monograph of the pill-millipedes (Zephroniidae) inhabiting India, Cylon and Burma. *Journal of Bombay Natural History Society* 12, 269-285.

Ponge, J.-F. 1999a. Heterogeneity in animal communities and the development of humus forms. In: Going Underground – Ecological Studies in Forest Soils (Ed. Rastin, N. and Bauhus, J.), Research Signpost, Trivandrum, 33-44.

Pramanik, R., Sraker, K. and Joy, V.C. 2001. Efficiency of detritivore soil arthropods in mobilizing nutrients from leaf litter. *Trop. Ecol.* 42, 51-58.

Prpic, N.-M. 2004. Homologs of wingless and decapentaplegic display a complex and dynamic expression profile during appendage development in the millipede Glomeris marginata (Myriapoda: Diplopoda). *Frontiers in Zoology* 1, 1-12.

Rawlins, A.J., Bull, I.D., Poirier, N., Ineson, P. and Evershed, R.P. 2006. The biochemical transformation of oak (Quercus robur) leaf litter consumed by the pill millipede (Glomeris marginata). *Soil Biol. Biochem.* 38, 1063-1076.

Rawlins, A.J., Bull, I.D., Ineson, P. and Evershed, R.P. 2007. Stabilisation of soil organic matter in invertebrate faecal pellets through leaf litter grazing. *Soil Biology and Biochemistry* 39, 1202-1205.

Redi, B.H., van Aarde, R.J. and Wassenaar, T.D. 2005. Coastal dune forest Development and the regeneration of millipede communities. *Restoration Ecol.* 13, 284-291.

Regier, J.C., Wilson, H.M. and Shultz, J.W. 2005. Phylogenetic analysis of Myriapoda using three nuclear protein-coding genes. Molecular Phylogenetics and Evolution 34, 147-158.

Reichle, D.E., Shanks, M.H. and Crossley, D.A. 1969. Calcium, potassium and sodium content of forest floor arthropods. *Ann. Environ. Soc. Amer.* 62, 57-62.

Rossi, W. and Weir, A. 1998. Triainomyces, a new genus of Laboulbeniales on the pill-millipede Procyliosoma tuberculatum from New Zealand. *Mycologia* 90, 282-289.

Sakwa, W.N. 1974. A consideration of the chemical basis of food preference in millipedes. Symposium of Zoological Society of London 32, 329-346.

Scheu, S. and Wolters, V. 1991. Influence of fragmentation and bioturbation on the decomposition of ^{14}C-labelled beech leaf litter. *Soil Biol. Biochem.* 23, 1029-1034.

Shear, W.A. 1969. A synopsis of the cave millipedes of the United States with an illustrated key to genera. *Psyche* 76, 126-143.

Shear, W.A. 1981. Two fossil millipedes from the Dominican amber (Diplopoda: Chytodesmidae, Siphonophoridae). *Myriapodologica* 1, 51-54.

Shear, W.A. 1984. Cave millipedes of the United States. III. Two new species from the western states (Diplopoda: Polydesmida, Chrodeumatida), *Myriapodologica* 351, 283-289.

Shelley, R.M. 2003. A revised, annotated, family-level classification of the Diplopoda. *Arthropoda Sel.* 11, 187-207.

Sibly, R.M. 1981. Strategies of digestion and defecation. In: Physiological Ecology: An Evolutionary Approach to Resource Use (Ed. Townsend C.R. and Calow, P.), Blackwell Scientific Publications, Oxford, 109-139.

Sierwald, P. and Bond, J.E. 2007. Current status of the Myriapod Class Diplopoda (millipedes): Taxonomic diversity and phylogeny. *Ann. Rev. Entomol.* 52, 401-420.

Sierwald, P. and Reft, A.J. 2004. The millipede collections of the world. Fieldiana N.S. 103, 1-100.

Sierwald, P., Shear, W.A., Shelley, R.M. and Bond, J.E. 2003. Phylogeny revisited in the light of the enigmatic order Siphoniulida. *J. Zool. Sys. Evol. Res.* 41, 87-99.

Smit, A.-M. and van Aarde, R.J. 2001. The influence of millipedes on selected sol elements: a microcosm study on three species occurring on coastal sand dunes. *Functional Ecology* 15, 51-59.

Sokoloff, J. 1914. Uber die spermatogenese bei Polyxenus sp. *Zool. Anz.* 44, 558-566.

Spiess, T. 1981. Structure and phylogenetic interpretation of diplopod eyes (Diplopoda). Zoomorphology 98, 241-260.

Stork, N.E. 1993. How many species are there? Biodiversity and Conservation 2, 215-232.

Striganova, B.R. and Prishutova, Z.G. 1990. Food requirements of diplopods in the dry steppe subzone of the USSR. *Pedobiologia* 34, 37-41.

Swift, M.J., Heal, O.W., Anderson, J.M. 1979. Decomposition in Terrestrial Ecosystems. Blackwell Scientific Publications, Oxford.

Tajovský, K. Šanůčtrková, H., Háněl, L., Balík and Lukešová. 1992. Decomposition of faecal pellets of the millipede Glomeris hexasticha (Diplopoda) in forest soil. *Pedobiologia* 36, 146-158.

Taylor, E.C. and Crawford, C.S. 1982. Microbial gut symbionts and desert detritivores. *Sci. Rev. Arid Zone Res.* 1, 37-52.

Teuben, A. and Verhoef, H.A. 1992. Direct contribution by soil arthropods to nutrient availability through body and faecal nutrient content. *Biol. Fert. Soils* 14, 71-75.

Urbasek, F. and Tajovský, K. 1991. The influence of food and temperature on enzymatic activities of the millipede Glomeris hexasticha (Diplopoda) *Rev. Ecol. Biol. Sol.* 28, 155-163.

van Aarde, R.J., Ferriera, S.M. and Kritzinger, J.J. 1996a. Millipede communities in rehabilitating coastal dune forests in northern KwaZulu-Natal, South Africa. *Journal of Zoology* (London) 238, 703-712.

van der Drift, J. 1951. Analysis of the animal community in a beech forest floor. Tijdschrift voor Entomologie 94, 1-168.

van der Drift, J. 1963. The disappearance of litter in mull and mor in connection with weather conditions and the activity of the macrofauna. In: Soil Organisms (Ed. Doeksen, J. and van der Drift, J.), North-Holland Publishing Company, Amsterdam, Netherlands, 125-133.

van der Drift, J. 1975. The significance of the millipede Glomeris marginata (Villers) for oak-litter decomposition and an approach of its part in energy flow. In: Progress in Soil Zoology (Ed. Vanek, J.), Academia, Prague, 293-298.

Verhoef, H.A. and Brussaard, L., 1990. Decomposition and nitrogen mineralization in natural and agroecosystems - the contribution of soil animals. *Biogeochem.* 11, 175-211.

Vohland, K. and Schroth, G. 1999. Distribution patterns of the litter macrofauna in agroforestry and monoculture plantations in central Amazonia as affected by plant species and management. *Appl. Soil Ecol.* 13, 57-68.

Wallwork, J.A. 1976. The Distribution and diversity of Soil Fauna. Academic Press, London.

Warchalowska-Sliva, E., Maryanska-Nadachowska, A and Kania, G. 2004. Cytogenetic studies of Glomeris (Diplopoda; Glomeridae). *Folia Biologica* (Krakow) 52, 1-2.

Warren, M.W. and Zou, X. 2002. Soil macrofauna and litter nutrients in three tropical tree plantations on a disturbed site in Puerto Rico. *Forest Ecol. Managem.* 170, 151-171.

WCMC (World Conservation Monitoring Centre) 1992. Global Biodiversity: Status of the Earth's Living Resources. Chapman and Hall, London.

Webb, D.P. 1977. Regulation of deciduous forest litter decomposition by soil arthropod faeces. In: The Role of Arthropods in Forest Ecosystem (Ed. Mattson, W.J.), Springer-Verlag, Heidelberg, 57-69.

Wesener. T. 2006. Giant pill millipede phylogeny and plate tectonics (Diplopoda, Sphaerotheriida). Ninety-ninth DZG Annual Meeting of the German Zoological Society, September 16-20, Münster, Germany, Abstract, 149.

Wesener, T. and Sierwald, P. 2005a. The gaint pill-millipedes of Madagascar. *Proc. California Acad of Sci.* 56, 557-599.

Wesener, T. and Sierwald, P. 2005b. New giant pill-millipede species from the littoral forest of Madagascar (Diplopoda, Sphaerotheriida, Zoosphaerium). *Zootaxa* 1097, 1-60.

Wilson, E.O. 1992. The Diversity of Life. Harvard University Press, Harvard.

Witkamp, M. and Crossley, D.A. 1966. The role of arthropods and microflora in the breakdown of white oak litter. *Pedobiologia* 6, 293-303.

Yeates, G.W., Tate, K.R. and Newton, P.C.D. 1997. Response of the fauna of a grassland soil to doubling of atmospheric carbon dioxide concentration. *Biol. Fert. Soils* 25, 307-315.

In: Organic Farming: Methods, Economics and Structure
Editors: Michael Nelson and Ivan Artamova
ISBN 978-1-60692-864-6

Chapter 4

METHODS FOR THE ANALYSIS OF STRUCTURE, PRODUCTION AND RESULTING ECONOMICAL TRAITS IN ORGANIC AND CONVENTIONAL FARMS: MEAT SHEEP EXAMPLES

***J. Cabaret*[1,1], *G. Laignel*[2] *and M. Benoit*[2]**
[1]INRA, IASP 213, 37380 Nouzilly-F and
[2]INRA Theix 63122 Saint Genès Champanelle-F

ABSTRACT

Farm structure and production traits are numerous, and it is a real difficulty to integrate such a large bulk of knowledge into simplified categorisations. One common tool is discriminant analysis, which requires an *a priori* categorisation. Conversely, clustering methods are available to identify homogeneous groups *a posteriori*. The difficulty of clustering methods is in the very large number of options. The discriminant analysis and the clustering methods intend to propose categories on traits which are evaluated at one moment without any attempt to relate and to suggest any hierarchy in the order or time of appearance of the traits. They are comparable to a snapshot at one moment. Phylogenetic analyses are intended to propose a sequence of appearances of the traits; they intend to reconstruct a history of events. In the present paper we used data from meat sheep farms in the centre of France. Some farms were organic and others were conventional; all farms studied were subsidised by the state. Firstly, we used cluster and phylogenetic analyses for categorisation of conventional (9) and organic (7) meat sheep farms. Secondly, the evolution on several years in seven organic farms was studied. The phylogenetic analysis (Hennig86 programme implemented with the freeware Tree Gardener) is performed on re-coded data into evoluted (derived) *versus* non-evoluted (ancestral) trait or level of the trait. The phylogenetic analysis we propose is a flexible tool that reconstructs the sequence of traits, and probably better corresponds to the strategy of the farmer in its environment.

1. INTRODUCTION

The first dilemma is that farm structure and production traits are numerous, and it is a real difficulty to integrate such a large bulk of knowledge into simplified categorisations. The second dilemma is that information can be obtained directly from farmers and then the number of cases will be limited or from census data, and then the number of farms is important but data may not be fully relevant for a specific investigation. This second dilemma is usually solved by tailored surveys for a particular point of view. Very few surveys rely on existing census data. The paper by Costas et al., 2004, is one rare example of census data utilisation. The analysis included 1252 livestock farms of Buenos Aires Province in Argentina. Data of the National Agricultural Survey of 1999 were used to construct the variables regarding cattle requirements and resource production. The multivariance analysis used allowed: (a) general characterisation of the farms, (b) creation of a typology and (c) description of groups and comparisons among them. Three technical levels were identified with regard to feeding systems: those that supplement self-produced resources, those that use grazing on improved pastures and those that rely on natural grassland grazing. The autonomy of the farm ranged from high (first system with complete cycle—brood cows, calves, replacement females, growing steers and fattening animals) to low in the third group—cow-calf operations with or without rearing. Since such census data are not repeated every year due to costs, researchers try to elaborate indexes that may represent a simplified tool for action (acceptability of technical proposals, prospective on viability (Sineiro-Garcia et al., 2004), building up of a general agriculture policy (Alvarez-Lopez, 2008), etc.). Costas et al. (2004) thus constructed two indicators: the Net Index of Forage Production for identifying the category of farm and the Self-produced Supplements that allowed separation of the type with the highest technical level. Zaroni and Carmo (2006) presented indicators on their own survey. Their study aimed to develop a scoring scale to estimate a farm's modernisation status. The methodology applied was the item response theory (IRT), which can be seen as a new and complex statistical methodology for the typology of small farmers.

Several researchers constructed their own survey, which found that the number of farms remains modest and can even be restricted to less than 10 farms (see Kaltoft, 1999 or McMahon, 2003). Those very low numbers are found mostly in sociological in-depth studies of farms and preclude the use of any statistical procedures. Moreover, the farms in these studies were not selected at random but were chosen for their exemplarity in one or several particular traits. However, topologies of farms are proposed based on ethics of agricultural practice (Barres et al, 1985), on values about nature (Kaltoft, 1999), on bearing witness in order to transform society (McMahon, 2003), or on relationships with stakeholders (Leguen and Jenot, 2007). The work of Castaldo et al. (2006) is a good example of the usual strategy: they intended to identify and classify the production systems for bovine fattening in pastoral conditions in the semiarid Argentina Pampa, taking into account the physical, productive and economic characteristics and its management. A cluster analysis was used to reduce the number of variables involved and to define farmer groups. The first subsystem was denominated by traditional grazing that responds to operations with low levels of productivity and daily average gain. The second subsystem is a transition between the traditional system and the technological system. The third subsystem is a technological grazing that utilises

1 Correspondence to: cabaret@tours.inra.fr

greater levels of feed and shows higher levels of productivity and average daily gain. Milan et al. (2006) used cluster analyses and multiple correspondence analyses: they identified three groups of farms: the first group has followed a strategy of intensification of the system and includes the smallest farms; the second group is constituted by farms of intermediate size, very specialised in beef production; and the third group includes large farms that have followed a strategy of extensification, complementing farm incomes by the exploitation of other livestock species. Srairi et al. (2003) used structural and economic parameters of the farms to collect data for descriptive analysis based on principal components and cluster analyses; five groups of farms were identified. Finally, most of the studies were dedicated to cattle and sheep farms need to be further investigated.

A wide array of multivariate analyses has been performed in the few available biotechnical descriptions of farms: discriminant analysis (1 use), K-means (2 uses), principal component analysis (5 uses), correspondence analysis (2 uses), and cluster analyses (4 uses). One of the most common tools is discriminant analysis, which requires an *a priori* categorisation. Clustering methods are available to identify homogeneous groups *a posteriori* as well as principal components (quantitative data) and correspondence analyses (qualitative data). The joint analysis of quantitative and qualitative data require transformation of data (all quantitative are transformed into qualitative ones, or more rarely qualitative data are transformed in quantitative ones). Some of the cluster analyses may incorporate jointly quantitative and qualitative data) (see example in Cabaret et al, 2008). All multivariate analyses aim at providing the "best" description of the data and do not perform statistical inference.

In phylogenetic analyses each character (coded as 0 when plesiomorphic or ancestral and 1 when apomorphic, e.g., derived or evoluted from the ancestral state) is polarised and is used directly in the analytic process. It is then completely different from other types of analyses since other analyses are established mostly on distance matrices and not on individual characters, which are not polarised. Few studies are available on the use of phylogenetics in fields other than systematics (ecology- Brooks, 1985 Lambshead and Paterson, 1986, Bellan-Santini, Dauvin and Bellan, 1994; economics-McCarty, 2005, Andersen, 2003; linguistics-Darlu et al, 1990; parasitology-Cabaret, 2003). In ecological and parasitological studies the phylogenetic analysis was more efficient than multivariate analyses to detect structures in the data.

The typologies of farms are still in the infancy for social, economic and technical descriptions. The aim of this paper is to propose a simple phylogenetical analysis based on parsimony of changes in binary character states of economical and production relevance, and to compare the results with the most popular cluster analyses. The studied farms are producing meat sheep either organically or conventionally, and we expect to have clear cut differences between those two types of animal farming. We studied the evolution of organic farms during several years, and we expected that the between-farms differences should be more important than the within farm variability (see Benoit and Laignel, 2004: on 15 years 50% of the farms did not change their typology).

2. Materials and Methods

2.1. Farm Economical and Production Characteristics

A limited set of actual farm results is presented in Annex 1 and is given for practice with few farms and characters.

Annex 1. File for Hennig86 (Out: outgroup with all ancestral state of characters and other lines corresponding to the 6 farms analysed; five columns are the states of the 5 characters in each farm)

```
xread
'farm characteristics'
5 7
Out 0    0    0    0    0
F1  0    1    1    0    1
F2  0    1    0    1    0
F3  1    0    0    0    1
F4  0    1    0    1    1
F5  1    0    0    0    0
F6  1    1    1    1    0
proc/;
```

The data were collected on one occasion in either organic (7) or conventional farms (9) and on several years (two to six) in seven organic farms. They were all meat sheep farms from the centre of France and are described in detail in Benoit and Laignel (2004) and Cabaret et al., 2008). The farms were in the framework of French subsidised agriculture.

The following criteria were used (Benoit and Laignel, 2006):

- the utilised total agricultural area (UTAA) is composed of grasslands (permanent, temporary and artificial pastures, maize for silage, etc.) and areas for cereals and proteaginous plants
- livestock units: 0.14 for ewes and 0.06-0.09 for grazing lambs
- most ratios (numerical productivity, gross margin, etc.) are expressed in relation to the yearly number of females over 12 months
- numerical productivity of the flock: number of lambs produced (sold and females kept on farm for reproduction) per ewe over 12 months.
- sheep gross margin: difference between gross product (sheep production sold, subsidies for sheep ...) and input for the flock (feeding, reproduction, veterinary costs, grassland fertilisers, services, fees, etc.)
- productivity of the work: [extended livestock units = livestock units + (Ha for cereals and proteaginous plants) /2] /full time worker on an annual basis.
- counterseason index: (Σ (no. of lambing of the period x coefficient)/(no. of lambing during the year). The coefficient has the following values: 0.8 (July), 1. (August to

15th of October), 0.9 (15/10 to 1/11), 0.7 (1/11 to 15/11), 0.6 (15/11 to 15/12), 0.5 (15/12 to 15/01), and 0 for all other periods.

2.2. Cluster Analyses

The term cluster analyses corresponds to a set of numerical techniques in which the main purpose is to divide the objects of study (here, farms) into discrete groups. The most commonly used cluster analyses are based on agglomerative hierarchical methods. The term agglomerative means that the dendrogram (tree-like diagram) is produced by starting with all the objects (farms) to be clustered and then separated successively, combining the most similar objects and/or until all are in a single hierarchical group. Average linkage techniques provide a more balanced approach to clustering than nearest and farthest neighbour techniques. There are two approaches, the unweighted pair group method using arithmetic averages (UPGMA) and the unweighted centroid clustering (UPGMC). The calculations were unweighted and gave equal weight to each point in each cluster. In general, the unweighted methods are used unless the data are expected to have some clusters that are much smaller than others. The centroid method replaces the clusters of objects (farms) with their centroids, that is with the points whose coordinates are averages of the objects in each group; this method is more appropriate when one is interested in chaining among clusters (groups of farms) rather than primary connections among objects at the lower level (Legendre and Legendre, 1998). When the data are transposed (variables become cases), it can be used to evaluate relationships between parameters used to characterise the farms.

2.2. Phylogenetic Analyses

A set of information on how to use the programs Hennig86 (Farris, 1989) and Tree Gardener (1997) is available at the following address:
http://wcentre.tours.inra.fr/sfpar/usehennig.htm. The programs can be downloaded from the same Internet site or are available on demand from one of us (J. Cabaret). Useful information on the use of Hennig86 is available (Lipscomb 1994,
http:/www.gwu.edu/~clade/faculty/lipscomb/web.pdf). Different commands (Annex 2) are available in Hennig86: h* command was the most used and constructs a simple cladogram by a single pass through the data and then applies branch-swapping to that tree. The cladograms in the shape of a tree-like graphic were constructed using Tree Gardener (1997).

Annex 2. Principal commands available in Hennig86 and used in this paper

hennig: Constructs a single tree by a single pass through the data. This is very rapid but may not find the shortest tree if there is much homoplasy (e.g., similarities not due to ancestral-derived states of characters) in the data.

hennig*: Constructs a single tree by a single pass through the data and then applies branch-swapping to the initial tree, retaining just one tree.

mhennig: Constructs several trees, each by a single pass but adding the taxa in a different sequence each time.

Annex 2. (Continued)

mhennig*: Constructs several trees, each by a single pass but adding the taxa in a different sequence each time and then applies branch-swapping to each of the trees, retaining just one tree for each initial one.
Nelsen: Calculates strict consensus tree

We use the term cladogram, as our analyses specified the relative degree of phylogenetic relationship between farms but did not attempt to specify direct ancestor-descendant relationships (like a pedigree) as in phylogenetic trees (Kitching et al, 1998). Cladograms were constructed using the principle of parsimony (and expressed for each cladogram by the length) that requires that *ad hoc* assumptions are minimised as far as possible, e.g., the number of changes in state of character from ancestral to derived and eventually back from derived to ancestral is kept to a minimum. We assume that the ancestral farm has all characters set to ancestral value. The overall consistency index (CI) indicates the degree of fit between the data and the constructed cladogram (it is a measure of homology: common state of characters found in a group, ancestral or derived); a CI of 100% is indicative of a perfect fit. The problems with consistency index are:

i) the inclusion of uninformative characters in the calculation will inflate the CI values,
ii) as the number of taxa increases, values of CI decreases, irrespective of the information content,
iii) the CI never reaches zero.

The retention index (RI, in %), is the measure of the degree of similarity of the state of characters (in fact, the rate of synapomorphy, e.g., derived characters in a cluster) in groups and higher values are expected in informative cladograms (Lipscomb, 1994). For example, a group of farms have the same state of derived characters (they may be large size of flock, high gross margin and high work productivity). Those indices are also available for each character. Very often several equally parsimonious cladograms are found and then a consensus cladogram is presented. Consensus trees (Nelsen procedure) are used to show the information about farms' relationships that all the equally parsimonious cladograms have in common. When a cladogram is elected we then concentrate on the state of characters that were influential in the construction. The CI and RI for each character give a first approach to detect the most important characters. For each influential character we detect their distribution among the group of farms is constructed in the cladogram. Statistical comparisons can be made using the Fisher exact test (Cabaret, 2003) or Mantel-Haenszel test, due to the fact that the Khi-square tests do not apply easily on small groups. For example, a derived character was found in 9 cases out of ten in one group and in one case out of six ; the test on such a two by two matrix (9, 1; 1, 5) provided a bilateral probability of 0.008, hence indicating an extremely significant relationship between the state of character and the constituted groups of farms. All probabilities in the rest of the text are calculated either with Fisher exact or Mantel-Haenzel tests (a freeware program is also available).

3. RESULTS

3.1. Comparative Analyses on Example Data

The dendrograms (fig.1) based on either UPGMA or centroid indicated that the veterinarian as a main counselor and lamb mortality were associated characteristics of interest—when mortality is high the veterinarian seems a resource for advice—whereas work productivity, gross margin per ewe and numerical productivity, indicating that the farm has good productivity—were a homogeneous group. The established groups of farms were different when using the UPGMA or centroid method. The following groups were evidenced when using UPGMA (fig.2 a): F6, then F3-F4-F5, then F2 and lastly F1. The centroid analysis (fig.2 b) indicated that the following farms were regrouped consistently: F3 and F5, F4 and F2, and lastly F6 and F1. The phylogenetic analysis (fig 3: Consensus cladogram) indicated that F3 and F5 were related (with high numerical productivity and low mortality of lambs), whereas a F1, F2, F4 and F6 group was characterised by low numerical productivity and high lamb mortality. The only common association is F3 and F5 in all analyses.

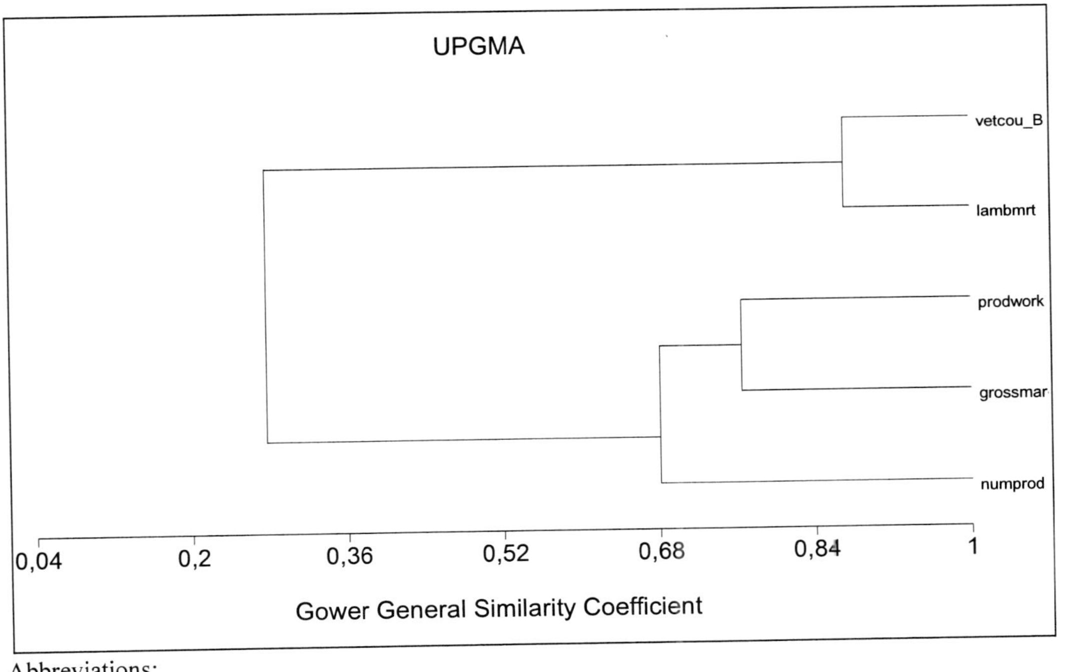

Abbreviations:

Vetcou_B:	veterinarian as a main counselor for sanitary problems (yes/No)
Grossmarg :	gross margin/year
Newe:	number of ewes in the flock
Numprod :	numerical productivity
Lambmrt:	lamb mortality
Prodwork:	work productivity

Figure 1. Dendrogram of characteristics of farm from the example file (Only UPGMA presented since centroid regrouping was nearly identical)

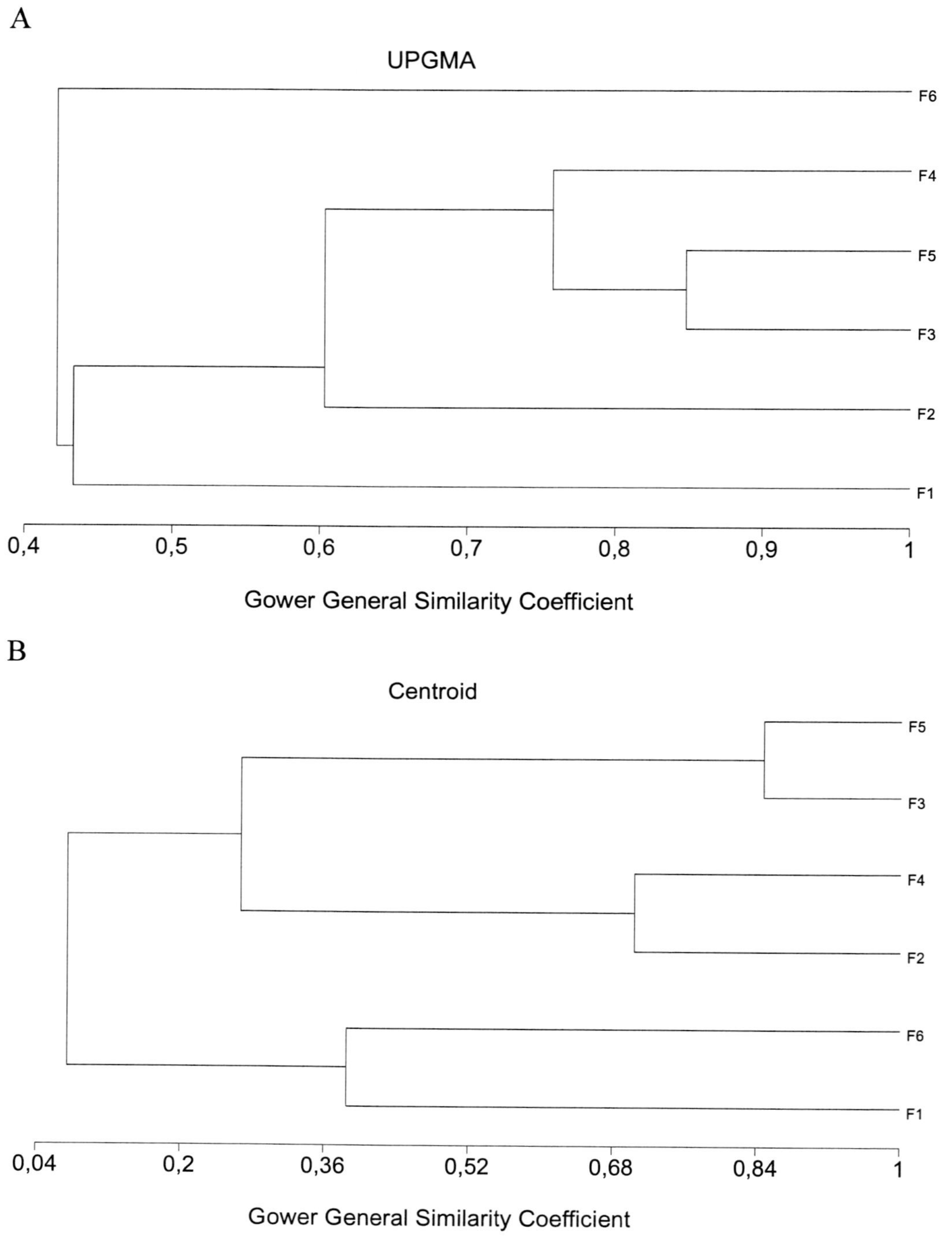

Figure 2. Dendrogram of characteristics of farm from the example file (A: UPGMA and B: centroid)

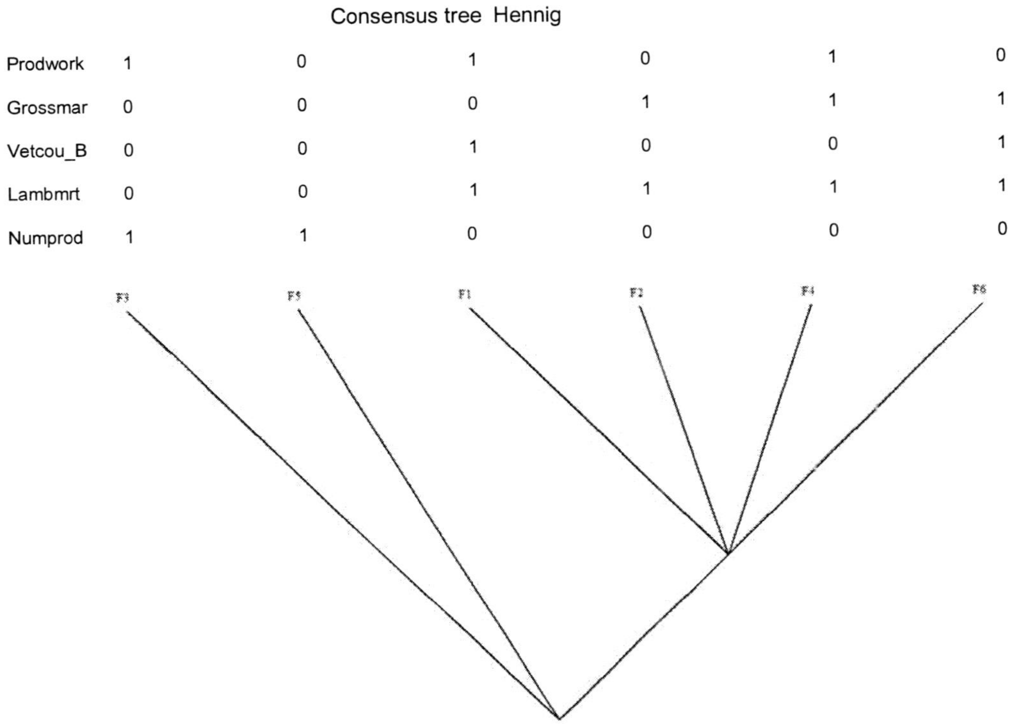

Figure 3. Consensus dendrogram obtained with Hennig86 on example data. Length 11, consistency index 45, retention index 33.

Table 1. Real flock data used for the example analyses

Cluster analysis	Actual value or presence 1 /absence 0				
Farm	Numerical productivity	Lamb mortality	Veterinarian as a main counselor	Gross margin	Productivity of the work
F1	108	23	1	58	105
F2	113	18	0	108	52
F3	149	9	0	85	99
F4	111	20	0	86	98
F5	123	11	0	79	92
F6	179	15	1	102	72
Phylogenetic analysis	State of character (0-ancestral and 1-evoluted)				
Farm	Numerical productivity	Lamb mortality	Veterinarian as a main counselor	Gross margin	Productivity of the work
F1	0	1	1	0	1
F2	0	1	0	1	0
F3	1	0	0	0	1
F4	0	1	0	1	1
F5	1	0	0	0	0
F6	1	1	1	1	0
State of character	0 when below median	0 when mortality is high	0 when veterinarian is not a counselor	0 when below median	0 when below median

3.2. Comparison of Organic and Conventional Meat Sheep Farms

The F9-conventional and F13-organic in the UPGMA dendrograms (figure 4 a) were atypical (although no single characteristic could explain their atypicality) and could not be regrouped with others. Two strongly identified groups (Gower distance over 0.6) were: F1-2-4-5-6-8-10-14 and F3-7-11-12-15-16, the first one being rather conventional and the second mostly organic. Two groups could be identified when using the centroid method (figure 4 b): F1-4-5-6-8-10-13-14 and F3-7-9-11-12-15-16, and could not be related to organic or conventional. Two main groups were identified using phylogenetic method (cladogram: length=24,CI=29 and RI=65) (figure 5): F1-3-10-11-12-13-15-16 based on high complementary drugs and minerals costs (p=0.13), high productivity of the work (p=0.13) - mostly organic and F4-5-6-7-8-9-14 based on high proportion of grassland (p=0.01) in their total agricultural area, higher number of ewes (p=0.13)—mostly conventional. Subgroups could be identified: high gross margin and numerical productivity in F12-15-16 (p=0.02) and high numerical productivity F4-6-7-9 (p=0.08). The cladogram provided information on groups and subgroups.

Table 2. Characteristics of 16 studied meat sheep farms in centre France (F1-F9: conventional and F10-F16: organic)

Farm	Veterinary costs	Complementary drugs and minerals	Percent of grassland of the total agricultural area	Number of ewes	Numerical productivity	Gross margin	Productivity of the work
F1	6.00	2.00	94	600	110	70	90
F2	6.30	2.10	82	870	108	58	105
F3	9.50	14.20	90	412	113	108	52
F4	12.00	0.90	95	465	149	85	99
F5	3.70	1.70	64	630	111	86	98
F6	4.40	0.90	85	522	123	79	92
F7	6.30	1.60	77	373	179	102	72
F8	7.80	3.20	70	472	113	81	105
F9	8.10	1.60	49	773	170	132	59
F10	3.40	3.60	88	213	88	58	58
F11	8.60	2.30	94	449	110	97	56
F12	3.20	1.80	89	348	149	112	49
F13	7.50	19.30	89	343	128	62	110
F14	3.30	1.50	60	221	96	66	92
F15	5.40	14.40	76	365	164	103	102
F16	5.90	3.30	88	807	152	94	51

3.3 Evolution in Organic Farms According to Year

The following groups were constructed: F5, F4-6-7, F1-2, F3 (UPGMA- figure 6A), F4-6-7, F3, F1-2-5 (Centroid –Figure 6b), F1-2-4-6, F3, F7 (Cladogram- Figure 7). The common regrouping is that F3 stands alone in all three analyses and that F4-6 are grouped; however, the groups are different for all other farms. The farms during their evolutions are fairly similar from one year to another, although F1 farm was distributed in several groups.

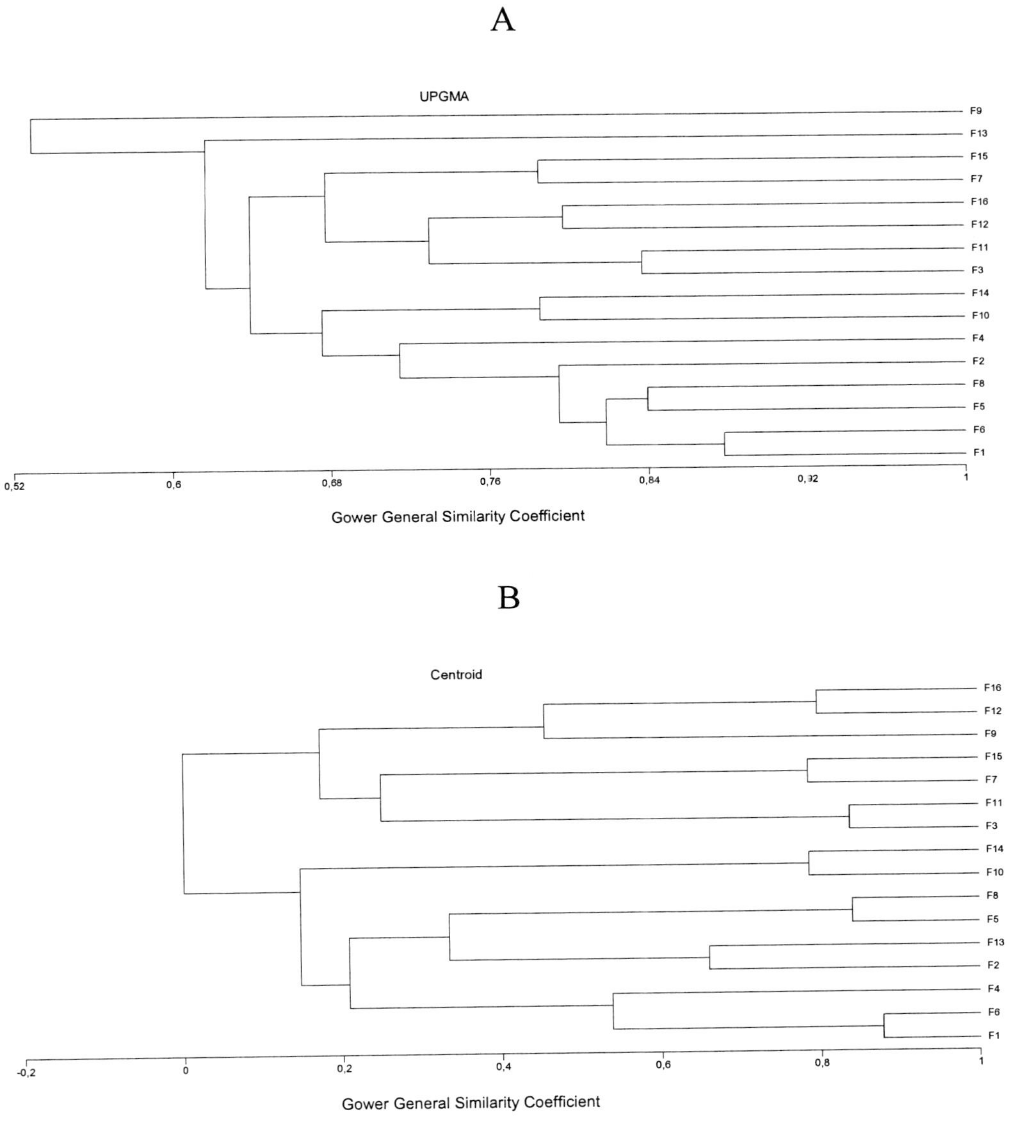

Figure 4. Organic (F10 to F16) and conventional farms (F1 to F9) dendrograms based on UPGMA-A or centroid-B (seven variables)

The first difference was based on the low autonomy of the farms regarding concentrate feed (high value of concentrate bought and low kg of concentrate produced on the farm) for the farms on the left of the cladogram (F7b to F4c) whereas the autonomy was high in the farms located on the right of the figure (F4c toF1a). In the first group, a first sub group comprised F4d to F3c characterised by high gross margin, large flocks (p=0.04) and an important production of heavy lamb type (p=0.04), whereas the second subgroup had a higher stocking rate.

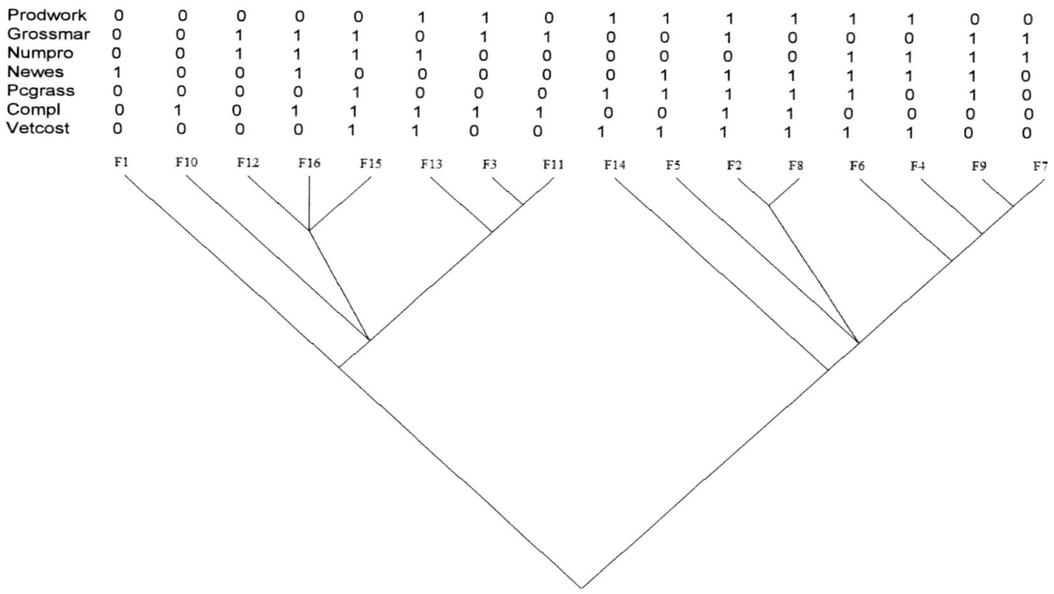

Figure 5. Cladogram (of the relationships between 16 organic (F10-F16) and conventional (F1-F9) farms. Abbreviations in table 3. Length 24, consistency index 29 and consistency index 65.

Table 3. Characteristics of seven organic meat sheep farms monitored during two to seven years monitoring

Code	Characteristics of the farm	Average	Standard deviation	Ancestral character (coded 0 when low and 1 when high)
Compl	Complementary drugs and minerals/year/ewe euros	5.4	2.3	low
Couseas	Counterseason index	1.26	0.24	low
Grmarg	Gross margin per ewe and per year euros	78.0	20.9	low
Kgcoba	Kg concentrate /ewe/year baught	74.1	42.4	low
Kgcofar	Kg concentrate /ewe/year from farm	53.7	24.8	high
Kgcofee	Kg concentrate feed/ewe	21.6	8.9	low
Kglam	Kg carcass /heavy lambs	18.1	1.1	low
Livsunit	Livestock unit (sheep)	62.4	26.0	low
Nempl	Number of employees	0.12	0.16	low
Newe	Number of ewes over 12 months	383	164	low
Nfamw	Number of family worker	1.13	0.29	high
Numprod	Numerical productivity	1.26	0.24	low
Org:year	Organic since (year)	5.9	4.7	high
Othprod	Other animal productions	Yes (coded 0) in 3 farms		
Pcautf	% of fodder autonomy	65.9	7.7	low
Pcautfd	% of food autonomy	76.7	9.6	low
Pchelam	% of heavy lambs sold as organic	62.7	40	high
Pcpastu	% of UTAA dedicated to pastures and forages	85.9	10.5	low
Powmilk	Powdermilk/ewe/year euros	0.9	0.7	low
Stocpas	Stocking rate per ha pasture	1.10	0.16	low
UTAA	Ha utilised from total agricultural area	83.0	21.4	low
Vetcost	Veterinary costs/ewe/year euros	5.6	0.5	low
Wool	Wool/per ewe/year euros	5.5	4.4	high
Workprod	Work productivity	69.2	21.4	low

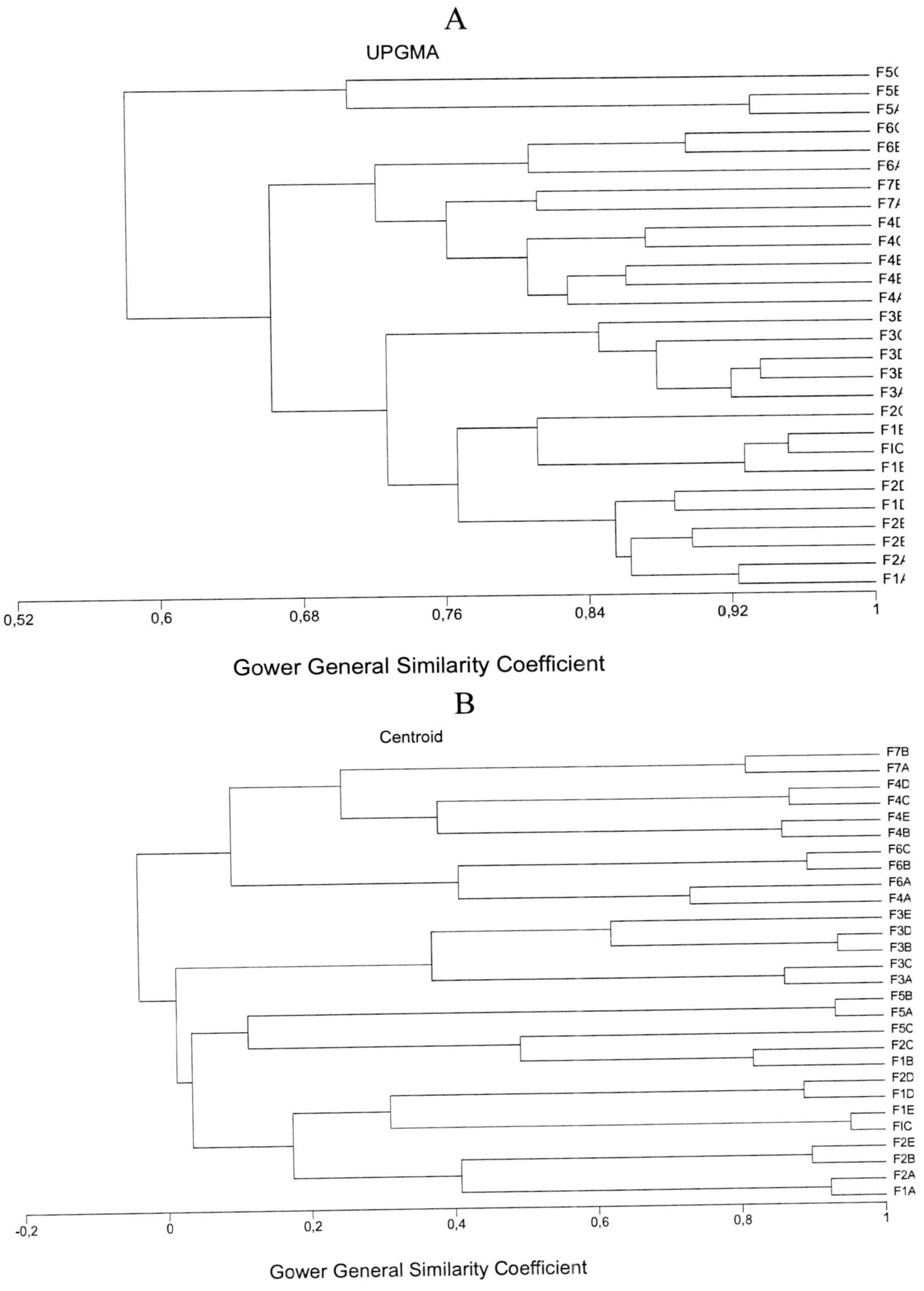

Figure 6. Dendrogram (UPGMA-A or Centroid-B) of seven organic farms evaluated on several years (A to E according to the year)

The second group (F4c to F1a) used a large amount of concentrate feed per ewe (p=0.005) and obtained good figures for numerical productivity (p=0.001). No subgroups could be identified and the farms had progressively acquired other characteristics, such as using counter season reproduction (p==0.000), then utilise largely complementary medicines, and finally, (F2b to F1a) was characterised by a low number of farm workers, had other animal productions and had large autonomy of fodder. Several cladograms (data not shown) were constructed using several commands of the program (mhennig*; hennig* and consensus when several trees were equally parsimonious). The conclusions did not change.

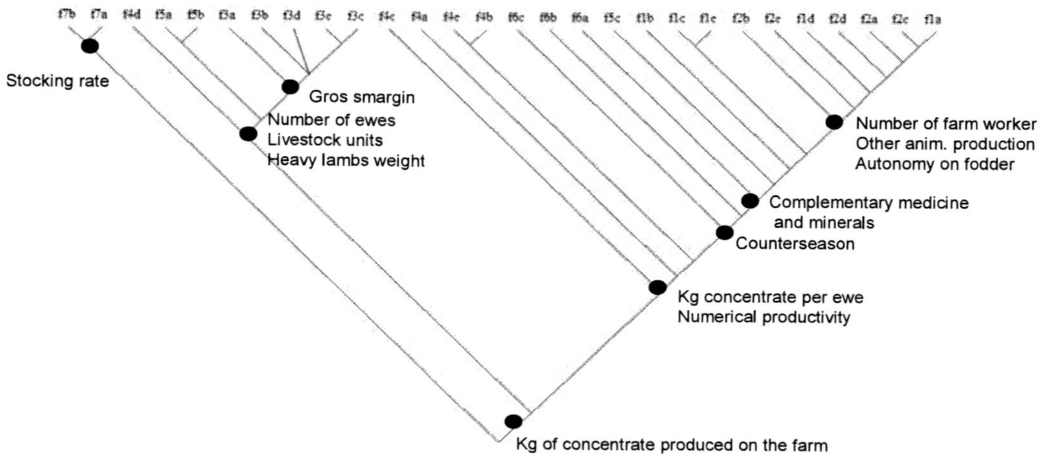

Figure 7. Cladogram of seven organic farms evaluated on several years (A to E according to the year). (f1a correspond to farm 1 and visited on year a). The length was 127, consistency index 18 and retention index 66.

4. Discussion

Choice of Ad Hoc Method: Phylogeny or Cluster Analyses?

The quality assessment of the grouping of farms into categories is a difficult task. It can be assessed from comparisons between results and hypotheses: do we distinguish better organic from conventional farmers with any of the method? In that case we suppose that our hypotheses are sound or highly probable. We can also compare our grouping with previous studies in similar areas (when available) and compare typologies; on which basis should we declare that one method is more efficient than another? We suggest that the method offering the best network of explanations for reconstruction should be favoured.

A typology of sheep meat farms was constructed by Benoit and Laignel (2004) in the centre of France. They described in 20 non-organic farms five typologies, based on principal components and cluster analyses:

i) *mostly* agriculture, with small flocks, low work productivity, high counter season index, large concentrate feed per ewe, low forage autonomy,

ii) extensive farming, with the *largest UTAA*, low stocking rate, low counter season index, limited use of concentrate per ewe, low numerical productivity,
iii) pasture farming, where *pastures represent 97% of the UTAA*, large flocks, low counter season index, limited use of concentrate per ewe, very large gross margin, very low debts for investment,
iv) mixed (cultures and animals) with large fodder autonomy, large flocks, large number of employees, fairly good gross margin per ewe, and *very high productivity of the work,*
v) mixed (cultures and animals), with *low food autonomy* (fodder, concentrate), high counter season index and they received the highest rate of state subsidies (213% of income).

We underlined the major trait that characterised the groups of farms; in fact, typology was nearly based on a single "causal" factor that was mostly external to the management of farmers (type of agricultural area dedicated automatically to pastures or cultures), even for the mixed typologies. The farms we studied in this paper did not belong to group 1 (mostly agriculture) or even 2, which are neatly delineated by their UTAA and the quality of soils; they were also chosen for having a clear meat sheep dominant activity. The example file was based on conventional meat sheep and presents a more contrasted typology (figure 3) based on numerical productivity (derived), gross margin (ancestral), veterinarian not being a main counselor for sanitary problems (ancestral) and lamb mortality (ancestral). The F3 and F5 farms were a monothetic group (both farms had a similar state for four characters) whereas the F1, F2, F4, and F6 were monothetic for numerical productivity and lamb mortality and polythetic for other characteristics (the same state of the character was not found in all four farms. The phylogenetic analysis gave a much clearer picture than the cluster analyses; the F3 and F5 farms were aggregated to the other farms and to different ones, depending on the analysis (UPGMA or Centroid). Using the Benoit and Laignel (2004) classification, when based on the productivity of the work, gross margin per ewe, and numerical productivity they should belong to mixed with food autonomy, mixed with low food autonomy and mixed with low (F5) or high (F3) food autonomy, respectively. It is indicative that the mixed (animal/culture) farms need a deepen approach than the one proposed by Benoit and Laignel (2004), probably more centred on the technical choices of farmers.

The fact that organic farmers were present in our sample also tended to select for farmer dependent choices rather than solely the available environment. It is clear that characters involved in differentiating between organic and conventional farms are mostly in the hand of the farmer (choice of using largely complementary medicines, numerical productivity as a result of technical capabilities, and productivity of the work). The distinction between organic and conventional was clearer when using phylogenetic analysis (3 out of 16 were not correctly identified: figure 5) rather than cluster analyses (five or six out of 16 were not correctly identified, see figure 4 a, b). The identified groups (organic versus conventional farms) were far from being monothetic (not a single character state was in common in all farms of the group), indicating again that typology may not rely mostly on a single character which was in part the rationale of the former typology (Benoit and Laignel, 2004) when applied to specialised production (here meat sheep).

The analysis of seven organic farms during several years necessitates a very sensitive method since it is expected that the same farm should look alike from one year to another, and possibly be different from other farms. The cluster analyses indicated that farms remained similar from one year to another (see F3, F4, F6 for example, in both analyses), although F1 and F2 were often not delineated. The phylogenic analysis provide similar information on the stability of F3, F4, F6, but delineated better the F1 and F2 farms. It was noticed that feed autonomy was the first determinant of farm classification and then many groups of farms were nearly monothetic: F7 with stocking rate, F3 and F5 with the size of the flock and quality of production with heavy weight lambs, F1-2-4-6 with the use of concentrate per ewe and numerical productivity. In fact most of the farm groups were monothetic for several states of characters. Thus between farms, the classification was frequently polythetic (see paragraph before on between farm comparisons), whereas within farms data are often monothetic. The phylogenetic analyses provided efficient tools for *a posteriori* constitution of groups of farms; it had the advantage of indicating also which state of character was influent in this classification and permitted statistical inference when associated with Fisher exact test. Our results were obtained on relatively small samples and it would be of interest to investigate larger data set to confirm the validity of the phylogenetic approach.

One or Several Phylogenetic Approaches?

The presented phylogenetic methodology remains fairly simple: only binary (and not multistate) data are used and parsimony principle was used in its simplest way. The use of other more sophisticated programs like Mega (Kumar et al., 2001) for economics (Andersen, 2003). However the utilisation of software dedicated to molecular data (Mega) with four bases (four states of character) needs adaptation to a set of other characters (presence/absence, multistate). The characters included in the present farm analysis are sometimes from biological (numerical productivity, for example, which integrate several life traits of sheep), economical (gross margin per ewe and per year) or cultural (choice of organic farming and use of complementary medicine) domains. They do not have the same rapidity of evolution (Mace and Holden, 2005) and in that case we should prefer the neighbour joining method to UPGMA (based on a similar rate of evolution for all characters) in phylogeny packages. We can question also the hypothesis of parsimony, since the policy on state or European agricultural subsidies has changed over the years, and farmers may adapt their practices to obtain these subsidies. The relative stability of farming systems (50% of the farms did not change their production systems) during 15 years (Benoit and Laignel, 2004) promotes partly the hypothesis of parsimony. We may regret that the coding for character states is restricted to ancestral/derived state since many characters are quantitative and the loss of information is large; multistate characters should be preferred in that case, although coding could be a difficult operation. It was surprising that using quantitative data in non-phylogenetic dendrograms were not as efficient as phylogenetic cladograms to discriminate between farms since the polarisation of characters may compensate the lost information due to coding. The accuracy or the usual variability from one year to another may overestimate differences between farms when quantitative data are used, which does not occur so much on coded data. The evolutionary unit is species in biology and we considered the farm as an evolutionary unit, which is not completely adequate when cultural traits are important, and we may think that it is the case in organic farming (Barres et al, 1985; Kaltoft, 1999; McMahon, 2005). Mace and Holden (2005) showed that for cultural traits replication was taught by parent-offspring transmission, peer groups, schools and by reading (instead of parent-offspring by means of mendelian inheritance) and that horizontal transfer is common through borrowing or imposition (it is fairly rare in species with sexual reproduction), which means that the evolutionary unit is difficult to delineate. The existence of horizontal transfer should promote the use of evolutionary networks (Huson and Bryant, 2006) instead of cladograms. Many of the proposals (parsimony, coding, horizontal transfer) presented here should be tested on a large database to assess robustness of cladistic analysis in different domains (economical, production, or mixed domains).

From Statistical Analysis to Scoring

These statistical analyses are in the hand of specialists and scoring is usually easier to use for stakeholders, farmers included. Zaroni and Carmo (2006) aimed to develop a scoring scale to estimate a farm's modernisation status. The methodology applied was the item response theory (IRT), which, in agricultural social sciences, can be seen as a new statistical methodology for the typology of small farmers. Their primary interest is focused on establishing the position of the individual farm along some latent dimension. The latent trait is not directly observable but the theory has no trouble in assuming it known. The most distinct feature of IRT is that it adopts explicit models for the probability of each possible response to a test IRT, which derives the probability of each response as a function of the latent trait and some item parameters. IRT models, however, are catching on very slowly in mainstream research. One of the main problems causing this is the lack of widely available, user-friendly software. Some scoring can be derived from phylogenetic analysis. Let us turn to figure 3 for simplicity. We suppose that we are interested in gross margin/ewe/year for scoring economically sound farms. We notice that F3, F5, F1 have a 0 code for gross margin (which is below average) and code 1 in F2, F4 and F6. We observe that this grouping on code 1 is somewhat inefficient because it does not match with the grouping based on the five characters. If we concentrate on a group identified with the five characters after phylogenetic analysis, F1-F2-F4-F6, we can use the sequence of state of character to detect high gross margin, for example for F2: from Prodwork to Numprod, 01010. It is thus similar to a barcode. It must be remembered that this barcode makes sense only in a phylogenetically identified group and not on the raw data that were analysed. Thus, on figure 5, in the organic group (left) high gross margin/ewe/year was found under the following codes (from Prodwork to Vetcost, and in column 3 to 5 from left): 0110000, 0111010, and 0110111. These codes are easily translated in characteristic states, giving thus, among organic farms, typology of farms with high gross margin. This may be a future tool for end users and policy makers.

Although phylogenetic analyses need to be further improved for full use by stakeholders, the analysis we proposed is of interest for grouping similar farms and addressing typology requirements in its present state.

Acknowledgments

The European Program Parasol and Comité interne de l'Agrobiologie (CIAB-INRA) funded this research. Discussions with colleagues of the steering committee of the INRA

Theix Organic platform (A. Haegelin, C. Nicourt, and S. Prache) improved the methodological proposal presented herein.

REFERENCES

Alvarez-Lopez, C. J,. Riveiro-Valino, J. A., Marey-Perez, M. F. (2008). Typology, classification and characterization of farms for agricultural production planning. *Spanish Journal of Agricultural Research,* 6, 125-136.

Andersen E.S. 2003. The Evolving tree of industrial life: an approach to the transformation of European industry. Paper for the the second workshop on the Economic Transformation of Europe, Torino, 31 Jan-2 Feb 2003. *DRUID and Department of Business Studies,* Aalborg University, 33 p.

Barres, D., Bonny, S., Lepape, Y., Remy, J. (1985). Une éthique de la pratique agricole. Agriculteurs biologiques du Nord-Drome. Economie et Sociologie rurale, *Grenoble-Grignon-* Paris. 109 p.

Benoit, M.,Laignel, G. (2002). Constraints under organic farming on French sheep meat productions: a legal and economic point of view with an emphasis on farming systems and veterinary aspects. *Veterinary Research*, 33, 613-624

Benoit, M., Laignel, G. (2004). Exploitations ovins viande en zone de plaine : évolution sur 16 années et évolution de la diversité. *Proceedings Société Française Economie rurale, Lille*. 15 p.

Benoit, M., Laignel G., 2006. Méthodologie d'élaboration de résultats technico-économiques en élevage ovin allaitant. Illustration en France, en zone de plaine et de montagne. Options méditerranéennes, A, 70, 57-65. *Séminaire méditerranéen, Analyse technico-économique des systèmes de production des ovins et des caprins : méthodologies et valorisation pour le développement et la prospective. FAO*, SIA et Univ. Sevilla

Borgerhoff Mulder M. 2008. Using phylogenetically based comparative methods in anthropology: More questions than answers. *Evolutionary Anthropology: Issues, News, and Reviews*, 10, 99-111.

Cabaret, J. (2003). Relating parasite communities to host environmental conditions using phylogenetic tools. *Parasite*, 10, 287-295.

Cabaret J., Benoit M., Laignel G., Nicourt, C. (2008). Situation and management of internal parasites for conventionnal and organic meat sheep farmers in central France : how does it fit for targeted selective treatments? *Veterinary Parasitology*, submitted.

Castaldo, A., Acero, R., Perea, J., Martos, J., Valerio, D., Pamio, J., Garcia, A. (2006). Typology of cattle fattening systems in the Pampa Region, Argentine. *Archivos de Zootecnia* 55, 83-193.

Darlu,P., Ruhlen, M., Cavalli-Sforza,L.L. (1990). A taxonomic analysis of linguistic families. In Wang N.S.Y(Ed.). *Language change and linguistic evolution*. London.

Fiorelli,C., Dedieu, B., Pailleux,J.Y. (2007). Explaining diversity of livestock-farming management strategies of multiple-job holders: importance of level of production objectives and role of farming in the household. *Animal, 1*, 1209-1218.

Funck Teixeira, JR., Marcelo Abreu da Silva, M. (2007). Tipologia de sistemas de produção bovina com ênfase na ocorrência de ectoparasitoses. *Revista Brasileira de Zootecnia*, 36, 2176-2183.

Gaspar, P., Mesıas ,F.J., Escribano, M., Rodriguez de Ledesma, A., Pulido F. (2007). Economic and management characterization of dehesa farms: implications for their sustainability. *Agroforest Systems* 71, 151–162.

Gaspar, P., Mesıas, F.J., Escribano, M., Rodriguez de Ledesma, A., Pulido F. (2008). Sheep farms in the Spanish rangelands (dehesas): typologies according to livestock management and economic indicators. *Small Ruminant Research*, 74, 52–63.

Girard, N., Duru, M., Hazard,L, Magda, D. (2008). Categorising farming practices to design sustainable land-use management in mountain areas. *Agron. Sustain. Dev.* 28, 333–343.

Hamadou, S., Marichatou, H., Kamuanga, M. Kanwe, A. B., Sidibe, A. G. (2003). Diagnostic of periurban dairy farms: typology of farms in the periphery of Bobo-Dioulasso (Burkina Faso). *Journal of Agriculture and Environment for International Development*, 97, 69-92.

Huson, D.H., Bryant, D. (2006). Application of Phylogenetic Networks in evolutionary studies. *Molecular Biology and Evolution*, 23, 254-267.

Kaltoft, P. (1999). Values about nature in organic farming practice and knowledge. *Sociologia Ruralis*, 39, 39-53.

Kumar, S., Tamura, K., Jakobsen, I. B. and Nei, M. (2001), MEGA2: Molecular Evolutionary Genetics.Analysis software, *Bioinformatics* 17, 1244–1245. See also the link to the software.*URL: http://www.megasoftware.net/*

Legendre, P., Legendre, L. (1998). Numerical ecology. *Development in environmental modelling 20.* Elsevier, Amsterdam, 853 p.

Leguen, R., Jenot, F. (2007). L'enquête sociologique, outil d'analyse du management d'une filière animale. Le cas de la filière caprine de Poitou-Charentes Vendée. *Rencontre. Recherche Ruminants* 14, 401-404.

McCarthy I.P. (2005). Toward a phylogenetic reconstruction of organizational life. *Journal of Bioeconomics*, 271–307.

McMahon, N. (2005). Biodynamic farmers in Ireland. Transforming society through purity, solitude and bearing witness. *Sociologia Ruralis*, 45, 98-114.

Maseda, F., Diaz, F., Alvarez, C. (2004). Family dairy farms in Galicia (N.W. Spain): classification by some family and farm factors relevant to quality of life. *Biosystems Engineering*, 87, 509–521.

Mace R., Holden, C.J. (2005).A phylogenetic approach to cultural evolution. *Trends in Ecology and evolution*, 20, 116-121.

Milan, M. J., Bartolome, J., Quintanilla, R., Garcia-Cachan, M. D., Espejo, M., Herraiz, P. L., Sanchez-Recio, J. M., Piedrafita, J. (2006). Structural characterisation and typology of beef cattle farms of Spanish wooded rangelands (dehesas). *Livestock Science* 99, 197-209.

Somda J., Kamuanga,M., , Tollens E. (2005). Characteristics and economic viability of milk production in the smallholder farming systems in The Gambia. *Agricultural Systems*, 85, 42–58.

Multivariate statistical package, (2001). MVSP. User' manual. Version 3.1. KCS, Pentraeth, Wales, UK.

Srairi, M. T., Benabdeljelil, K., Toure, A. (2003). Farm types in mountain areas in Morocco through the analysis of animal husbandry activities. *New Medit* 2, 15-22

Srairi, M. T.; Kiade, N. (2005). Typology of dairy cattle farming systems in the Gharb irrigated perimeter, Morocco. *Livestock Research for Rural Development, 17*, paper 12.

Costas, A. M., Abbiati, N. N., Pereyra, A. M. (2004). A typology for classifying beef cattle - producing farms in Buenos Aires Province of Argentina *Archivos Latinoamericanos de Produccion Animal 12*, 28-41.

Sineiro Garcia, F., Lopez Iglesias, E., Lorenzana Fernandez, R., Valdes Pacos, B. (2004). Typology of farms based on their economic and demographic viability; application to dairy and cattle farms in Galicia (Spain). *Economia Agraria y Recursos Naturales*, 4, 63-85.

Tamura K, Dudley J, Nei M., Kumar S. (2007) MEGA4: Molecular Evolutionary Genetics Analysis (MEGA) software version 4.0. *Molecular Biology and Evolution* 24: 1596-1599.

Tooby J., Cosmides L. (1989). Adaptation versus phylogeny: the role of animal psychology in the study of human behaviour. *The International Journal of Comparative Psychology*, 3, 175-

Zaroni, M. M. H., Carmo, M. S. do. (2006). Family farmer typology: developing a scoring scale for ranking agriculture modernization. *Agricultura em Sao Paulo*, 53, 33-61.

In: Organic Farming: Methods, Economics and Structure
Editors: Michael Nelson and Ivan Artamova
ISBN 978-1-60692-864-6

Chapter 5

ORGANIC MEAT: MARKET DEVELOPMENT AND CONSUMER WILLINGNESS TO PAY

Fabio Napolitano[1], Antonio Girolami and Ada Braghieri
Dipartimento di Scienze delle Produzioni Animali - Università degli Studi della Basilicata – Potenza, Italy.

ABSTRACT

This chapter provides an overview on the evolution of global market demand for organic meat and factors affecting it. Future prospects for development of organic meat consumptions and possible problems concerning trading are analysed. The price that people are willing to pay is the major determinant of the market share of organic meat. Although intent to purchase depends upon the interactions of quality attributes such as appearance and colour, it has been hypothesized a consistent effect of organic labelling as compared to some sensory characteristics on the price offered by consumers for organic meat. Previous experiments have shown that expectations induced by the information can affect the quality perception. Therefore, consumers' behaviour and attitudes toward organic meat are tested through the study of the effect of information about organic farming practices on product liking and consumer willingness to buy organic beef and pay the extra production costs. Our results indicate that this information, if given to the consumers, can be a major determinant of beef liking, thus providing a potential tool for meat differentiation to traditional farms where husbandry is based on extensive rearing systems and high animal welfare standards. Accordingly, consumers are prepared to spend more for organic beef, thus indicating that reliable information about the organic farming system may markedly increase consumer willingness to pay.

1 Correspondence to: ph: +39 0971 205078; fax: +39 0971 205099; E-mail: fabio.napolitano@unibas.it; ada.braghieri@unibas.it

INTRODUCTION

The global market for organic food was valued at USD (United States dollars) 25 billion in 2003, which can be considered a niche (Willer and Yussefi, 2004). Although production of organic products is increasing across the globe, sales are concentrated in the industrialised parts of the world. North America and Western Europe account for 95 % of the world retail sales of organic food products with USD 13 and 10.4 billion, respectively. Although consumer interest is growing in other regions, the demand is confined to the industrialised world largely because of the price premium of organic products. Many developing countries have large sections of their populations below the poverty line, and this makes it difficult for a market for organic products to develop. A typical consumer of organic products has the following attributes: lives in urban areas, usually in a big city, in a medium to high-income household with relatively high purchasing power, belongs to middle-high social classes and is sensitive to factors like quality, provenance and production methods (Willer and Yussefi, 2004). The industrialised nations have a sizeable middle-class, and this is the reason why most organic food sales are concentrated in these countries. As more countries develop economically and as their populations become more affluent, demand for organic products is likely to rise. This will cause sales of organic products to become less concentrated in the world. Rapid economic growth in countries like China, Brazil, and South Africa is causing the upper social classes to expand, and this is creating a market for organic food. In other regions, there is an increase in organic farm land because farmers are attracted to the export benefits of organic production. Although most production in Asian and African countries will be for export markets, it is also creating regional markets to develop in which organic farmers market their organic crops to consumers in their region. Sales of organic products are slowing in certain countries, especially in Western Europe; however, the market is becoming increasingly global.

THE MARKET OF ORGANIC MEAT

During 2005, while beef trade was growing at an average of 3.0 %, global pork export was led by the European Union, followed by USA, China, Brazil, Mexico and Canada, whereas Japan was the largest consumer market for pork followed by Russia, Mexico, Korea and USA (Organic Monitor, 2006).

The Western European market for organic meat was traditionally the largest in the world, however it has now been equalled by North America. The European market has faced rapid expansion since the mid 1990s and is now reporting slowing growth rates as certain sectors approach maturity. Supply-demand imbalances have become a characteristic of the European organic food industry. A number of countries are showing oversupplies in sectors like organic meat whereas other sectors like organic cereals continue to suffer from product shortages. A reason for this is that much of the early converts to organic agriculture were dairy and beef cattle farmers.

The average European expenditure rate on organic products is USD 27.2 per annum. There is much variation in the expenditure rate between European countries, ranging from USD 7.3 (Spain) to USD 105 (Switzerland) per capita.

In 2003, EU, including 15 countries, certified non-dairy cattle (mainly suckler cows) amounted to about 1 million heads corresponding to 1.7 % of total non-dairy cattle herd (European Commission, 2005). Austria ranked first with 25 % of EU, followed by Italy with 15 %, the United Kingdom with 13 % and Germany with 12 %. Certified pigs amounted to 450,000 heads or about 0.4 % of total pig herd. Only a few member states, such as Germany with 8 % and France with 4 %, had a significant share of certified animals. The number of certified pigs is still relative low. The relative slow development may be linked with higher production costs in organic pig meat production. Certified broiler number amounted to more than 9 million, France ranked first with 5.1 million. Sheep and goats amounted to 2.4 million heads or 2.4 % of total sheep and goats herd. Of the about 2 million certified sheep, 35 % were located in the United Kingdom and 21 % in Italy. According to Hamm et al. (2002) beef had the highest market share (1.7 %) followed by sheep and goat meat (0.7 %), pork (0.3 %) and poultry (0.3 %).

Only little and partial information is available on prices for organic meat and meat products as prices diverge depending on the country and on the product. For instance, farmer price for organic pork in 2002 was about 2.46 €/kg on EU-15 average, only Italy reported a price 20 % below this average, but some countries had significantly higher prices like Greece (5.00 €/kg), Ireland (3.49 €/kg), Luxembourg (3.27 €/kg) and the United Kingdom (3.22 €/kg). Price premiums for organic pork ranged from 45 % in Germany and Austria to 132 % in the Netherlands, while EU average was about 62 %. Consumer price for organic pork cutlet was about 13.17 €/kg but differed considerably between the 15 member states: the Danish price was about the double of the price in Spain, Portugal and Finland. Consumer price premium for organic pork cutlet (81 %) was higher than average farm gate price premium and ranged extremely from 0 % in Portugal to 165 % in Greece. The high divergence in consumer prices and price premiums reflect often the different sales chains used, i.e. consumer prices are usually much higher in organic food shops than in supermarket chains. There was a considerable variability among the 15 Member States in farm gate prices for beef as well as in consumer prices for minced beef. As for beef, farm gate prices for organic and conventional were very high in Greece resulting in a high consumer price for minced beef. Organic price premium at farm gate level varied from 17 % in Denmark to 190 % in Spain and was on average about 49 % (European Commission, 2005).

The North American market for organic products is reporting the highest growth worldwide. The meat sector is the fastest growing organic food industry with sales expanding by 51 % in 2005 (Organic Monitor, 2006). The USA organic beef market has been showing the highest growth with sales doubling each year since BSE was reported in 2003. High market growth rates have led to organic meat supply shortages with producers unable to meet growing demand from retailers. The organic beef and pork markets are the most affected because of low production levels. American farmers have shown little interest in producing these organic meats due to high production costs and the lack of distribution infrastructure. Although production has increased since 2004, supply is expected to lag demand for a number of years. As a consequence prices are rising for a range of organic meat products and an increasing number of North American processors are looking overseas for supplies. In particular, organic meat products imports are mainly from Latin America, Australasia and Canada. Although organic beef has the fastest growing market, organic poultry is the most widely available in North America (nearly two-thirds of this sector). USA retail sales of organic poultry were USD 161 million in 2005, well under 1 % of conventional poultry sales.

However, retail sales of organic poultry have almost quadrupled since 2003, and estimates of annual growth rates range from 23 to 38 % through the end of the decade, with annual sales reaching almost USD 600 million by 2010 (Nutrition Business Journal, 2006). Approximately half (51 %) of organic poultry sales were in natural food stores in 2003, 45 % in mass market grocery stores (including conventional grocery, mass merchandiser, and club stores), and 4 % through direct sales and other distribution channels (Nutrition Business Journal, 2006). As in the conventional food sector, sales of broilers account for the majority of organic poultry sales. The relatively short production cycle and low price premium are responsible for organic chicken to be the most popular organic meat with consumers. Prices for organic broilers were steady at an average of USD 2.17 per pound from July 2004 through early 2006, with a range of USD 1.89 to USD 2.45. Average prices for conventional broilers, on the other hand, ranged from USD 0.59 to USD 0.82 per pound during the same period (Oberholtzer et al., 2006). Organic beef in comparison is more rarely found in retailers because of small-scale production, and inadequate distribution infrastructure, although the success of competing products like “natural” beef is decreasing. USDA (United States Department of Agriculture) roles for “natural” beef are less strict than those for organic products: antibiotics, hormones and animal by-products are banned but conventional feedstuffs are allowed.

The number of certified animals in 2005 were: over 36,000 beef cows, 10,000 hogs and pigs, 4,400 sheep and lambs and over 10,000,000 broilers. In comparison with 2002 the increase ranged from 9 % (sheep and lambs) to 264 % (hogs and pigs).

The Argentine government has established national standards for organic products. These are at least as stringent as those of the International Foundation for Organic Agriculture (IFOAM) and the European Union. The organic industry in Argentina is mainly export oriented (85 % by value is exported with an estimated value of USD 32 million). Organic meat production, and beef in particular, is growing as a consequence of climatic suitability and soil fertility and can be cited as an example of a developing country supplying external markets. However, a domestic market is being developed in Buenos Aires. In 2000, out of a total production of 40,000 tons of organic produce, meat and livestock products represented 8 %. Of this total, 35,000 tons were exported, of which meat and animal products accounted for 3 %. This would imply that a substantial proportion of organically produced meat was consumed in the domestic market. It is not known what proportion was sold as organic meat and what proportion went into conventional sales. The European Union is the principal destination of Argentina's organic beef exports. Meat exporting began in the mid 90s with beef, and more recently Patagonian lamb became the predominant export for international markets. In 2002 there were 754,000 sheep and 122,000 head of cattle certified in Argentina (Willer and Yussefi, 2004).

Elsewhere, both Brazil and Uruguay are seeking to develop exports of organically produced meat (Willer and Yussefi, 2004). In the case of Brazil, organic beef production systems are being developed in the centre of the country. The authorities in Brazil see the organic production system as a way of boosting local cattle producers' income, while reducing the environmental damage that traditional methods of cattle raising have caused. While 90 % of Brazil's overall organic production is exported, Brazil's internal market for organic foods is growing at around 25 % a year.

Although the Australasian continent comprises almost 40 % of global organic farmland with 12.1 million hectares, the market represents a fraction of the global total (Willer and Yussefi, 2004). Sales of organic products were estimated at about USD 200 million in 2002 with Australia comprising the most. Beef cattle farmers use much of the organic farmland in Australia as pastureland. The Australian and New Zealand organic food industry is export-oriented with significant quantities of primary products, including beef and lamb, going to other northern hemisphere countries and relatively low amounts sold in the domestic market. However, sales of organic products within Australia are growing at about 15 to 20 % per year.

Exporters are finding increasingly difficult to meet supply gaps because of the differences in organic standards between regions. For instance National Organic Program standards developed in USA for organic meat products differ from those of the EU, and, although trade liberalisation may be removing tariff barriers, standards are becoming the major impediment to free trade in the global organic meat industry.

CONSUMER WILLINGNESS TO BUY ORGANIC MEAT

Food quality is not an objective feature of the products as it is affected by consumer perception, and it is often referred to as perceived quality (Cardello, 1995). Many aspects can be used by consumers to perform their food choices. Intrinsic (e.g. cut, colour, fat rim) and extrinsic cues (price, origin, stamp of quality, production and nutritional information) are used to form expectations about product quality attributes, the latter can be classified in two categories: those experienced before or during consumption (experience quality attributes: e.g. price and sensory properties) and those not experienced directly, such as healthiness, naturalness, ethical aspects, etc., which should be communicated to be perceived as they are credence characteristics that can not be confirmed either before or after purchase (Grunert et al., 2004). Tangible aspects can markedly affect most purchasing decisions, independently from the knowledge and awareness of the consumer, whereas intangible attributes are important for high-involvement consumers possessing enough beliefs and attitudes (McEachern and Schroder, 2002). For these consumers, such intangible attributes play a central role in defining product quality.

According to a recent on line survey (ACNielsen, 2005), organic alternatives are purchased mainly for health reasons. Over two thirds of survey respondents think that organic foods are healthier for them and their children. This healthy perception is consistently strong across Asia Pacific. European people seem to be more conscious of the wider benefits of organic foods, such as protecting the environment and animals. Nearly a fifth think that organic foods preserve the environment, and 12 % cites kindness to animals as the key reason for purchasing organic alternatives. However, in other regions, less than a fifth of consumers are aware of the environmental benefits of organic foods, and even less mentions animal protection. As to the barriers to purchasing organic alternatives, high prices are perceived as the biggest problem for one third of respondents in Asia Pacific and over 40 % of European and North American people. Across Latin America and in South Africa, availability is considered a problem. Interestingly, lack of availability prevails over lack of credibility as a reason for not purchasing organic foods globally. An element favouring the spread of organic meat into the general retail sector is that purchasers of organic food tend to be in the higher income segment (F.A.O., 2002). Consequently, supermarkets seek to attract such customers by providing a wide range of food, including organic meats. The increased involvement of supermarkets, with their centralised systems of purchasing and distribution, may result in pressure to reduce the current price differential between organic and conventional products. As an extension of this trend, some supermarkets have introduced "own brand" organic meat products, with such products being priced below those of competing brands.

The general sense among consumers who would not buy organic products is that they are usually more costly. Manufacturers and retailers should take into account the geographical differences in terms of acceptance and perception of organic food. For example, in Thailand and China credibility is more an issue, while in Asia Pacific perceptions of price of organic foods is the major constraint in eight out of 13 markets studied.

Meeting organic certification requirements usually implies higher production costs. For example, it is reported that the cost of producing organic beef in the United Kingdom is 20 % higher than under conventional methods. In some cases, the high cost of converting to organic meat and dairy production has led to subsidies being paid to the farmer. However, this is not the case in many countries (e.g. USA and developing countries). Thus the extra production costs have to be paid by consumers.

Study of the Effect of Information on Organic Beef Acceptability and WTP

For consumers from western countries, price is not the only determinant behind animal-food purchases as they are acquiring an increasing interest in farming practices and the related organic standards. Consumers do not seek the cheapest food but the best value for money, i.e. the maximum benefit for what they are prepared to spend (McInerney, 2004). Information is a key consumer need. Consumers should be provided with the information they require, which should be presented in a simple form in order to enable them to make informed choices. In fact, it has been observed that consumers are willing to access to more information about organic food (Brennan et al., 2003).

Numerous studies have been conducted on the effect of information on food liking (Aaron et al., 1994; Daillant and Issanchou, 1993; Kähkönen and Tuorila, 1995; Solheim, 1992; Westcombe and Wardle, 1997). In particular, information about the manufacturing process (organic vs. OGM) can affect product acceptability (Caporale and Monteleone, 2004). Recent studies have been conducted on the effect of information about animal welfare on lamb (Napolitano et al., 2007a) and beef liking (Napolitano et al., 2007b). All these experiments have shown that expectations induced by the information can affect the quality perception. Thus, if expectations are either positively (the liking score of the product tasted without external information is higher than expected) or negatively disconfirmed (the product is worse than expected), the assimilation model is generally applicable, which means that hedonic ratings move towards the expectations when an external information is given compared to tasting without external information (Anderson, 1973; Cardello and Sawyer, 1992). However, there is generally a significant difference between liking score with information and expected liking score, i.e. consumers do not completely assimilate (Lange et al., 1999; Siret and Issanchou, 2000).

The price that people are willing to pay is the major determinant of the market share of organic meat. Although intent to purchase depends upon the interactions of quality attributes such as appearance and colour (Brewer and McKeith, 1999), it has been found a more consistent effect of organic labelling as compared to some sensory characteristics on the price offered by consumers for organic pork (Dransfield et al., 2005). Using questionnaires on organic foods in Spain, consumers appeared to be prepared to pay about 12 % more for organic red meats and chicken (Gil et al., 2000). In France and The Netherlands, questionnaire responses suggested that almost half of consumers would pay 20 % more for pork from pigs raised outdoors (Carpentier and Latouche, 2005). Dranfield et al. (2005) stated that those studies over-estimated the premiums consumers would be willing to pay, while their results suggest that people would offer 5 % extra, with about one-fifth of consumers willing to pay 20 % extra, for organic pork. However, hedonic and purchase intent measurements may be not representative of the real behaviour of consumers. They may declare high preferences and purchase intent for products with high-perceived quality, albeit not buying them under economic constraints (Lange et al., 1999).

Recently, Lange et al. (2002) and Napolitano et al. (2008) observed that, as for food liking, the assimilation model is applicable also to consumer willingness to pay. As a consequence, it can be hypothesized that information about the organic manufacturing process can increase both meat acceptability and willingness to pay thus providing a means to cover the extra production costs sustained by organic farmers.

Little is known on the effect of the information about organic production on the real consumer willingness to pay for meat. Recent studies demonstrated that auctions are able to place consumers in real situations where they can show their true preferences. In particular, the Vickrey second price auction is widely used to assess consumer willingness to pay real goods (e.g. Melton et al., 1996), including foods (Lange et al., 2002), and the value consumers give to food safety (e.g. Hayes et al., 1995) and animal welfare (Napolitano et al., 2008). According to this specific type of auction consumers are individually asked to submit a sealed bid corresponding to the highest price they would agree to pay for a particular product. The highest bidder (i.e. the winner), by paying the second highest price, has the opportunity to buy a product at a price equal to or, more often, lower than the value he assigns to the product (Vickrey, 1961).

The present study was aimed to verify whether consumers confirm their higher liking for organic meat by willing to pay extra costs in a situation where a potential purchase performed by consumers, such as the Vickrey auction, is included.

Methodology

Beef was obtained from twelve Podolian young bulls aged 16 months and reared either in conventional (CB) or organic local farms (OB). Carcasses were aged 15 days in order to increase meat tenderness and make the product acceptable from a sensory point of view (Braghieri et al., 2005). Both products were obtained from the muscle *Longissimus dorsi*. Meat samples (10 x 10 x 1 cm) were grilled at 300°C to an internal temperature of 75°C assessed using a thermocouple probe inserted into the meat. Mean cooking time was 6 min. Samples were offered to the subjects immediately after cooking in booths where salt was available *ad libitum*.

Subjects were recruited in Potenza (main town in the region of Basilicata, southern Italy). The consumer panel consisted of fifty subjects recruited on the basis of age and level of education. In addition, subjects were selected using predetermined screening criteria based on consumption frequency of beef and organic products.

Sixty-five subjects were interviewed and were asked their frequency of consumption of beef at home (1 = never; 2 = once a year or less; 3 = 3-5 times a year; 4 = less than once a month; 5 = 1-2 times a month; 6 = more than twice a month; 7 = at least once a week).

The selected consumer panel included subjects who reported to consume beef at least "1-2 times a month" and organic products occasionally (at least "once a year or less" but "less than once a week"). Subjects had a mean age of 49 years and were almost equally distributed for sex. The main features of the subjects participating to the consumer panel are depicted in Table 1.

Table 1. Socio-demographic features of the subjects participating to the consumer test

		Number	*Percentage*
Age	20-39 years	14	*28*
	40-59 years	17	*34*
	> 59	19	*38*
Sex	Female	22	*44*
	Male	28	*56*
Education level	Primary school	3	*6*
	Secondary school	6	*12*
	High school	20	*40*
	Graduated	17	*34*
	Post-graduate	4	*8*

The experiment was planned in four tests (Table 2). In the first test the consumers were offered both CB and OB in a balanced order of presentation. They were asked to taste the meat and rate their liking receiving no information on the products (Perceived liking). In the second test the subjects received two sheets with the information concerning the farming systems (conventional or organic). They were asked to read carefully the information and give their liking expectation for that product (Expected liking). First and second tests were performed in the same day. The day after the third test was performed: the consumers were given OB only along with the information sheet. They were instructed to read the information before tasting the sample and express their liking score (Actual liking).

Table 2. Summary of the experimental design for the assessment of consumer liking (L) and willingness to pay (WTP)

Test	Day	Stimulus presentation	Type of evaluation	Type of rating
1	1	Beef	Tasting without information	Perceived L
2	1	Information	Expectation	Expected L
3	2	Beef + information	Tasting with information	Actual L
4	2	Product display	Auction	WTP

Consumers rated their liking on a 9-point hedonic scale labelled at the left end with "extremely unpleasant", at the right end with "extremely pleasant" and at the central point with "neither pleasant nor unpleasant" (Kähkönen, Tuorila and Rita, 1996).

In tests 2 (expectations produced by information) and 3 (acceptability generated by information and tasting of the product) the following information concerning the farming systems were given to consumers:

1) Conventional beef: conventional beef cattle farming ensure standards of animal welfare as set by the current legislation; the administration of pharmaceuticals is allowed within the suspension limits, as well as the use of GMO and chemicals for the production of animal feeds, in accordance with the current legislation; a high stocking density per hectare is allowed.
2) Organic beef: organic beef cattle farming practices ensure standards of animal welfare higher than those set by the current legislation by promoting grazing systems and the expression of species-specific natural behaviour; the use of pharmaceuticals is markedly reduced; the use of GMO and chemicals for the production of animal feeds is banned; stocking density per hectare is low in order to reduce the impact of faming on the environment.

The second price Vickrey auction (Vickrey, 1961) was used to assess consumer willingness to pay organic beef. Participants attended a short presentation explaining the procedure to be followed for the auction. It was made clear that the submission of bids implied a commitment to buy the product. Participants agreeing to the procedure signed a consent and received 10 Euro in cash. Subsequently, a formal training on the use of the Vickrey second price auction was conducted. It was explained that the maximum price accepted to pay for 100 g of organic beef had to be written on paper, separately by each participant. The bid had to be comprised between 0.00 and 5.00 €. On each form provided to consumers there was a suggested price (ranging from 0.29 to 2.71 €; mean = 1.54 €, corresponding to the actual commercial value of organic beef), however it was explained that they could offer lower or higher prices. The participant submitting the highest price (winner) had to buy the product, not at the submitted price, but at the second highest price (i.e. the second highest submitted bid). In case of more consumers offering the same highest bid only one participant, randomly chosen by another consumer, would be selected as winner. This procedure allowed one of the participants to buy organic beef at a price lower than or equal to the price they would normally accept to pay. It was also explained that the study aimed to know the value that the product had for the consumers, not its commercial value, and that the best option for them was the submission of their real reservation price. In order to ascertain that all participants correctly interpreted the procedure, some practice was conducted using snacks. Subsequently, an auction was conducted where consumers submitted bids after the display of a 100 g slice of organic beef.

The Student's paired t-tests were used to evaluate differences between mean scores either obtained for the two products (OB and CB), or obtained for the same product under different conditions (tasting only, information only, tasting with information). The same test was used to compare the suggested price for organic beef and the actual bids offered by consumers.

Results and Discussion

Ratings given by consumers to OB and CB are summarised in Table 3. Consumers rated both products at scores above the central point (5 = neither pleasant nor unpleasant) for perceived liking. These results indicate that the meat from both organically and conventionally reared beef cattle was characterised by a good eating quality. However, mean scores of perceived liking were higher for OB as compared to CB (P<0.001). No information is available on the effect of organic farming on beef acceptability, although previous studies reported minor effects on flavour (Marino et al., 2006) and tenderness (Braghieri et al., 2005). In this study organic farming practices induced a much higher acceptability as compared with conventional techniques (P<0.001) with scores above 7 (pleasant).

Table 3. Rating (± S.E.) given by the consumer panel during the three hedonic tests

Type of rating	Organic beef	Conventional beef
Perceived liking	7.22 ± 0.16a	5.94 ± 0.20b
Expected liking	7.88 ± 0.12a	5.18 ± 0.24b
Actual liking	7.72 ± 0.15	NR
P-E	-0.66*** Negative disconfirmation[1]	0.76*** Positive disconfirmation[2]
A-P	0.50* Assimilation[3]	-
A-E	-0.16 Complete[4]	-

P = perceived liking mean scores (blind condition); E = expected liking mean scores (only information); A = actual liking mean scores (with information); NR = not recorded.

a,b = P<0.001; * = P<0.05; *** = P<0.001.

[1] The product is worse than expected

[2] The product is better than expected

[3] Actual liking moves towards the expectations

[4] Assimilation occurs, and actual liking is not different from expectations

Expected liking scores were higher for OB than for CB animals (P<0.001). These results indicate that consumers are aware of the possible positive effects of organic farming on product quality and safety. Our study also confirms previous reports stating that consumers use the application of organic standards as an indicator of product attributes such as food safety, food quality and food healthiness (Brennan et al., 2003; Michaelidou and Hassan, 2008). Fresh meat is usually commercialised as undifferentiated product. For instance, Grunert (1997) suggests that the butcher is considered as an expert by consumers and, therefore, reliable to predict meat quality and characteristics. As long as fresh meat is mainly sold as a commodity, there is also only a limited incentive for meat producers to differentiate their product (Grunert et al., 2004). Any form for improved or otherwise differentiated meat quality requires new ways to signal the quality to the consumer. Thus, meat product differentiation needs a constant and reliable signalling of quality through appropriate information given to consumers in order to motivate them and increase their willingness to buy and pay for meat (Bredahl, 2004). This differentiation can be based on both product or process characteristics. For animal-based products process characteristics may be represented by the farming practices and the related organic standards. In fact, consumers showed a WTP for OB higher than the suggested price (P<0.001), the latter corresponding to the commercial value for organic beef (Table 4).

Table 4. Mean ratings (±S.E.) of WTP

	WTP
Suggested price	1.54 ± 0.12a
Consumer bid	2.65 ± 0.13b

a,b = P<0.001

Results concerning the effect of information on expected and actual liking of beef are shown in Table 3. For both products (OB and CB) the expected acceptability was significantly different from the perceived liking expressed in blind conditions (P<0.001), thus indicating that a disconfirmation occurred. In particular, the consumers found OB worse than expected (negative disconfirmation), whereas CB was considered better than expected (positive disconfirmation). These results indicate that information about farming practice can have a marked impact on consumer expectancy with organic standards associated with high expected product quality and conventional farming associated with lower expected product quality. Accordingly, previous studies revealed a marked effect of positive information about farming practices and animal welfare on actual liking of meat (Napolitano et al., 2007a,b) and willingness to pay for yogurt (Napolitano et al., 2008).

A significant difference between perceived and actual liking was observed for OB as the former was lower than the latter (P<0.001). Therefore, the information given about organic farming was able to affect the actual liking of beef. In this case, the effect of information can be explained on the basis of the assimilation model, which can be observed when the actual liking of the product moves in the direction of the expectations. In particular, the information concerning the use of organic farming techniques characterised by high levels of animal welfare, product safety and low impact on the environment had a positive impact on actual liking. Consumers completely assimilated their liking in the direction of expectations, as indicated by the fact that expectancy was not significantly different from actual liking (P>0.05), which in turn showed scores close to 8 (very pleasant). The complete assimilation observed for this product is likely to be due to the important role played by the information in the determination of actual liking of organic beef. In addition, the complete assimilation was possibly facilitated by the good eating quality of the organic product, as indicated by the high value of perceived liking observed in blind conditions.

CONCLUSION

The main limit to purchasing organic meat remains price because of high production costs, which are affected by organic rules (higher space allowance, origin of feedstuffs, etc.) and small-scale production systems. Two strategies to overcome this constraint are: the spread of organic meat in supermarkets, which is likely to induce a reduction of current price (this approach may be suitable for meat which is organic but otherwise undifferentiated from conventional products); the induction of increased willingness to pay by constant and reliable quality signalling systems capable to provide an ethical value to the product, which may become even higher if associated to traditional farming systems and typical meat productions.

The results concerning the study of the effect of information on beef liking show that consumers are influenced by information about organic production and move their actual acceptability in the direction of expected liking, possibly because consumers are aware of the ethical value of organic farming and its effects on product safety. In particular, the hedonic discrepancy was totally assimilated indicating that actual liking of organic beef was markedly affected by information. Conversely, the information concerning conventional production had a detrimental effect on expectancy. Therefore, the information about organic farming, if given to the consumers, can be a major determinant of beef liking, thus providing a potential tool for meat differentiation to traditional farms where husbandry is based on extensive rearing systems and high animal welfare standards. Accordingly, consumers are prepared to spend more for organic beef as compared to the suggested price, thus indicating that reliable information about the organic farming system may markedly increase consumer willingness to pay.

Acknowledgment

We are grateful to the Regione Marche for supporting the program E.QU.I.ZOO.BIO. "Efficienza, Qualità e Innovazione nella Zootecnia Biologica" (SP 11), and thereby the effort of the authors. The authors thank A. M. Riviezzi, G. Impemba and N. Piazzolla for assistance in conducting the experimental trial.

References

Aaron, J. I., Mela, D. J. & Evans, R. E. (1994). The influences of attitudes, beliefs and label information on perceptions of reduced-fat spread. *Appetite, 22,* 25-37.

Anderson, R. E. (1973). Consumer dissatisfaction: the effect of disconfirmed expectancy on perceived product performance. *Journal of Marketing Research, 10,* 38-44.

ACNielsen (2005). Organic and functional foods have plenty of room to grow according to new ACNielsen global study. *http://usacnielsencom/news*

Braghieri, A., Cifuni, G. F., Girolami, A., Riviezzi, A. M., Marsico, I. & Napolitano, F., (2005). Chemical physical and sensory properties of meat from pure and crossbred Podolian bulls at different ageing times. *Meat Science, 69,* 681–689.

Bredahl, L. (2004). Cue utilisation and quality perception with regard to branded beef. *Food Quality & Preference, 15*, 65-75.

Brennan, C., Gallagher, K. & McEachern, M. (2003). A review of the "consumer interest" in organic meat. *International Journal of Consumer Studies, 27,* 381-394.

Brewer, M. S. & McKeith, F. K. (1999). Consumer related quality characteristics as related to purchase intent of fresh pork. *Journal of Food Science, 64,* 171–174.

Caporale, G. & Monteleone, E. (2004). Influence of information about manufacturing process on beer aceptability. *Food Quality & Preference, 15,* 271-278.

Cardello, A. V. & Sawyer, F. M. (1992). Effects of disconfirmed consumer expectations on food acceptability. *Journal of Sensory Studies, 7,* 253-277.

Cardello, A. V. (1995). Food quality: relativity, context and consumer expectations. *Food Quality & Preference, 6,* 163-170.

Carpentier, A. & Latouche, K. (2005). French consumers' and citizens' concern: which willingness to pay for "green pork"? International workshop on green pork production, Paris, May 25-27, 2005. pp17-18.

Daillant, B. & Issanchou, S. (1993). Influence of sugar and fat contents on preference for cream cheese. A preliminary study. *Food Quality & Preference, 4*, 133-139.

Dransfield, E., Ngapo, T.M., Nielsen, N.A., Bredahl, L., Sjödén, P.O., Magnusson, M., Campo, M.M. & Nute, G.R. (2005). Consumer choice and suggested price for pork as influenced by its appearance, taste and information concerning country of origin and organic pig production. *Meat Science, 69,* 61-70.

European Commission. 2005. Attitudes of consumers towards the welfare of farmed animals. Special Eurobarometer 229. Wave 63.2.TNS Opinion and Social *http://ec.europa.eu/public_opinion/archives/ebs/ebs_229_en.pdf*

FAO (2002). Market developments for organic meat and dairy products: implications for developing countries. Committee on commodity problems, Rome 27-29 August (2002). *http://www.fao.org/DOCREP/MEETING/004/y7029e.htm*

Gil, J. M., Gracia, A. & Sanchez, M. (2000). Market segmentation and willingness to pay for organic products in Spain. *International Food and Agribusiness Management Review, 3,* 207–226.

Grunert, K. G. (1997). What's in a steak? A cross-cultural study of the quality perception of beef. *Food Quality & Preference, 8,* 157-174.

Grunert, K.G., Bredahl, L. & Brunsø, K. (2004). Consumer perception of meat quality and implications for product development in the meat sector – a review. *Meat Science, 66,* 259-272.

Hamm, U., Gironefeld, F. & Halpin, D. (2002). Analysis of the European market for organic food - *Organic Marketing Initiatives and Rural Development* (Vol 1). School of Management and Business, University of Wales, Aberystwyth, Wales, UK.

Hayes, D. J., Shogren, J. F., Shin, S. Y. & Kliebenstein, J. B. (1995). Valuing food safety in experimental auction markets. *American Journal of Agricultural Economics 77,* 40–53.

Kähkönen, P. & Tuorila, H. (1995). The role of expectations and information in sensory perception of low-fat and regular-fat in sausages. *Appetite, 24*, 298-299.

Kähkönen, P., Tuorila, H. & Rita, H. (1996). How information enhances acceptability of a low-fat spread. *Food Quality & Preference, 7,* 87-94.

Lange, C., Rousseau, F. & Issanchou, S. (1999). Expectation, liking and purchase behaviour under economical constraint. *Food Quality & Preference 10*, 31-39.

Lange, C., Martin, F., Chabanet, C., Combris, P. & Issanchou, S. (2002). Impact of the information provided to consumers on their willingness to pay for champagne: comparison with hedonic scores. *Food Quality & Preference 13*, 597-608.

Marino, R., Albenzio, M., Girolami, A., Muscio, A., Sevi, A. & Braghieri, A. (2006). Effect of forage to concentrate ratio on growth performance and on carcass and meat quality of Podolian young bulls. *Meat Science, 38,* 269–277.

McEachern, M. G. & Schröder, M. J. A. (2002). The role of livestock production ethics in consumer values towards meat. *Journal of Agricultural and Environmental Ethics, 15,* 221-237.

McInerney, J. (2004). Animal welfare, economics and policy. Report prepared for DEFRA. *http://statistics.defra.gov.uk/esg/reports/animalwelfare.pdf*

Melton, B., Huffman, W., Shogren, J. & Fox, J. (1996). Consumer preferences for fresh food items with multiple quality attributes: evidence from an experimental auction of pork chops. *American Journal of Agricultural Economics, 78,* 916–923.

Michaelidou, N. & Hassan, L. M. (2008). The role of health consciusness, food safety concern and ethical identità on attitudes and intentions towards organic food. *International Journal of Consumer Studies, 32,* 163-170.

Napolitano, F., Braghieri, A., Caroprese, M., Marino, R., Girolami, A. & Sevi, A. (2007a). Effect of information about animal welfare, expressed in terms of rearing conditions, on lamb acceptability. *Meat Science, 77*, 431-436.

Napolitano, F., Caporale, G., Carlucci, A. & Monteleone, E. (2007b). Effect of information about animal welfare and product nutritional properties on acceptability of meat from Podolian cattle. *Food Quality & Preference, 18,* 305-312.

Napolitano, F., Pacelli C., Girolami A. & Braghieri, A. (2008). Effect of information about animal welfare on consumer willingness to pay for yogurt. *Journal of Dairy Science, 91*, 910–917.

Nutrition Business Journal (2006). Organic Foods Report, Penton Media Inc., San Diego, USA.

Oberholtzer, L., Greene, C. & Lopez, E. (2006). Organic poultry and eggs capture high price premiums and share of specialty markets. USDA Outlook Report No LDPM-15001. *http://wwwersusdagov/Publications*

Organic Monitor (2006). The North American Market for Organic Meat Products. Organic Monitor Ltd.

Siret, F. & Issanchou, S. (2000). Traditional process: influence on sensory properties and on consumers' expectation and liking. Application to 'pâté de campagne'. *Food Quality & Preference, 11,* 217-228.

Solheim, R. (1992). Consumer liking for sausages affected by sensory quality and information on fat content. *Appetite, 19,* 285-292.

Vickrey, W. (1961). Counterspeculation, auctions, and competitive sealed tenders. *Journal of Finance, 16,* 8–37.

Westcombe, A. & Wardle, J. (1997). Influence of relative fat content information on responses to three foods. *Appetite, 28,* 49-62.

Willer, H. & Yussefi, M. (2004). The World of Organic Agriculture - Statistics and Emerging Trends. International Federation of Organic Agriculture Movements, Bonn, Germany.

In; Organic Farming: Methods, Economics and Structure ISBN 978-1-60692-864-6
Editors: Michael Nelson and Ivan Artamova

Chapter 6

ORGANIC STEER PRODUCTION BASED ON DAIRY BRED BULL CALVES – A FARM STUDY IN DENMARK

***Bea Klinzing Nielsen*[1,1]*, Troels Kristensen*[2] *and Stig Milan Thamsborg*[3]**

[1]Department of Large Animal Sciences, Faculty of Life Sciences, University of Copenhagen, Frederiksberg, Denmark.

[2] Department of Agricultural Systems, Faculty of Agricultural Sciences, University of Aarhus, Tjele, Denmark

[3]Danish Centre for Experimental Parasitology, Department of Veterinary Pathobiology, Faculty of Life Sciences, University of Copenhagen,Frederiksberg, Denmark

ABSTRACT

The objective of this study was to optimize organic steer production by analyzing the effect of production factors as feed level, grazing strategy, and finishing and welfare strategies on production objectives as liveweight gain, carcass conformation and health. Six Danish organic farms were part of the experiment that endured 2½ years. Registrations included liveweight, feed consumption, welfare-related observations and slaughter results with special emphasis on the finishing period. Daily gain of steers varied between the farms from 533 to 823 g and from 562 to 867 g on pasture and during the stable period respectively. Average feed consumption per steer per year was 1770 Scandinavian Feed Units (SFU) ± 186 based on 4 farms. Feed consumption in summer accounted 0.47 to 0.57 of total feed consumption per year. Finishing strategies showed great variations between farms with liveweight at slaughter from 458 to 595 kg, age at slaughter from 20 to 25 months and length of finishing period from 2.3 to 3.4 months. The total amount of cereals consumed in the finishing period was 100-357 kg/steer. Slaughter results show a variation between farms in carcass conformation from 3.5 to 4.9 and fatness from 2.5 to 3.2. Liveweight at slaughter was the only variable with general effect on carcass conformation (0.6 point on the EUROP-scale with increase in 100 kg liveweight at slaughter). The welfare registration scheme indicates good conditions both in the barn and on pasture, low disease rate and minor problems concerning welfare

1 Corresponding author: Tel.: +45 36885842; fax: + 45 36770266, E-mail address: *b.nielsen@norfeed.dk*

aspects in organic steer production. It is concluded that finishing strategy should focus on liveweight at slaughter. Length of finishing period as well as daily gain during the finishing period in the range practiced on the farms seemed to be of minor importance with regard to carcass conformation.

Keywords: beef production, finishing, organic farming, production system, welfare

1. INTRODUCTION

A Danish survey has shown that the majority of bull calves born on organic dairy farms are sold to conventional farms for fattening whereas a smaller proportion is reared as steers on the farm of origin [Nielsen & Thamsborg, 2002]. Similar situations are experienced in other European countries [B. Henriksen, M. Hovi, A. Martini, I. Rubelowski, W. Zollitsch, personal communications 2002]. Primary explanations are low expected economy and lack of resources as own feedstuff and stall capacities [Nielsen & Thamsborg, 2002]. Organic steer production in Europe is based on grazing in the summertime, feeding of minimum 60% of roughage during whole the year and a ban of GMO-products and hormones. Thus, the majority of the existent literature concerning steer production cannot be applied to organic production systems. The knowledge concerning steer production based on grazing and feeding with high amounts of roughage is limited as dairy breed bull calves in many conventional production systems typically are not castrated and are slaughtered at an age of 8-14 months after fattening indoor with high amounts of cereals. In Great Britain and Ireland, steers based on dairy breeds and crossbreds are produced in systems based on grazing [Baker et al., 1992]. Studies with grazing steers have documented acceptable gains in the grazing season, although problems with parasitic infections are quite common [Younie, 1990; Andersen et al., 2002; Andersen et al., 2003]. However many farmers reject steer production due to expected bad economy [Nielsen & Thamsborg, 2002]. A Californian investigation concluded that a beef system based on grazing appears to be more cost effective than a system based on intensive feeding in a feedlot (Paganini, 2004; Sainz and Paganini, 2004). The steers in this experiment were implanted and treated with dewormers and results are thus difficult to compare with organic production systems. Knowledge concerning the on-farm production system may give valuable information and help in the translation of results from experimental studies to the real situation on a farm. The objective of this study was to optimize organic steer production on farm level by analyzing the effect of production factors as feed level, grazing strategy, finishing and welfare strategies on production objectives as liveweight gain, carcass conformation and health with special emphasis on the finishing period.

2. MATERIALS AND METHODS

2.1. Farms

Three organic dairy farmers and three crop farmers buying dairy bred bull calves from organic dairy farmers were included in the study (Table 1).

Table 1. Presentation of the 6 farms in the registration program (mean of 2 year)

Farm Number	Main production	Area, ha		Animals[1]		Production, per ha			Breed [2]
		Rotation	Permanent grasslands	Dairy cows	Steers	Milk, kg	Meat, kg	Roughage, SFU [4]	
149	Crop	276	36	0	59	---	---	---	RDDB
309	Dairy	214	18	148	132	4953	356	39	DH
319	Dairy	272	14	134	25	3026	106	26	DH
409	Crop	48	31	0	98	0	306	25	DH /RDDB
419[3]	Crop	146	8	0	169	---	---	---	DH
649	Dairy	206	20	97	87	2791	252	30	DH

[1] estimated from feeding days / 365
[2] RDDB= Red Danish Dairy Breed; DH=Danish Holstein
[3] based on 2000 only
[4] Scandinavian Feed Units

The farms participated in a larger project including other types of organic production systems with focus on – how to produce organic products with high use of on farm resources. In this paper only the items related to steer production are highlighted. The farms were selected after an initial farm visit based on advisors knowledge of the farm, the farmers intention to participate in the registration programme and the farms perceived ability to carry through the registrations. All farmers had a considerable steer production.

2.2. Registrations

The study lasted 2 ½ years from January 2000 to April 2002. Registrations done on the farms, source of registration and frequency were listed in Table 2. A welfare assessment scheme was developed to describe welfare during winter when stabled in the barn and during summer on pasture. Welfare recordings were measured on the system as well on the animals and were done on group level per box or paddock. The running of management and decisions were taken by the farmer based on feeding plans, feed budgets and plans for use of pastures. Additional registrations on the farms with regard to crop production and economics were part of a national agricultural production programme [Anon. 2000; 2001].

Table 2. Registrations done concerning all steers, as well as steers in the finishing period; source and frequency of registrations.

Source	Registrations concerning	
	All steers	Finishing steers
National cattle database	Farm of origin, Birth date, Date of arrival to the farm, - of departure from the farm, breed, veterinary treatments	
Interviews of the farmer	Castration age and method	
	Management practices concerning inspection and litter	
		Time of start and end of finishing
Farm technicians		

Table 2. (Continued)

Source	Registrations concerning All steers	Finishing steers
Liveweight	At arrival to farm, turn-out, turn-in, end of the year	Monthly in the winter period, at start of finishing period and slaughter
Feed consumption	Monthly; group level	Monthly; group level
Welfare	Monthly; group level	Monthly; group level
Condition score	Monthly; group level	At start and end of finishing period
Feed quality [1]	Monthly on pasture, 2-3 times per winter	2-3 times per winter

[1] Sample taken by farm technicians, analysed by private laboratory

2.3. Estimations and Statistical Analyses

All data were analysed on farm level. Production data and turn over was established from herd data covering a period of typical one year based on total feeding days and total feed consumption per year as well as number and average liveweight of steers at start, at end, imported and slaughtered. Average daily gain per steer was estimated from total gain per period (difference between liveweight at start and end of the period) divided by total feeding days. Feed consumption per steer during housing was estimated from total feed consumption per period for the group and total feeding days. Feed consumption in summer was estimated from daily gain during grazing and expected feed utilization based on standards [Strudsholm et al., 1999]. Feed quality was primarily taken from feed analysis. When feed analyses were not available, values from feeding tables were used [Strudsholm et al., 1999].

In the finishing period feed utilization was estimated on group level based on mean feed consumption by the group and individual daily gains. For steers registered in the finishing period, gain from birth to slaughter was calculated from liveweight at birth, at slaughter and age at slaughter. If slaughter weight was not registered it was calculated by carcass weight and a theoretic dressing percentage of 50. The effect of liveweight at slaughter, daily gain in the finishing period, daily gain from birth to slaughter, age at start of finishing, the length of the finishing period, sire and condition score at slaughter on carcass conformation (EUROP-scale) were analysed by analysis of variance. The effect of liveweight at start of finishing, daily gain from birth to slaughter, age at start of finishing, the length of the finishing period, sire and condition score at start of finishing period on daily gain in the finishing period were analysed by an analysis of variance. Statistical analyses were done on farm level as well as for all steers. Statistical analyses were performed using SAS (SAS, 1999). Results are presented as mean ± standard deviation. Levels of significance are presented as * for $P<0.05$, ** for $P<0.01$ and *** for $P<0.001$.

3. Results

3.1. Production Results

The farms had an annual number of 25-202 steers and 37-132 steers were slaughtered during the trial period (Table 3). Farms 319 and 419 stopped the steer production in 2001 and 2002 therefore some data are missing from these farms. Farm 149, 319 and 419 had recently started the steer production, as can be seen from the low mean liveweight at the beginning of the experiment in 2000 with a large increase to 2002 (Table 3). Thus, results per year were biased on these farms especially on Farm 319, as the farmer started with a group bull calves in the beginning of 2000 without starting new animals further on. Therefore, focus will be on results per season or period. Liveweight of imported animals on Farm 309 and 649 is based on theoretical values for newborn calves from their own farm. The age at time of importing the calves differed between the farms explaining the varying liveweight between the farms. Average daily gain from birth to slaughter was between 0.64 and 0.74 kg/day (Table 3).

Table 3. Turnover of steers on the farms and daily gain from birth to slaughter.

	Farm No.					
	149	**309**	**319**	**409**	**419**	**649**
Annual number of steers	59	132	25	98	202	87
At start (1/1/00)						
- Number	29	120	32	59	86	73
- Liveweight (kg)	200	293 ± 159	190 ± 178	358 ± 117	139 ± 74	272 ± 150
Imported						
- Number	94	105	14	202	226	120
- Liveweight (kg)	128 ± 41	42 ± 0	122 ± 203	160 ± 73	55 ± 10	42 ± 0
Slaughtered						
- Number	45	122	37	113	132	79
- Liveweight (kg)	596 ± 69	645 ± 77	578 ± 189	572 ± 58	604 ± 68	561 ± 62
Dead						
- Number	5	11	7	4	34	16
- Liveweight (kg)	230	111	77	217	163	87
At end (31/4/02)						
- Number	73	92	2	144	146	98
- Liveweight (kg)	360 ± 141	459 ± 98	450 ± 177	289 ± 135	---	317 ± 166
Average daily gain, kg:	0.73	0.74	0.71	0.64	0.36 0.82 [1]	0.70

[1] based on 2000 data

3.2. Feeding and Gain per Season

The mean feed consumption per steer per year was 1770 SFU (Scandinavian Feed Units) ± 186 based on the results on 4 farms (Figure 1). The feed consumption in summer accounted 0.47 to 0.57 of total DM per year, and pasture and total roughages accounted for 0.22 to 0.56 and 0.64 to 0.84 of the total DM consumption per year respectively (Figure 1). Concentrates accounted for 0.16 to 0.37 of the total DM consumption per year. The consumption of cereals was 319 to 809 kg DM per steer per year.

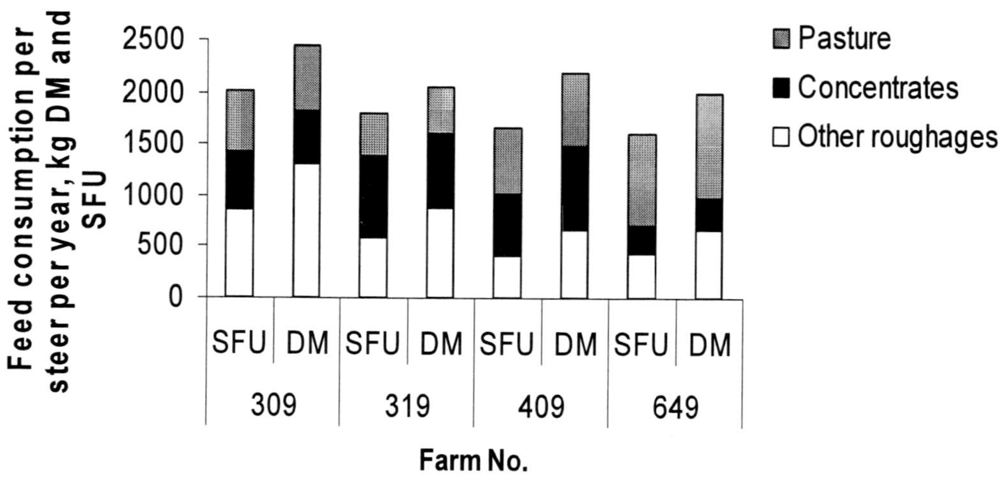

Figure 1. Feed consumption per steer per year on the 4 organic farms divided on pasture, other roughages and cereals and given as Scandinavian Feed Units (SFU) and kg dry matter (DM)

Table 4. Feed conversion ratio and days on pasture.

	Farm No.					
	149	**309**	**319**	**409**	**419**	**649**
Average daily gain, g:						
- On pasture [1]	533 ± 45	710 ± 88	760 ± 2	742 ± 88	823	808 ± 41
- During housing [1]	867 ± 128	793 ± 91	654 ± 198	649 ± 296	725 ± 23	562 ± 45
Feed conversion ratio (SFU/ kg gain) in winter[2]						
- 2000	---	6.1	3.6	7.8	---	8.0
- 2001	---	9.5	14.2	---	---	5.9
Days on pasture[3]	169 ± 6	192 ± 7	170 ± 47	141 ± 14	204	154 ± 9

[1] Standard deviation is given between animals.
[2] No standard deviation are given to feed conversion ratio as these values are estimated on group level. Anon. 2000; 2001
[3] Standard deviation is given between years 2000-2002.

The daily gain on pasture varied from 533 to 823 g between farms (Table 4). On most of the farms steers grazed both ryegrass/clover pastures as well as marginal grazing areas with the exception on Farm 419. Here only ryegrass /clover pastures were used for grazing. The daily gain for steers with a liveweight less than 120 kg at start of the period was lower compared to the 140 kg liveweight group, both for steers in the barn ($P<0.05$) and on pasture ($P<0.001$) (Figure 2). Gains during the stable period varied from 562 to 867 g/day between the farms. Feed conversion ratio on Farm 319 was biased due to an unbalanced steer production on that farm. There were also great variations in feed conversion ratio between years on the other farms. Days on pasture varied between 141 and 204 days.

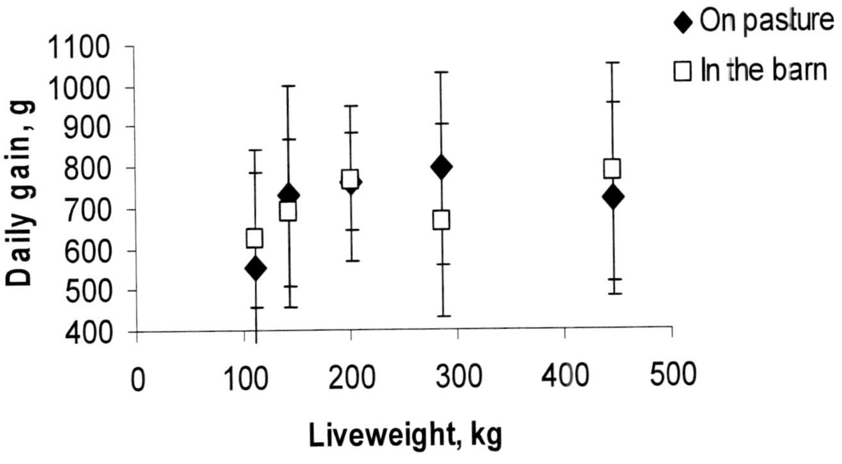

Figure 2. Average daily gain on pasture and in the barn for all farms according to liveweight (average for 2000-2001) expressed as mean ± standard variation

3.3. Management Practices, Welfare and Health

In winter all steers were housed in loose housing systems with deep litter. The groups of steers in every pen were established according to age, liveweight and pen size and remixed if needed according to the system. Average group size was 14, but with large variation both within and between farms (Table 5). The variation in liveweight within each group was on average 76 kg but on some farms up to 178 kg. The lying area was dry on 43%, moist on 35% of the area and very wet in 15% of the area (Figure 3). Feed hygiene and quality was generally good (Table 5). Carbon dioxide and ammonia was tested directly at the beginning of the registration period, but the values were very low, and the registration was stopped. Air quality was normal. The atmospheric humidity as well as temperature was slightly above the values outside. Most of the farmers informed that straw was given as an average less than every second day. Animals were inspected daily. The method used for castration was Burdizzo (crushing of the spermatic cord) on 3 farms and surgical removing of testes on 2 farms. Castration was done at an age of 2-5 months. On the crop farms the steers were moved to the farm at an age of 2 weeks up to 4 months.

On pasture, the variation in liveweight within groups on paddocks was on average 151 kg. The majority of the pasture was dry (93%) (Figure 3). Water supply was based on both natural water sources as well as drinking vats and water supply on pasture was primarily continuous. There were 1.2 ± 0.5 water places per pasture. On all farms supplementary feed was given in periods on the pasture, primarily straw, that was fed on the ground or in a feeding trough. On pasture, minerals were given on most farms ad libitum but there were also periods without mineral supply on pasture. Inspection of the animals on pasture was as average done 1 times a day every second day, depending on age and pasture. Rotation between pastures was used a lot on all farms from July until housing in October / November, primarily depending on forage supply.

Table 5. Description of housing facilities (during winter) and the conditions on pasture (during summer) by observations using a welfare assessment scheme.

	Winter [2)]	Summer [2)]
Number of observations [1)]	555-650	200
Number of animals per pen /paddock	14 ± 12	38 ± 27
Slipperiness (1= good grip; 4= very slippery)		
- Lying area	1.1 ± 0.3	---
- Feeding area	2.1 ± 1.2	---
Feed hygiene (1=clean; 3=contaminated with faeces, stones etc.)	1.1 ± 0.4	---
Feed quality (1=good; 4=bad (mould))	1.3 ± 0.5	---
Water supply (1=continuous; 2=daily; 3=less than daily)	---	1.0 ± 0.3
Cleanliness of drinking trough (1=no faeces, straw or others; 3=a lot of faeces, straw or others)	1.5 ± 0.7	1.3 ± 0.6
Air quality (1=normal; 3=pungent smell of ammonia)	1.3 ± 0.4	---
Daylight (1=light; 4=medium)	2.1 ± 1.0	---
Atmospheric humidity (relative)		
- Inside	83.2 ± 7.8	---
- Outside	74.1 ± 16.7	
Temperature, °C		
- Inside	12.4 ± 5.4	---
- Outside	7.9 ± 6.5	

1) The number of observations describes total observations done on all pasture and boxes on all farms, i.e. a total of 555 and 650 boxes dependent on observation and 200 paddocks were observed in the registration period.

2) Standard deviations were given between all observations and were rather high, because the values are dependent and at some observations it might be ranked 0% for "total dry" and 90% for "moist" and in another case the rankings were 90 and 0% respectively.

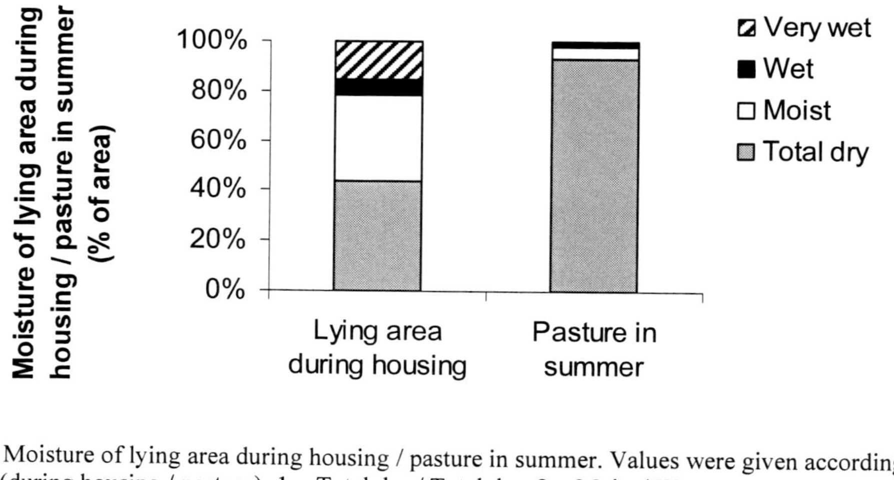

Figure 3. Moisture of lying area during housing / pasture in summer. Values were given according to the scale (during housing / pasture): 1 = Total dry / Total dry; 2 = Moist / Water covers a part of the hooves; 3 = Wet / Animals walk in water to the knees; 4 = Very wet / ---

Observations concerning welfare-related issues on animal level are presented in Table 6. A total of 765 boxes and 146 pastures were observed during the registration period. The steers were without or with little faecal contamination both on pasture and in the barn. The majority of steers had a dry hair coat when stabled in the barn and the hair coat was either nice or not nice nor ruffled on the majority of steers observed. Most of the steers had a normal body condition score. Most of the steers were supposed to be completely dehorned, however about 10% of the observed steers were not dehorned at time of the registration. Faeces consistency was higher when stabled in the barn than when grazing on pasture (Figure 4). Distinct swelling or lack of hair on hock, distinct swelling, abscess or skin lesions, ringworm (dermatomycosis), inappropriate long claws and sore feet were rarely observed on the farms whether during the summer nor the winter period (mean 1.0-1.6). Respiratory symptoms (frequent or laboured respiration) were not observed (1.1 ± 0.4) and coughing was observed in less than 10% of animals (1.8 ± 0.9) during whole the year.

Table 6. Welfare-related observations on steers during housing and on pasture. Values are given according to scale: 1=0%; 2=<10%; 3=10-50%; 4=50-90%; 5=>90% of the animals [1].

	During housing	On pasture
Number of observations	765	146
Cleanliness of the animals		
- No faecal contamination	2.5 ± 1.5	3.1 ± 1.3
- Little contamination	3.4 ± 1.4	2.8 ± 1.1
- Strong contamination	1.2 ± 0.6	1.1 ± 0.4
Wetness of the hair coat		
- Dry	3.2 ± 1.6	---
- Moist	2.5 ± 1.4	---
- Wet overall	2.0 ± 1.3	---
Length of the hair coat		
- Trimmed	1.1 ± 0.6	---
- Not trimmed but short	3.7 ± 1.5	---
- Long	2.0 ± 1.3	---
Hair coat		
- Nice	2.7 ± 1.5	3.4 ± 1.3
- Not nice nor ruffled	3.0 ± 1.4	3.0 ± 1.2
- Little ruffled	1.8 ± 1.1	1.5 ± 0.9
- Much ruffled	1.1 ± 0.4	1.0 ± 0.1
Body condition score		
- Very poor	1.2 ± 0.6	1.1 ± 0.3
- Poor	2.3 ± 1.1	2.4 ± 0.9
- Normal	3.6 ± 1.3	3.9 ± 0.8
- Fat	1.4 ± 0.9	1.7 ± 0.9
- Grossly fat	1.1 ± 0.5	1.0 ± 0.2

[1] The registrations were done on a scale from 1-5, according to the proportion of animals in the observed groups, as i.e. some steers in the box may be strong contaminated with faeces and some steers may be without any contamination. Every single observation thus sum to 100%, however in the table average values of all observations are given and thus the total sum may not exactly sum to 100%.

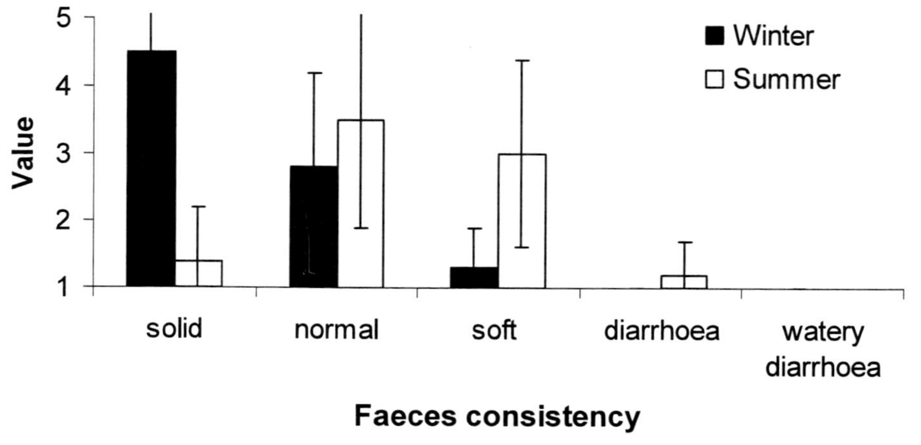

Figure 4 Faeces consistency from steers on pasture (summer) and in the barn (winter). Values were given according to scale: 1=0%; 2=<10%; 3=10-50%; 4=50-90%; 5=>90% of the animals.

3.4. Finishing Strategies

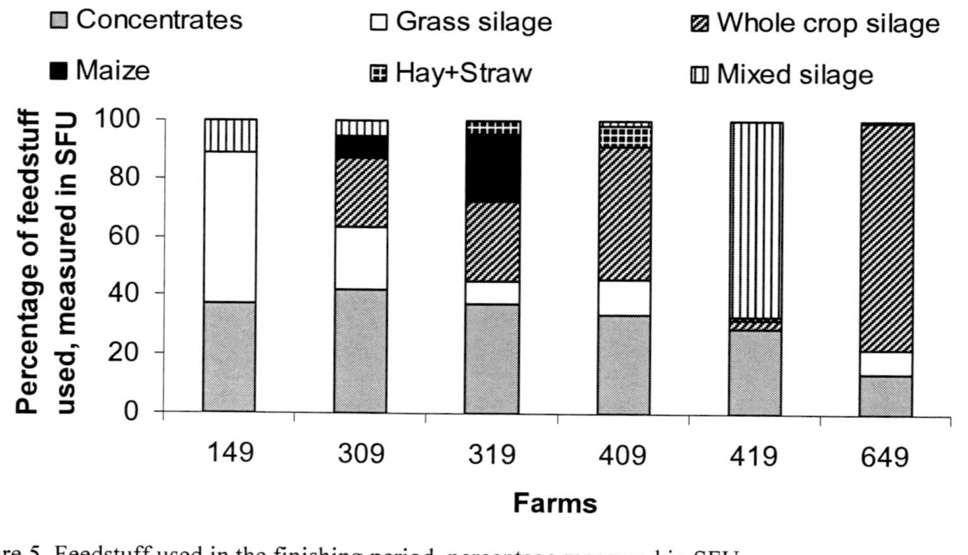

Figure 5. Feedstuff used in the finishing period, percentage measured in SFU

Management strategies concerning the finishing of steers were different between the 6 farms (Table 7). Factors as liveweight (458-595 kg), age at start of finishing (20-25 months) and at slaughter (565-667 kg and 23-28 months respectively) and length of finishing period (2.3-4.0 months) varied. Feeding during the finishing period was based on 1.3 to 4.7 SFU concentrates per steer per day, primary cereals, together with ad lib feeding of silage (Table 8). On Farm 419 and Farm 649, no cereals were given at all in a period. The total amount of concentrates consumed in the finishing period was 100-357 kg/steer. Daily weight gain varied from 0.58 to 1.25 kg (Table 8). Great variations in feed conversion ratio on some farms were due to weight losses in some steers. Only few farmers finished the steers on the pasture. During the finishing period, cereals were used less than or equal to 42% of the feedstuff used, measured in SFU (Figure 5). Whole crop silage was the most common type of silage in the finishing period. Slaughter results showed a variation between farms in dressing percentage from 49.4 to 53.8, carcass conformation from 3.5 to 4.9 and fatness from 2.5 to 3.2 (EUROP-scale) (Table 9). Most of the slaughtered steers achieved a carcass conformation of 3-5 (Figure 6).

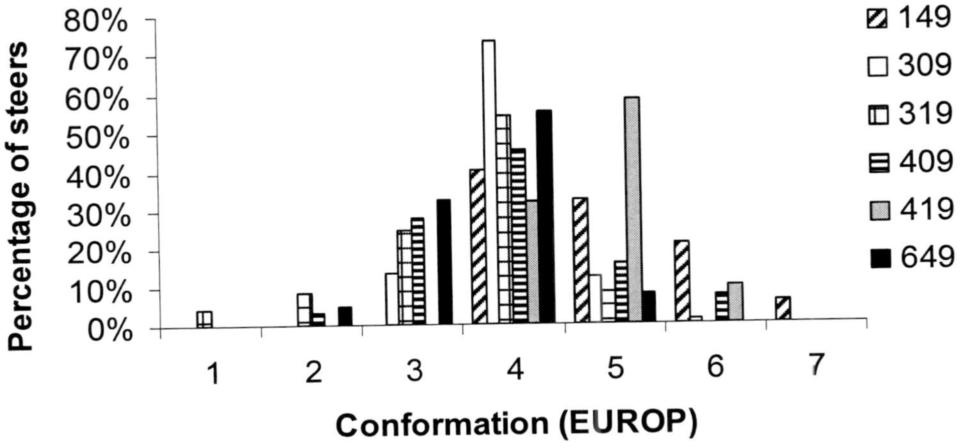

Figure 6. Carcass conformation (EUROP) in slaughtered organic dairy bred steers

Table 7. Characteristics of steers at start and end of the finishing period and length of finishing period.

	Farm No.					
	149	**309**	**319**	**409**	**419**	**649**
Number of steers	55	99	24	78	31	67
Liveweight, kg						
- at start of finishing	496 ± 57	595 ± 43	560 ± 35	458 ± 49	516 ± 51	485 ± 81
- at slaughter	586 ± 44	667 ± 42	646 ± 64	577 ± 37	617 ± 53	565 ± 63
Age, months						
- at start of finishing	22.5 ± 3.4	25.1 ± 1.7	24.8 ± 1.0	20.1 ± 2.2	19.4 ± 1.5	20.8 ± 3.1
- at slaughter	25.5 ± 2.7	27.4 ± 1.6	28.1 ± 1.3	25.0 ± 1.8	22.8 ± 0.5	24.3 ± 1.0
Length of finishing period, days	89 ± 44	69 ± 35	102 ± 14	119 ± 44	105 ± 47	80 ± 61
Body Condition Score						
- at start	3.4 ± 0.5 [1]	3.1 ± 0.3	2.8 ± 0.2	2.8 ± 0.5 [1]	2.7 ± 0.4	3.0 ± 0.3
- at slaughter	4.4 ± 0.4 [1]	3.2 ± 0.3	3.2 ± 0.6	---	3.5 ± 0.4	3.2 ± 0.4

1) The value is based on 23, 17 and 17 steers for condition score at start (Farm 149 and Farm 409) and at slaughter (Farm 149) respectively

Table 8. Feed consumption, feed conversion ratio and gain in the finishing period

	FARM NO.					
	149	**309**	**319**	**409**	**419**	**649**
Feed consumption, SFU/steer/day	9.7	11.2	9.5	7.9	9.8	9.0
Feed consumption, SFU/steer	863	776	969	946	1029	716
Gain, kg/day:						
- 2000/01	1.06 ± 0.40	0.97 ± 0.51	---	1.14 ± 0.23	1.02 ± 0.43	0.58 ± 0.23
- 2001/02	0.98 ± 0.36	1.25 ± 0.54	0.86 ± 0.44	0.75 ± 0.19	---	1.09 ± 0.33
- whole period	1.01 ± 0.37	1.16 ± 0.54	---	0.86 ± 0.27	---	0.81 ± 0.38
Total gain in the finishing period	90 ± 49	71 ± 31	85 ± 46	102 ± 52	99 ± 59	61 ± 50
Feed conversion, SFU/kg gain						
- 2000/01	11.3 ± 6.0	16.2 ± 11.1	---	8.1 ± 1.7	13.0 ± 12.1	13.5 ± 8.2
- 2001/02	---	10.0 ± 29.4	12.1 ± 14.5	10.2 ± 3.3	---	11.6 ± 5.3

Table 9. Carcass characteristics of organic dairy bred steers

	FARM NO.					
	149	**309**	**319**	**409**	**419**	**649**
Number of steers	52	98	24	69	31	65
Carcass weight, kg	300 ± 23	334 ± 20	319 ± 35	287 ± 20	332 ± 26	282 ± 34
Dressing percentage[1]	51.2	50.0	49.4	49.7	53.8	49.9
Carcass classification:						
- Conformation	4.9 ± 0.9	4.0 ± 0.5	3.5 ± 0.9	4.1 ± 1.2	4.8 ± 0.6	3.7 ± 0.7
- Fatness	3.0 ± 0.1	3.1 ± 0.4	2.8 ± 0.7	2.6 ± 0.6	3.2 ± 0.5	2.5 ± 0.6
- Colour	3.0 ± 0	3.1 ± 0.2	3.1 ± 0.3	3.1 ± 0.3	3.0 ± 0	3.2 ± 0.4

[1] estimated from liveweight at slaughter and carcass weight

Table 10. Estimates from the analysis of variance with effect on carcass conformation (EUROP-classification)

	Effect of					
Farm	**LW at slaughter, kg**	**Daily gain during the finishing period, kg**	**Daily gain from birth to slaughter, kg**	**Age at slaughter, months**	**Length of finishing period, days**	**R^2**
149	0.0083**	NS	NS	NS	NS	0.17
309	0.042*	0.363**	-34.5*	-0.960*	-0.029*	0.15
319	NS	NS	9.39***	NS	0.029*	0.60
409	0.004*	NS	NS	NS	NS	0.07
419	0.006**	NS	NS	NS	NS	0.23
649	0.009***	-0.701*	NS	NS	0.002*	0.44

Results from the analysis of variance showed a positive effect of LW at slaughter (5 farms) on carcass conformation (Table 10) with an average of 0.6 point increase in EUROP scale with 100 kg increase in liveweight at slaughter. There was only limited effect of daily gain during the finishing period and daily gain from birth to slaughter, body condition score at slaughter and sire on carcass conformation. There was significant effect of liveweight and age at start of finishing and daily gain from birth to slaughter on daily gain during the finishing period (Table 11). There was effect of sire on daily gain on Farm 319 ($P<0.01$) and

Farm 649 (P<0.05). In Table 12 characteristics are given for steers with low and high liveweight at start of finishing and for steers with carcass conformation less, equal and more than 4. The groups of low and high liveweight at start of finishing were created by two groups of equal size. Liveweight at slaughter, daily gain during the finishing period and condition score at slaughter increased with increased carcass conformation.

Table 11. Estimates from the analysis of variance with effect on daily gain in the finishing period

Farm	LW at start of finishing	Daily gain from birth to slaughter	Effect of Age at start of finishing	Length of finishing period	Body condition score at start of finishing	R^2
149	-0.009***	7.74***	0.230***	NS	NS	0.79
309	-0.0123***	10.17***	0.293***	-0.0055***	NS	0.65
319	-0.0098***	8.32***	0.199***	-0.0020**	NS	0.99
409	-0.0079***	6.20***	0.163***	-0.0016*	---	0.89
419	-0.0064***	5.36***	0.221***	NS	NS	0.81
649	-0.014***	10.02***	0.188***	-0.0040*	-0.330**	0.77

Table 12. Characteristics for groups for low and high liveweight (LW) at start of finishing and carcass conformation (< 4, equal to 4 and > 4).

	Carcass Conformation	LW at slaughter	Daily gain	Length of finishing period, days	Condition Score At start	Condition Score At slaughter
LW at start of finishing						
- LOW	4.0	---	0.92	118	2.9	3.4
- HIGH	4.2	---	1	69	3.0	3.3
Conformation						
- Less than 4	---	578	0.78	94	2.8	3.0
- Equal to 4	---	615	0.99	82	3.0	3.4
- More than 4	---	627	1.09	96	3.0	3.6

4. Discussion

4.1. Farms

The dairy farms participating in the present experiment had in average less livestock units per ha compared to other organic dairy farms in Denmark (0.56 and 0.79 cows/ha respectively) [Nielsen & Thamsborg, 2002]. Thus, the large area available on these farms results in high forage resource for a steer production. Shortage of own feedstuff is one of the primary reasons mentioned by farmers for not finishing bull calves on their own farm

[Nielsen & Thamsborg, 2002] and production of home-grown feed is a condition for a good economy in organic dairy and beef production [Tvedegaard, 2002].

4.2. Feeding and Gain

Gains during the housing periods varied from 526 to 867 g/day between the farms. The high daily gain during winter is not generally due to a good feed conversion ratio but is accompanied by a high feed consumption. The decisions of the farmers to feed at high feeding level during winter is astonishing, as low daily gain during winter could save resources and promote compensatory growth in the following grazing season [Wright et al., 1986]. The daily gains during observed summer and winter is in accordance with other experiments [Ainslie et al., 1992; Grenet et al., 1997; Andersen et al., 2003]. However, average gain from birth to slaughter was lower compared to gains in steers grazing ryegrass / white clover pastures (0.80-0.90 kg/day) [Andersen et al., 2003]. This may be explained by the periodic use of marginal grazing areas on the private farms resulting in a lower average daily gain.

Feed consumption in summer accounted for about half of total feed consumption per year, primarily from pasture, indicating that summer grazing play an important role in organic steer production. Experiments with steers grazing exclusively marginal pastures indicate that about 30% of the total feed consumption can be based on marginal pastures which are often related to low growth rates [Andersen et al., 2002]. The farmers typically have the possibility to change between ryegrass/clover and marginal pastures, thus utilizing these pastures optimal and in this way increasing the total feed consumption from pasture. The variations between farms also indicate possibilities of increasing feed consumption from pasture up to 56% of total feed consumption per year. Grazing strategies as more days on pasture may decrease feed costs. Variations between farms up to 2 months indicate possibilities to improve management at this point.

It should be mentioned that the estimation of feed consumption in summer is based on daily gain and expected feed utilization. If daily gain on pasture is low, as i.e. by grazing marginal grazing areas, estimated feed consumption would be low as well. The total amount of cereals used per produced steer (1183 kg DM at slaughter at 24 months) markedly exceeded the amount used in other experiments with steer production based on grazing in the summer with a consumption of 400-500 kg cereals [Flynn, 1985; Jans & Troxler, 1996; Grenet et al., 1997]. Cereals used for finishing steers accounted for about 18 to 30% of the total consumption of cereals when slaughtered at 24 months. The finishing period should be the time of using the highest amounts of cereals, thus it seems that especially cereal consumption before finishing was higher than in other experiments. The mean feed conversion ratio was 7.5 SFU/kg gain and is in accordance with other experiments (8.3 SFU/kg gain) [Andersen et al., 2003]. However, lower feed conversion ratio was found in steers following grazing marginal pastures (5.7 SFU/kg gain) [Andersen et al., 2002]. This may be explained by a lower liveweight in the given period as feed conversion ratio increase with liveweight.

4.3. Finishing

Some farmers started finishing in the winter period, then the steers were turned-out to pasture because steers were not finished for slaughter and then finishing continued after turn-in. These animals were not included in the analysis.

It seems from the feed recordings in the finishing period that the demand of 60% roughage according to European standards for organic farming is met.

Feed consumption on the farms was in-between the observations made by Andersen et al. (2002; 2003). However in the present experiment feed conversion ratio was much higher (11.8 SFU/kg gain) than in experiments by Andersen et al. (2002) (8.3-8.5 SFU/kg gain). This may be explained by low gain in the period before housing in the experiments from Andersen et al. (2002) and following compensatory growth in the barn. No data are available from the farms concerning daily gain in the last month before housing, but it seems probable that the farmers were aware of low daily gain in the period before housing and adjusted management according to circumstances.

Liveweight at slaughter was the only general factor with a major effect on carcass conformation on several farms with an increase in carcass conformation of 0.6 point (EUROP-scale) with increase of 100 kg liveweight at slaughter. The regression coefficient was very low suggesting a low validity of the model. Parameters dependent on each other as liveweight and age may explain the low values. However similar to the present results, increase of 100 kg liveweight resulted in increased conformation of 0.5 point (scale of 1-7, 1-5) and 1.0 point (EUROP-scale) [O'Ferrall & Keane, 1990; Steen & Kilpatrick, 1995; Andersen et al., 2002]. In the present experiment, there was no effect of sire on carcass conformation, contrary to other experiments showing high effect of sire. This may be explained by the high number of sires used on the farms with one farm using 20 different sires to 67 steers.

Daily gain during the finishing period had no effect on carcass conformation in the present experiment. Contrary to this, higher feed level during finishing resulted in improvement in carcass conformation [Keane & Allen, 2002]. Similar, carcass conformation seemed to be affected by daily gain, length of finishing period and condition score, although no significant effect was found. In another experiment effect of the length of finishing period on carcass conformation was found [Andersen et al, 2002]. However, length of finishing period was confounded with liveweight, as steers finished for at longer time also had a higher liveweight at slaughter. Bulls slaughtered from pasture without finishing had a lower dressing percentage, a higher muscularity and a lower fat content compared with bulls at same liveweight fed on high feed levels in the barn [Therkildsen et al., 1995], however no differences in carcass conformation were found between these 2 groups.

The study clearly showed that liveweight at slaughter is the primary factor with effect on carcass conformation. Length of finishing, daily gain and condition score at slaughter had no significant effects on carcass conformation, thus the effect of these factors seemed to be of minor importance in the range practiced on the farms. In the present registration program no comparison was done between finishing and not finishing therefore, no conclusion can be drawn whether finishing is important or not with reference to the results found by Nielsen et al. [2005] and Nielsen & Kristensen [2007] showing that finishing of dairy bred steers did not improve the economic output. Further on carcass conformation tended to increase with daily gain during the finishing period and condition score at slaughter and these factors might be important tools in management decisions as well (Table 12).

4.4. Welfare and Health

Health and welfare problems appear to be limited in organic steer production. The registrations confirm this although the mortality is high on some of the farms. In other experiments parasite problems especially when grazing marginal areas [Thamsborg, 2000], and liver abscesses in animals fed periodically with high amounts of concentrates [Andersen et al., 2003] were found.

Similar a Danish survey among organic farmers with a steer production reported low level of health and welfare problems [Nielsen & Thamsborg, 2002], however internal parasites especially lungworm and ringworm were mentioned by the farmers. The present experiment did not give any information about the clinical level of parasite infections, but normal faeces consistency and low level of respiratory symptoms indicated no untreated infections. Health and welfare problems seem not to be the case, although steer production is an extensive production and reduced management could be expected on dairy and crop farms having dairy and crop production respectively as their primary interest.

Conclusion

The study clarified that feed consumption from pasture as well as high amounts of roughages during winter feeding and finishing in the barn plays a major role in organic steer production systems. It is concluded that large variation between farms concerning management strategies shows the possibilities for improving the production system. This is especially obvious regarding days on pasture which should be increased on several farms as well as feed consumption of cereals in the time before finishing that should be reduced to a minimum. Liveweight at slaughter seemed to be the most important factor with effect on carcass conformation being a key factor for the production economy. It is suggested that factors as length of finishing period and daily gain during the finishing period has minor importance, as none of these factors had effect on carcass conformation.

Acknowledgments

The study received financial support from the Danish Research Centre for Organic Farming (DARCOF). The authors wish to thank the farmers and technicians being part of the registration programme.

REFERENCES

Anonymous 2000; 2001. *Gårdrapport Studielandbrug*. Danish Agricultural Advisory Centre, Århus. (In Danish).

Ainslie, S.J., Fox, D.G. & Perry, T.C. 1992. Management systems for Holstein steers that utilize alfalfa silage and improve carcass value. *J. Anim. Sci. 70* (9), 2643-51.

Andersen, H.R., Kristensen, T., Bligaard, H.B. & Thamsborg, S.M. 2002. Studeproduktion ved afgræsning af ferske enge. *Report No. 40 (Husbandry)*. Danish Institute of Agricultural Science. Foulum, Denmark. (In Danish).

Andersen, H.R., Kristensen, T., Bliggaard, H.B. & Nielsen, B. 2003. Forskellige produktionssystemer for stude på kløvergræs. *Report No. 40 (Husbandry)*. Danish Institute of Agricultural Science. Foulum, Denmark. (In Danish).

Baker. 1992. Pasture fattening in the humid temperate zone. In: Jarrige, R., Beranger, C. (Eds.), *Beef cattle production. Elsevier*, Amsterdam, 259-270

Flynn, V. 1985. Beef production from Friesians calves. *Veterinary Update* : 34-36.

Grenet, N., Micol, D., Haurez, Ph., Dozias, D. & Pflimlin, A.1997 Production of high quality beef extensive conditions in France. Effects of extensification on animal performance, carcass composition and product quality. *Proceedings of a workshop*, Belgium. 248-267.

Jans, F., Troxler, J. 1996. Ochsenmast auf ungeduengten Weiden in Hoehenlagen [Schweiz]. *Agrarforschung* (Switzerland) 3 (4): 169-72. (In German)

Keane, M.G. & Allen, P. 2002. A comparison of Friesian-Holstein, Piemontese X Friesian-Holstein and Romagnola X Friesian-Holstein steers for beef production and carcass traits. *Livest. Prod. Sci. 78*, 143-58.

Nielsen, B. & Thamsborg, S.M. 2002. Dairy bull calves as a resource for organic beef production: A farm survey in Denmark. *Livest. Prod. Sci. 75* (3), 245-55.

Nielsen, B.K., Kristensen, A.R. & Thamsborg, S.M. 2005. Optimal decisions in organic steer production – a model including winter feeding level, grazing strategy and slaughtering policy. *Livestock Production Science 88*, 239-250.

Nielsen, B.K. & Kristensen, A.R. 2007. Optimal decisions in organic beef production from steers - Effects of criterion of optimality and price changes. *Livestock Science 110*, 25-32.

O'Ferrall, G.J.M. & Keane, M.G. 1990. A comparison for liveweight and carcass production of Charolais, Hereford and Friesian steer progeny from Friesians cows finished on two energy levels and serially slaughtered. *Anim. Prod. 50*, 19-28.

Pananini, R.V. 2004. An alternative natural beef production system: A differentiation strategy for California producers and packers. Giannini Foundation of Agricultural Economics. *ARE Update 8* (1), 9-11.

Sainz, R & Pananini, R.V. 2004. Effects of different grazing and feeding periods on performance and carcass traits for beef steers. *Journal of Animal Science 82*, 292-297.

SAS Institute Inc. 1999. Version 8, Cary, NC, USA.

Steen, R.W.J. & Kilpatrick, D.J. 1995. Effects of plane of nutrition and slaughter weight on the carcass composition of serially slaughtered bulls, steers and heifers of three breed crosses. *Livest. Prod. Sci. 43*, 205-13.

Strudsholm, F., Aaes, O., Madsen, J., Kristensen, V.F., Andersen, H.R., Hvelplund, T. & Østergaard, S. 1999. Danske fodernormer til kvæg. *Report No. 84. Danish Agricultural Advisory Service*. Skeiby.

Thamsborg, S.M. 2000. Uptake of nematode infections in steers reared at different grazing intensities. In: Grassland farming - balancing environmental and economic demands. (Eds. Søegaard et al.), *Proceedings of the 18th EGF*, Aalborg, Grassland Science in Europe. 561-63.

Therkildsen, M., Vestergaard, M., Ramsgaard Jensen, L. & Andersen, H.R. 1995. Betydningen af fodringsintensitet, afgræsning og slutfedning på produktion og slagtekvalitet af SDM ungtyre. *Research Report No. 35, Statens Husdyrbrugsforsøg*. Frederiksberg. pp 36.

Tvedegaard, N. 2002. Økologisk mælkeproduktion. *Fødevareøkonomisk Institut. Report No. 137*. Copenhagen. (In Danish)

Wright, I.A., Russel, A.J.F. & Hunter, E.A. 1986. The effect of winter food level on compensatory growth of weaned, suckled calves grazed at two sward heights. *Anim. Prod 43*, 211-23.

Younie, D. 1990. Eighteen-month beef production: Organic and intensive systems compared. In: *Organic Meat Production*. Chamberlain, A.T; Walsingham, J.M. & Stark, B.A. (editors), Chalcombe publications. 41-53.

In; Organic Farming: Methods, Economics and Structure
Editors: Michael Nelson and Ivan Artamova
ISBN 978-1-60692-864-6

Chapter 7

FERTILIZATION LOADS OF LESS COMMON ELEMENTS TO ARABLE SOILS IN AUSTRIA

Manfred Sager [1]

Austrian Agency for Health and Food Safety Competence Centre of Elements
Spargelfeldstraße 191 A – 1220 Vienna, Austria

ABSTRACT

All kinds of fertilizers available on the Austrian market have been analyzed for their nutrient element und micro-element contents. From a total of 2759 samples, taken in 2002 - 2006, 2391 samples were assigned as solid and nutritional, and classified due to their main element characteristics N-P-K-C-S. Commercial organic fertilizers, manures of any kind, and composts have been summarized in one category each. The elements (in alphabetical order) Al, As, B, Ba, Be, Ca, Cd, Co, Cr, Cu, Fe, K, Li, Mg, Mn, Mo, Na, Ni, P, Pb, Sr, V, and Zn were determined by ICP-OES after acid digestion. Because all datasets are asymmetric, with many low and some high values, medians and 5-95% concentration ranges are given, for each micro-element in 17 tables. Concentrations in dry mass are compared with mean soil and mean crust values. Loads for a fertilization rate of 100 kg N and 100 kg P were compared with atmospheric deposition data and presumable crop uptake. At usual fertilization rates for N and P, the input of most micro elements was maximum for composts. Manures and commercial organics also supplied micro elements at higher levels than mineral fertilizers, except Cd-As-Cr. In addition, manures, some K-salts, and PK-fertilizers supply a lot of Na. On a long term, phosphate fertilizers will contribute to the enrichment of arable soils with Ba, Be, Cd, Cr, and V, because they supply more than atmospheric perrecipitation and crop outputs. Composts will probably enrich soils with As, Ba, Be, Cu, Cr, Li, Mn, Ni, Pb, V, and Zn.

[1] e-mail: manfred.sager@ages.at.

1. INTRODUCTION

Soils have various functions. They are the most important substrate to grow crops for human nutrition, for domestic animals, or maybe for the supply of energy in the future. In addition, they have important functions of filtering and storage of groundwater. For urban civilizations, they give space for recreation areas like parks, playgrounds, hiking trails etc.. Threats to arable soils arise from urbanization, contamination, salinization, compaction and erosion.

Fertilization means the input of plant nutrient compounds to the soil in order to increase crop yield. Sufficient nutrient supply at a favorable soil pH, as well as to stable humics and soil life should be achieved. In Austria, "Guidelines of Adequate Fertilization" (Richtlinien für die sachgerechte Düngung) have been issued to recommend ecologically and economically feasible amounts of nutrient loads. The term "fertilizer" covers a wide range of mixtures of inorganic salts as well as organic compounds and wastes. Apart from main constituents N-P-K-S, their composition is highly variable among fertilizer types. This includes micro-elements necessary for plant growth (e.g. Zn, Mo, Co, B), unwanted and largely toxic elements (e.g. Cd, Pb), and elements of low physiological activity (e.g. Li). Quite often, a surely too simple question is put forward: what is better, mineral fertilization or organic amendments?

Indicators for sustainable ecological conditions are the nitrogen budget, the phosphorus budget, the humics budget, biodiversity, energy input, soil compaction, soil erosion, and the emission of gases (Kehres 2007). All fertilizing procedures should sustain favourable humics contents and soil pH. The goal of fertilization is to supply the crop with sufficient essential nutrients. Above all it has to be considered that nitrogen is much slower available from plants, dungs and crop residues, than from mineral fertilizers and manures (Peszt, Claupein et al. 1997). Undoubtedly, a certain amount of organic amendment is necessary to ensure proper microbial activities and an organic carbon pool to favour the activity of the soil fauna (e.g. earthworms), which prevents too much compaction and erosion.

Application of fertilizers in inappropriate amounts and element proportions may influence the quality of crops as well as the environment.

Whereas mineral fertilizers are industrial or mining products, wastes from farmed elements and from plant production are recycled to the soil as organic fertilizers. This recycling of nutrients is necessary to protect surface waters from eutrophication, and to maintain nitrate in potable waters at low levels.

When we look at current fertilization practice, fertilizer dosage is not simply done just by fertilizer weight, but with respect to the input of a certain nitrogen or phosphorus load per area. Thus, the load of all accessory elements has to be considered per nutrient rather, in order to achieve a realistic classification of wanted or unwanted enrichments or deficiencies. Whereas the farmer can influence the loads to his field via the fertilization pathway, additional inputs occur from atmospheric deposition. Therefore, it is worth while to compare total deposition data with median loads put to the fields together with 100 kg N or 100 kg P per hectare, and obtained from various types of fetilizers. Main pathways of output are the crops, groundwater leaching, and soil erosion. From this, possible enrichments or depletion can be concluded in the long term.

Numerous investigations about nitrogen, carbon, phosphorus, sulfur, and potassium cycles have been made. Some trace metals levels have been of concern also, because their toxicity effects have been widely known, like Pb and Cd. Other elements which may be beneficial for plant growth, but also toxic at higher levels, are even added to fertilizers, like Co, Mo and B. Usually, their level in fertilizers is controlled by authorities, if their contents has been put on the label.

The dataset compiled within this context should help to estimate threats of contamination and salinization from fertilizer application, and the selection of suitable fertilizer types.

2. Material and Methods

2.1. Laboratory Procedure

Commercial organic and mineral fertilizers were delivered dry, they were homogenized by milling, and taken as such. Composts, garden moulds, dungs, and above all the manures had appreciable water contents. They were mixed by stirring, weighed as wet samples, and water was determined in aliquots. The presented data refer to the dry mass.

5 g of mineral fertilizer samples were dissolved on the boiling water bath in 3M hydrochloric acid, and filtered into a 250 ml volumetric flask with paper filter. Together with the filter, this residue was ashed in a muffle furnace at 560°C, and the ash dissolved with 3M hydrochloric acid at the boiling water bath for half an hour, filtered into the previously used volumetric flask, and made up to the mark. Organo- mineral fertilizers were dry ashed at first, extracted with acid, filtered, passed the muffle furnace a second time, extracted and filtered again (see also Sager 2006).

Most manures, dungs and composts were delivered fresh from the farmers. Means from at least double digests were used. Manures were oven dried overnight in the digestion beaker, and dry- ashed thereafter. In case, manures, dungs, composts or other organics had to analyzed for sulfur and/or selenium also, they were dried and subsequently ashed in presence of magnesium nitrate.

The resultant sample solutions were measured by ICP-OES (Perkin-Elmer Optima 3000XL), with the plasma torch in the axial mode, usually from 1+24, 1+4, and undiluted sample solutions, in order to meet an adequate calibration range, and to recognize matrix effects. In the calibrant solutions, main and trace elements were simultaneously present to imitate a probable matrix. Together with each batch, 2 blanks and 2 digests of a control sample (NPK fertilizer) was run, to indicate long term precision. For e.g. triple phosphates, K-salts and limestones, more dilution was necessary to obtain correct results for the main elements. The accuracy was checked by control measurements with atomic absorption (Perkin Elmer AAS300), wherever possible, as well as the certified reference material NIST 1400 Bone Ash and BCR Nr. 32 Phosphate Rock (Natural Moroccan Phosphorite). A mineral phosphate sample was run within every batch as a control for day-to-day variations. We also take part successfully in regular ring tests (MARSEP program) organized by Wageningen Agricultural University (NL).

Usually, the elements (in alphabetical order) Al, As, B, Ba, Be, Ca, Cd, Co, Cr, Cu, Fe, K, Li, Mg, Mn, Mo, Na, Ni, P, Pb, Sr, V, and Zn were read. Due to variable blanks from the glass, the boron data were read, but not evaluated. Detection limits were taken from the repeatability (2 s) of 11 pairs of blank digests. Na and Ba levels were sufficiently high to cope with interactions from the glass.

2.2. Data Processing

All compilations and calculations were done in EXCEL. At first, the resultant data sets were copied to a general result file, and given a code in order to group them according to sample types. Data of each sample type were stored within a proper file for further treatments. In addition to the own analytical data, water contents, total nitrogen, ammonium and nitrate nitrogen, as well as the post code of the sampling site were imported from other sources within our institution. Within each file, the samples were grouped according to the postcode of the sampling location, which go along the main railway lines, but roughly represent the province and the region. Just small corrections had to be made for the exact geographical assignment of their origin resp. sampling site.

Finally, box- plots were done utilizing the SPSS statistic program.

2.3. Calculation of Annual Extraction by Crops

For calculations of annual extraction by crops, median data obtained from European samples supplied from the International Plant Exchange Program (Wageningen Agricultural University) were multiplied with an average output (Houba, Uittenbogaard 1994). As these samples have been analyzed by many labs, data are very safe, and they refer to dry weight. Wheat was assumed to deliver an output of 4,5 tons of grains + 3,6 tons of straw per hectare. The yield of potatoes was estimated to be 30 tons fresh weight, equivalent to 6 g dry weight. Maize was assumed to yield 8 tons per hectare. Of course, the output will be lower than given in the tables, if some straw remains on the field.

2.4. Atmospheric Deposition

Unfortunately, much more data about atmospheric concentrations than about atmospheric deposition were found for the central European region. As the east of Austria is too dry for lichens, the respective data are often extrapolated, or from lonely mountains in the plains. Therefore it makes sense to look at older data for comparison of atmospheric deposition with fertilizer input.

In 1984/85, the author determined some main and trace element in total atmospheric deposition sampled at 40 sites in the Marchfeld, northeast of Vienna (results still unpublished). Glass beakers of 100 cm^2 area were exposed about 1 month, their contents digested with nitric acid, and made up to 50 ml. This solution was measured for main and trace elements on a simultaneously reading ICP, and in addition, Pb-Cd-Ni-V were determined by graphite furnace AAS, arsenic by hydride AAS, and Zn by flame AAS. Data from outliers high in aluminum were discarded because of presumable contamination from soils, and data from outliers high in phosphorus were discarded because of presumable input of birds′ excrements. The corrected median values of atmospheric deposition (g per hectare and year) are compiled in the respective data tables.

In 1997/98 a revival of the total deposition measurements started with the same ICP, upon which the fertilizer samples were run. The total deposition samples were digested with conc. HNO_3, made up to 25 ml, and measured. Median data from 11 sites sampled in triplicate on experimental fields in the east of Austria during four monthly periods, are given in the tables.

2.5. Mean Crust and Soil Data

Mean crust data have been taken from Bowen (1979). For soils, the median data of the most frequent soil types of the Austrian soil inventory have been considered. This soil inventory refers to the aqua regia -extractable fraction (digested without hydrofluoric acid), which may be more similar to the procedure applied to the fertilizer samples, than a total digestion. In case, Austrian data were not available, mean soil data have been taken from Ure and Berrow (1982), who give total amounts.

3. Results

3.1. Frequency of Samples

The frequency of samples among various fertilizer types reflects the somehow the frequency of emergence on the market. However, a lot of N-only compounds, K-salts, and limes have not been run for at the ICP, because analysis for "heavy metals" was not ordered; within these samples, unwanted traces are usually very low.

From a total of 2759 samples, less common sample types as well as liquid samples were omitted for evaluation. Sewage sludges are largely analyzed elsewhere. All samples containing substantial organic carbon were classified as organic fertilizers, unless labelled as manures, dungs, or composts. Garden moulds are subject to the Fertilizer Act as well, though they are not an additive to the soil but a substrate of their own. This led to a final evaluation of 2391 samples (table 1).

For stastistical treatment of the obtained sample concentrations, all were equally treated, irrespective of their amount sold on the market, or the bag size for big farmers and hobby gardeners. It can be assumed, however, they have been mixed from the same components.

Table 1. Median of nutrient contents and number of investigated samples of the current evaluation (mainly 2002-2006)

	% N	% P P_2O_5= 2,29* %P	Main elements	Type	Number of samples
One main nutrient	-	-	**K**; Mg	K-salts	22
	-	0,01	Ca, Mg	Lime, dolomite	90
	17,1	-	**N**; Mg, S	Mg - sulfates/nitrates	18

Table 1. (Continued)

	% N	% P P_2O_5= 2,29* %P	Main elements	Type	Number of samples
	26,7	-	**N**; Ca	Ammonium nitrate lime	20
	20,9	-	**N**; S	Ammonium sulfate	9
	21,2	-	**N**; Ca	Calcium nitrate	4
	-	12,0	**P**	Rock phosphate	57
	-	10,3	**P**	Super phosphate	19
	-	20,9	**P**	Triple phosphate	93
	-	8,2	**P**; Ca, Mg	Ca-Mg-phosphates	43
	-	7,2	**P**; Ca,Mg,Fe	Basic slag	6
Two main nutrients	-	4,62	**P**; **K**, Ca,Fe	Basic slag potash	11
	23,5	7,1	**N, P**	NP 24/12	14
	18,0	20,2	**N, P**	D-ammonium phosphate	123
	Var	Var	**N, P**	Other N-P	12
	-	6,22	**P, K**	P-K fertilizers	125
Three main nutrients	15,0	4,60	**N, P, K**	N-P-K all proportions	475
	15,0	3,79	**N, P, K**	N-P-K with sulfate, all proportions	471
Organics	6,80 *	1,90 *	**N, P, K, C**	Commercial products with organic carbon	417
	7,74 *	1,39 *	**N, P, K, C**	Manures and dungs	219
	1,23 *	0,56 *	**N, P, K, C**	Composts	65
	**	0,11	**N, P, K, C**	Garden moulds	94
Not evaluated				Sewage sludges	5
				Powdered rocks	36
				Inorganic soil additives	14
				Peat	9
	None			Trace element additives	53
				Commercial liquid fertilizers	235

* regional variations

** Total nitrogen in garden moulds was determined per litre; not available per dry mass

3.2. General

Within this compilation, substances added to arable soils in order to increase their levels of available nutrient elements N-P-K-S, have been treated. Liquid products and products containing soluble trace element combinations without nutrients NPK have been excluded. At first, elemental contents of fertilizers are compared with mean crust and mean soil values, and then with respect to their N and their P contents. Mineral fertilizers are usually sold in dry pellets, and the concentrations were taken as such. Data from composts, manures and dungs were transferred to dry mass data.

For those types of fertilizers, which are supplied to increase soil nitrogen, a fertilization load referring to an equivalent of 100 kg per hectare were compared, which is within the scope of practical farming. Accessory elements in ammonium nitrate lime were always lowest, and in composts largely highest. The elements Ba, Mn, Mo and Pb strictly followed the ascending order ammonium nitrate lime < mineral fertilizers < organics < manures < composts. The group of Al, Li, Co, Ni followed this order also, but had more in the diammonium phosphates than in the organics. Sr and V equivalents for 100 kg N were higher in NPK with sulfates than in those without, and also higher than in manures. Just for 3 elements, Cd-As-Cr, mineral fertilizer contained a higher load for 100 kg N than manures and organics, but still lower than the composts. Just Na contributed its highest load from manures instead from composts.

Another aspect is to compare the accessory elements added along with nutrient phosphates. The loads equivalent to 100 kg P (= 229 kg P_2O_5) were much higher than equivalent to 100 kg N throughout, but in current agricultural practice, 100 kg P per hectare is unusually high. For a lot of concomitant elements to 100 kg P, the load decreased from garden moulds to composts to manures to mineral fertilizers (Al-Ba-Be-Co-Cu-Mn-Mo-Ni-Pb-Zn), partly because garden moulds and composts were lower in P. Additionally, top values from garden moulds followed by composts came from As-Cr-Sr-V. Remarkable exceptions from this sequence were noted for Cd and Na. Cd load for 100 kg P from garden moulds remained highest, but it was lower from organics and manures than from mineral fertilizers. Maximum amounts of Na came from manures instead from garden moulds. For Li, composts were slightly higher than garden moulds. Loads from mineral phosphates and rock phosphates were largely overlapping, except for Cu-Zn, which were lower, and Mn-Pb-Sr-V, which were higher than in processed mineral mixtures. Among the group of mineral fertilizers, Cu/P was larger in NPK, NPK with sulfate and NP 24/12, and Zn was larger in NP 24/12. It is obvious that Na input from ammonium salts should be lower than from others, but also PK-type products bring more Na than NPKs.

Use of basic slags as a P-source is history for Austria, the last samples were taken in 2005. Contrary to a former data compilation (1986-93), the Cd/P was like for other phosphates, whereas in former times it was much lower. Basic slags contained more Co-Cr-Li-Mn-Pb-Sr-V than other mineral phosphates, whereas Al-As-Ba-Be-Cu-Mo-Na-Ni-Zn were in the same range.

3.3. Aluminum

Total aluminum is a main component of the earth crust, its mean crust concentration being 8,2%, and its mean soil concentration 6,7 %. Because silicates and refractory oxides remain largely insoluble, the data obtained by the dry- ashing / acid leaching method do not reflect total contents, but largely the amounts bound to soil humics and hydroxides. Plant - available Al (at about 200 mg/kg) significantly limits plant growth in acid soils, causing P-deficiency or Ca- deficiency symptoms (Sager 1986).

Fortunately, in P-less fertilizers, Al was found at a median of about 100 mg/kg, but for the highest values there has to be paid some attention on their application to acid mineral soils. In presence of excess phosphates, low solubilities of $AlPO_4$, resp. $Al(OH)_2(H_2PO_4)$ prevents its toxicity towards plant roots. Rock phosphates contained a median of about 200 mg/kg Al. In organics, Al was found from below below detection limit up to 1,45 %, and highest levels were found in garden moulds and composts. But as complexes of Al^{3+} with humics are second strongest after Cu^{2+}, the organic matrix counteracts any action of Al.

Table 2. Data for aluminum

Al: Mean crust = 82 g/kg Mean soil = 67 g/kg						
Type	g/kg	range	kg for 100kg N	range	kg for 100 kg P	range
K-salts	0,039	< 0,002 - 0,82				
Ammonium sulfates	0,082	0,003 - 1,98				
Ammonium nitrate limes	0,096	0,063 - 0,355	0,035	0,02 - 0,06		
Magnesium sulfates/nitrates	0,108	< 0,002 - 1,14				
Commercial organics	0,57	0,005 - 14,5	0,80	<0,02 - 103	2,77	0,04 - 453
Limes, dolomites	0,59	0,033 - 17,5				
N-P-K fertilizers	0,66	0,002 - 2,90	0,43	<0,02 - 2,12	1,44	<0,02 - 14,1
P-K fertilizers	0,71	0,16 - 2,23			1,16	0,70 - 2,66
N-P-K + sulfate	0,72	0,02 - 1,89	0,49	0,02 - 1,51	1,68	0,09 - 8,59
Basic slags	1,07	0,84 - 4,11			1,51	1,30 - 7,14
NP 24/12	1,07	0,035 - 2,67	0,58	0,43 - 0,91	1,61	0,79 - 2,44
Super phosphates	1,12	0,86 - 2,30			1,11	0,87 - 1,86
Ca-Mg-phosphates	1,47	0,92 - 2,61				
Manures and dungs	1,89	0,17 - 6,17	2,23	0,40 - 182	13,5	2,08 - 67,2
Rock phosphates	2,02	1,44 - 3,01			1,63	1,24 - 2,94
Triple phosphates	2,74	1,18 - 4,50			1,17	0,60 - 2,04
Di-ammonium phosphates	3,39	1,14 - 5,25	1,84	0,59 - 2,82	1,63	0,67 - 2,70
Basic slag potash	3,74	2,86 - 4,60				
Garden moulds	6,58	1,07 - 13,73			321	87,3 - 1417
Composts	14,39	9,62 - 19,83	80,5	15,8 - 284	275	87,2 - 1047

The equivalent for 100 kg N was highly variable and thus nearly unpredictable for organics (0,006-103 kg Al(100 kg N) ands manures (0,4 - 318 kg Al/100 kg N, whereas ranges for the ammonium phosphates were rather narrow.

Mineral fertilizers contained about 1-2 kg (acid extractable) Al equivalent to 100 kg P. Composts (275 kg Al/ 100 kg P) and garden moulds (321 kg Al/100 kg P) had much more, but did not approach the proportion of the mean soil values Al/P = 84 (table 2).

3.4. Arsenic

When grouped to geological units, background values in Austrian soils range from 2,1-14,7 mg/kg (Danneberg 1999). Most fertilizer samples were below this range, just DAP and composts might have reached soil levels. Manures, commercial organics, and samples without P main component were lowest.

The amount of arsenic equivalent to 100 kg N was lower than the atmospheric deposition for mineral fertilizers, but sometimes higher for organics and composts. Similarly, fertilization of 100 kg P per hectare yielded lower arsenic input than atmospheric deposition. However, As/P from composts was about 20 times higher than from mineral sources, and clearly higher than atmospheric input (table 3).

Table 3. Data for arsenic

As: Mean crust = 1,5 mg/kg Soil medians (Austria) 4-15 mg/kg Mean soil = 11,3 mg/kg	Crop uptake: wheat 0,27 + 0,62 g maize 2,2 g	Atmospheric deposition: Marchfeld 1985: 6,0 g/ha.a East of Austria 1997/98: 2,1 g/ha.a

	´mg/kg	range	´g for 100 kg N	range	´g for 100 kg P	range
Manures and dungs	0,68	< 0,1* - 3,54	0,83	< 0,2 * - 82	4,3	< 0,5 - 38,7
Commercial organics	0,73	< 0,5 - 5,27	1,02	< 0,5 - 45	4,3	< 0,5 - 247
Ammonium sulfates	1,04	< 0,5 - 1,83				
Ammonium nitrate limes	1,30	< 0,5 - 2,03	0,36	< 0,5 - 0,75		
K-salts	1,56	0,37 - 2,55				
P-K fertilizers	1,79	< 0,5 - 6,2			3,05	< 0,5 - 7,4
Basic slag potash	1,94	1,47 - 2,30				
Garden moulds	2,13	< 0,5 - 8,0			180	< 0,5 - 668
Super phosphates	2,18	0,81 - 9,7			2,05	1,74 - 15,8
Limes, dolomites	2,48	< 0,5 - 6,8				
N-P-K fertilizers	2,54	< 0,5 - 15,5	1,51	< 0,5 - 12,9	5,4	< 0,5 - 42,5
Rock phosphates	2,82	< 0,5 - 11,3			2,52	< 0,5 - 9,8
N-P-K +sulfate	3,07	< 0,5 - 11,2	2,01	< 0,5 - 8,3	8,2	< 0,5 - 33,5
Magnesium sulfates/nitrates	3,23	0,91 - 4,09				
Ca-Mg-phosphates	3,55	0,68 - 18,7				
Basic slags	4,00	1,55 - 7,3			5,7	3,26 - 10,2
Triple phosphates	4,50	< 0,5 - 17,2			2,10	< 0,5 - 6,5
Composts	5,97	2,27 - 10,9	31	7,8 - 127	107	24 - 461
NP 24/12	6,81	1,63 - 15,9	2,84	1,09 - 4,47	9,2	1,86 - 16,9
Di-ammonium phosphates	9,04	0,86 - 29,8	5,5	< 0,5 - 18,4	4,9	< 0,5 - 16,5

* ... determination with hydride-AAS

Table 4. Data for barium

Ba: Mean crust = 500 mg/kg mean soil = 568 mg/kg	Crop uptake: maize 27 g	Atmospheric deposition: East of Austria 1997/98: 20 g/ha.a

Type	´mg/kg	range	´g for 100 kg N	range	´g for 100 kg P	range
K-salts	0,86	0,09 - 2,3				
Ammonium sulfates	1,64	0,29 - 11,3				
Ammonium nitrate limes	3,53	0,18 - 16,2	0,82	< 0,05 - 6,16		
Limes, dolomites	5,70	0,43 - 136				
Magnesium sulfates/nitrates	8,07	< 0,05 - 25,1				
Di-ammonium phosphates	11,5	1,26 - 234	5,5	0,76 - 128	4,8	0,61 - 107
N-P-K fertilizers	13,5	< 0,05 - 118	8,2	< 0,05 - 97	31	< 0,05 - 291
Commercial organics	14,4	0,48 - 108	22,6	0,54 - 620	93	3,0 - 3173
N-P-K + sulfate	24,1	1,2 - 102	16,5	1,1 - 89	60	4,1 - 291
Manures and dungs	37,4	10,0 - 114	58	11,8 - 2606	273	84 - 1032
NP 24/12	44,8	15,8 - 84	19,0	8,3 - 31	64	18,0 - 96
Super phosphates	46,9	34,4 - 212			51	36,3 - 120
Garden moulds	55,5	9,7 - 145			4328	1101 - 15753
P-K fertilizers	59,1	6,9 - 196			109	23,5 - 213
Basic slags	72	43 - 212			121	60 - 304
Ca-Mg-phosphates	75	22,2 - 212				
Basic slag potash	93	42 - 142				
Triple phosphates	99	13 - 148			46,6	6,9 - 67
Rock phosphates	149	93 - 274			134	73 - 204
Composts	169	84 - 307	907	215 - 3213	3507	1072 - 17696

3.5. Barium

All barium concentrations found in the entire dataset were below mean crust (500 mg/kg) and mean soil (568 mg/kg) levels. However, barium present in feldspars and micas has not been recovered here. Samples without phosphate main component were largely below 10 mg/kg, but NPK and organic fertilizers were also at a median level of just 14 mg/kg. Significantly more barium (10-fold) was found in rock phosphates and composts, manures were in the middle between NPK and composts. Variations were larger for NPKs, DAPs and commercial organics.

Fertilization of 100 kg N from mineral fertilizers contained less barium than from organics, manures and composts. However, Ba-loads from organics and manures varied more than 3 orders of magnitude.

Though data about uptake of barium into crops are variable, 100 kg P presumably cover annual plant uptake from any source. Composts contained significantly more Ba/P than others (table 4).

3.6. Beryllium

Beryllium is known to be accessory in silicates, and the mean soil (1,5 mg/kg) is below the mean crust level (2,6 mg/kg). Among fertilizers, products without P main component hardly reached 0,1 mg/kg. Organics of any kind as well as NPKs, PKs, and garden moulds were in the range of arable soils or below. DAP and triple phosphates may exceed mean soil levels to about double.

Together with 100 kg N, the beryllium input from all phosphates and organics exceeded the annual input from atmospheric deposition. The load from composts was significantly the highest. The inputs from manures and organics referring to 100 kg N covered a broad range of about 3 orders of magnitude.

100 kg P from all kinds of mineral fertilizers, manures and organic fertilizers contained about 10 times more beryllium than the annual atmospheric precipitation, and 100 kg P from compost about 200 times more. Be/P for basic slags could be higher than for other mineral fertilizers (table 5).

Table 5. Data for beryllium

Be: Mean crust = 2,6 mg/kg mean soil = 1,5 mg/kg		Crop uptake: wheat 0,03 g potatoes 0,02 g		Atmospheric deposition below detection limit		
	´mg/kg	range	´g for 100 kg N	range	´g for 100 kg P	range
K-salts	< 0,04	< 0,04 - 0,13				
Magnesium sulfates/nitrates	< 0,04	< 0,04 - 0,26				
Ammonium nitrate lime	0,05	< 0,04 - 0,17	< 0,04	< 0,04 - 0,06		
Ammonium sulfates	0,05	< 0,04 - 0,38				
Limes, dolomites	0,12	< 0,04 - 1,72				
Commercial organics	0,17	< 0,04 - 1,73	0,27	< 0,04 - 15,4	0,99	0,07 - 59,9
Manures and dungs	0,24	< 0,04 - 0,97	0,33	0,05 - 27,7	1,83	0,35 - 7,70
N-P-K all proportions	0,35	< 0,04 - 1,27	0,24	< 0,04 - 0,93	0,84	0,07 - 3,76
N-P-K + sulfate	0,37	0,06 - 1,04	0,27	0,04 - 0,69	0,96	0,19 - 1,94
P-K fertilizers	0,53	0,21 - 1,40			0,86	0,55 - 1,67

Table 5. (Continued)

Be: Mean crust = 2,6 mg/kg mean soil = 1,5 mg/kg	Crop uptake: wheat 0,03 g potatoes 0,02 g	Atmospheric deposition: below detection limit

	´mg/kg	range	´g for 100 kg N	range	´g for 100 kg P	range
Garden moulds	0,63	0,13 - 1,97			46,2	10,3 - 204
Super phosphates	0,77	0,31 - 1,43			0,89	0,72 - 1,62
Basic slags	0,88	0,55 - 1,49			1,25	0,85 - 3,13
NP 24/12	0,89	0,18 - 1,72	0,38	0,34 - 0,86	1,24	0,02 - 1,62
Ca-Mg-phosphates	0,92	0,60 - 1,29				
Rock phosphates	1,18	0,89 - 1,76			0,98	0,73 - 1,52
Basic slag potash	1,35	0,38 - 2,55				
Composts	1,84	0,96 - 2,74	11,1	1,6 - 39,6	31,0	9,4 - 182
Triple phosphates	1,87	1,30 - 3,15			0,83	0,57 - 1,81
Di-ammonium phosphates	2,85	0,60 - 3,83	1,58	0,32 - 2,12	1,41	0,30 - 1,96

3.7 Cadmium

Analysis for cadmium in fertilizers is very popular, though toxicity effects of many other elements are known as well. Due to substitution of Ca in the apatite lattice, which is the main phosphate mineral, Cd in fertilizers strongly correlates with P. Medians of organics of any kind were found at about soil level, whereas mineral phosphates were higher throughout. Thus, most Cd was found in (increasing order) rock phosphates, Ca-Mg phosphates, triple phosphates, and diammonium phosphates. Top values may exceed background concentrations in arable soils at about 100-fold, which justifies regular controls and legal thresholds.

As most of the cadmium comes from phosphate sources, a 100 kg N- fertilization from products without P as main component lead to inputs clearly below atmospheric deposition and plant uptake, whereas composts and diammonium phosphates were above. Organics and NPKs were medium. The level of concern = 10g/ha for Cd, was never reached for a fertilization of 100 kg N.

However, 100 kg P per hectare from all kinds of mineral fertilizers might exceed this level of concern. Cd/P ranges were largely overlapping, but mind that ammonium phosphates contained about double Cd/P than rock phosphates. Contrary to most other elements, Cd/P from organics and manures was lower than from minerals (table 6).

Table 6. Data for cadmium

Cd: Mean crust = 0,11 mg/kg Soil medians (Austria) 0,12-0,34 mg/kg		Crop uptake: wheat var. potatoes 0,46 g maize 2,1 g		Atmospheric deposition: Marchfeld 1985: 1,48 g/ha a East of Austria 1997/98: 1,21 g/ha.a		
	´mg/kg	´range	´g for 100 kg N	´range	´g for 100 kg P	range
K-salts	0,05	< 0,05 - 0,16				
Ammonium nitrate limes	0,09	< 0,05 - 0,21	0,03	< 0,02 - 0,08		
Ammonium sulfates	0,13	< 0,05 - 1,42				
Magnesium sulfates/nitrates	0,15	< 0,05 - 0,67				
Limes, dolomites	0,19	0,05 - 0,82				
Commercial organics	0,20	< 0,05 - 4,60	0,38	< 0,05 - 7,82	1,7	< 0,1 - 18,3
Manures and dungs	0,38	< 0,05 - 1,22	0,44	< 0,05 - 12,1	2,5	< 0,1 - 11,5
Garden moulds	0,42	0,10 - 0,80			25,5	6,7 - 59
Composts	0,43	0,24 - 0,99	2,34	0,56 - 8,4	7,9	3,5 - 56
N-P-K fertilizers	0,98	< 0,05 - 10,17	0,60	< 0,05 - 7,18	2,74	< 0,1 - 18,4
N-P-K + sulfate	2,56	< 0,05 - 6,95	1,75	< 0,05 - 5,09	6,53	< 0,1 - 18,8
Basic slag potash	2,99	0,14 - 4,35				
P-K fertilizers	3,89	0,48 - 9,84			5,85	2,56 - 12,8
Super phosphates	5,40	2,77 - 11,8			5,63	4,24 - 13,1
NP 24/12	6,57	0,24 - 13,5	2,78	1,33 - 4,50	9,3	2,9 - 12,7
Basic slags	6,63	0,70 - 12,1			9,95	0,78 - 18,9
Rock phosphates	7,28	2,78 - 13,3			5,74	2,40 - 12,5
Ca-Mg-phosphates	7,35	3,14 - 13,4				
Triple phosphates	13,5	8,60 - 37,1			6,5	3,3 - 22,2
Di-ammonium phosphates	19,3	0,08 - 37,1	10,51	0,07 - 21,3	10,7	< 0,1 - 18,7

3.8. Cobalt

Though cobalt is essential for plant growth, soil background values (in Austria) are lower than mean crust. In fertilizers, Co was found to occur generally very low, and all phosphates were also below soil background. In NPK and commercial organics, significantly asymmetric frequency distribution of cobalt concentrations indicate occasional additions. Highest median concentrations of the cobalt, however, were found in garden moulds and composts.

An annual fertilization of 100 kg N together with atmospheric deposition carries along enough cobalt to balance plant uptake from main crops wheat and potatoes, whereas the input from di-ammonium phosphates, organics, and above all from composts, was clearly more.

A 100 kg P fertilization per hectare from basic slags, manures, and above all from composts, contained much more cobalt than the needs of wheat and potatoes, whereas input from other sources approximately balanced the crop output (table 7).

Table 7. Data for cobalt

Co: Mean crust = 20 mg/kg Soil medians (Austria) 6,5-13,7 mg/kg Mean soil = 12 mg/kg		Crop uptake: wheat 0,14 + 0,15 g potatoes 0,53 g maize 1,1 g		Atmospheric deposition: below detection limit		
	´mg/kg	´range	´mg/kg	´range	´mg/kg	´range
K-salts	0,11	< 0,1 - 0,27				
Ammonium nitrate limes	0,12	< 0,1 - 0,91	< 0,1	< 0,1 - 0,34		
Ammonium sulfates	0,26	< 0,1 - 1,33				
N-P-K + sulfate	0,29	< 0,1 - 1,95	0,21	0,03 - 1,42	0,69	< 0,1 - 5,24
Magnesium sulfates/nitrates	0,31	< 0,1 - 1,01				
P-K fertilizers	0,32	< 0,1 - 2,50			0,48	< 0,1 - 3,29
NP 24/12	0,46	< 0,1 - 1,66	0,21	0,18 - 0,44	0,70	0,31 - 1,90
Lime, dolomites	0,48	< 0,1 - 16,2				
Ca-Mg-phosphates	0,49	0,26 - 2,20				
N-P-K fertilizers	0,51	< 0,1 - 81,5	0,33	< 0,1 - 46,8	1,07	< 0,1 - 228
Commercial organics	0,59	< 0,1 - 9,55	0,81	< 0,1 - 65,8	3,72	0,17 - 260
Super phosphates	0,71	0,30 - 2,19			0,65	0,36 - 2,31
Rock phosphates	0,75	0,31 - 2,45			0,60	0,26 - 1,53
Triple phosphates	0,97	0,47 - 2,04			0,42	0,23 - 1,15
Basic slags	1,08	0,32 - 37,0			1,51	0,46 - 58
Di-ammonium phosphates	1,71	0,84 - 4,44	1,04	0,44 - 3,50	0,91	0,37 - 2,70
Manures and dungs	1,82	0,20 - 6,49	2,02	0,69 - 70	12,5	4,0 - 77
Basic slag potash	2,00	0,39 - 3,11				
Garden moulds	3,23	0,34 - 6,89			209,3	20,6 - 545
Composts	6,88	3,02 - 9,70	34,1	6,7 - 159	135,4	34,3- 525

3. 9. Chromium

In soils and river sediments, significant parts of total chromium (about 2/3) have been surely not recovered by the method applied, which are bound to refractory oxides and silicates. In composts and garden moulds, this recovery is much higher. Thus, the mean crust level is estimated to be 100 mg/kg, whereas the background value of acid leachable chromium in arable soils was found at 39 mg/kg (Danneberg 1999), and more than 100 mg/kg acid leachable chromium in soils indicates contamination.

In fertilizers, products without P main component and organics of any kind were lower than P-rich products and soil levels, but the concentration range covered 2 orders of magnitude in commercial organics and NPKs. Besides the known high values in basic slags which finally led to the implementation of threshold values, also DAP and triple phosphates often exceeded the level of concern of 100 mg/kg.

Table 8. Data for chromium

Cr: Mean crust = 100 mg/kg Soil medians (Austria) 29 - 49 mg/kg Mean soil = 84 mg/kg	Crop uptake: wheat 0,9 + 1,1 g potatoes 3,3 g		Atmospheric deposition: Marchfeld 1985: 2,3 g/ha.a East of Austria 1997/98: 2,3 g/ha.a			
	´mg/kg	´range	´g for 100kg N	´range	´g for 100 kg P	range
K-salts	0,91	<0,25 - 5,3				
Ammonium nitrate limes	0,97	<0,25 - 3,60	0,35	< 0,1 - 1,34		
Magnesium sulfates/nitrates	1,20	<0,25 - 15,0				
Limes, dolomites	2,24	<0,25 - 31,7				
Ammonium sulfates	5,42	0,37 - 12,5				
Commercial organics	7,89	<0,25 - 81,7	12,7	< 0,3 - 435	53	0,7 - 1916
Manures and dungs	8,68	0,88 - 96	10,0	1,00 - 154	56	7,6 - 762
Garden moulds	12,9	0,85 - 34,8			706	0,5 - 2688
N-P-K fertilizers	18,0	<0,25 - 161	11,1	< 0,2 - 101	45	< 0,2 - 461
Composts	30,8	17,3 - 114	163	67 - 1230	629	159 - 4710
N-P-K + sulfate	40,8	0,54 - 118	31	0,51 - 78	101	1,78 - 227
P-K fertilizers	56,6	4,06 - 104			89	48 - 147
Ca-Mg-phosphates	87,6	53,8 - 123				
NP 24/12	92,7	2,9 - 132	50	29 - 76	137	62 - 213
Super phosphates	107	60 - 156			94	75 - 152
Rock phosphates	108	64 - 138			89	59 - 115
Triple phosphates	180	109 - 281			83	45 - 186

The annual atmospheric precipitation approximately equalled crop extraction. At a fertilization rate of 100 kg N/ha resp. 100 kg P/ha, all kinds of fertilizers supplied (at the median values) additional chromium, except ammonium nitrate lime. Fertilization of 100 kg N never reached the level of concern for 2 years, of 625 g Cr/ha, except for some diammonium phosphates. However, top inputs came from basic slags, and for these the level of concern has to be well considered, apart from current thresholds expressed in terms of concentration (table 8).

Table 9. Data for copper

Cu: Mean crust = 50 mg/kg Soil median (Austria) 16 - 25 mg/kg Mean soil = 26 mg/kg			Crop uptake: wheat 18 + 9 g potatoes 29 g maize 60 g		Atmospheric deposition: Marchfeld 1985: 19,6 g/ha.a East of Austria 1997/98: 12,3 g/ha.a	
	mg/kg	range	g for 100 kg N	range	g for 100 kg P	range
K-salts	0,80	0,27 - 3,0				
Ammonium nitrate limes	0,90	0,11 - 10,0	0,29	< 0,07 - 3,81		
Ammonium sulfates	1,48	0,02 - 15,1				
Limes, dolomites	2,02	< 0,2 - 33,7				
Magnesium sulfates/nitrates	2,77	< 0,2 - 54,6				
P-K fertilizers	6,61	2,09 - 16,5			10,6	2,92 - 23,8
N-P-K + sulfate	8,8	0,59 - 594	5,88	0,41 - 478	22,4	2,20 - 2480
Basic slag potash	8,8	3,8 - 31,5				
Ca-Mg-phosphates	8,9	4,7 - 64,0				
N-P-K fertilizers	10,3	0,48 - 915	7,08	0,31 - 634	20,5	1,21 - 2344
Rock phosphates	11,8	7,6 - 38,6			9,8	6,1 - 21,2
Di-ammonium phosphates	12,4	0,26 - 54,7	7,24	< 0,1 - 29,8	6,70	< 0,1 - 27,1
Super phosphates	13,0	8,0 - 21,0			11,5	9,2 - 18,0
Basic slags	16,8	4,56 - 74,0			17,2	7,1 - 62,9
Commercial organics	19,2	2,76 - 137	28,0	2,23 - 598	123	9,9 - 2315
Garden moulds	22,2	13,0 - 193			2008	639 - 20508
Triple phosphates	23,2	2,69 - 35,4			11,2	1,46 - 18,0
Manures and dungs	64,7	8,5 - 568	88	19 - 723	547	134 - 2884
Composts	84,2	28,4 - 176	415	68 - 2238	1136	542 - 15880
NP 24/12	233	0,2 - 719	224	6,3 - 347	206	10 - 1085

3.10. Copper

For copper, a mean crust level of 50 mg/kg and a mean soil background level of 21 mg/kg have been established. For mineral fertilizers without P as a main component, just 1-2 mg/kg can be expected, whereas phosphates of any kind and garden moulds range like soils. Top values appeared in manures (above all in pig manures). Some commercial organics, garden moulds, manures and NP 24/12 exceeded the level of concern for soils, which is 100 mg/kg. Asymmetric frequency distribution in NPK fertilizers indicates occasional additions (or contaminations?).

Fertilization rates of 100 kg N per hectare from mineral fertilizers did not approach plant uptake alone, but well together with atmospheric deposition. Contrary to this, manures and composts supplied clearly more. 100 kg N given via manures from pigs, chickens and and turkeys easily superseded the level of concern for contaminations, which is 625 g Cu per hectare within 2 years, like some NPKs and organics which may contain added Cu-salts. These intentional additions should have been marked on the respective label.

Atmospheric deposition together with 100 kg P per hectare satisfied the needs of main crops wheat and potatoes. 100 kg P from composts, however, add a median of 1,1 kg Cu to the fertilized area, and also manures (above all from pig farms) easily contain the level of concern for contaminated soil, which is 625g/ha for 2 years (table 9).

3.11. Lithium

Lithium is largely bound to silicates (mean crust = 20 mg/kg), and its transfer to green plants is low. No soil data from Austria are currently available. In fertilizers, lithium was found to be uniformly low, about 1 mg/kg in mineral fertilizers and commercial organics, and about 2 mg/kg in limes, dolomites, Ca-Mg.phosphates, rock phosphates, PKs and NPKs. Contrary to Na (see there), there was no enrichment in K-salts. Just composts were significantly higher than others, corresponding to soil levels.

Median lithium input from composts along with 100 kg N was 70 g, and about 100 times lower from organics and manures. The Li-input for 100 kg P was 1-3 g from mineral fertilizers, 27 g from manures, and about 350 g from composts. Li/P in composts and garden moulds was smaller than the proportion for mean soil values (Li/P = 0,02). Data for plant uptake and atmospheric deposition are still unknown (table 10).

3.12. Manganese

In case of manganese, most fertilizer samples were below the mean crust level of 950 mg/kg and the mean soil level of 760 mg/kg, just basic slag and basic slag potash were clearly above. The lowest Mn-contents were found in K-salts, ammonium sulfates and NP 24/12. NPKs, NPKs with sulfate, ammonium nitrate lime and commercial organics showed a bimodal frequency distribution of their Mn- concentrations. The first frequency maximum at 11 mg/kg for ammonium nitrate limes and NPKs with sulfate, as well at 58 mg/kg for NPKs and organics may reflect the natural abundance. A second frequency maximum of about 400

Table 10. Data for lithium

Li: Mean crust = 20 mg/kg mean soil = 31 mg/kg		Crop uptake: wheat 0,08 + 0,53 g		Atmospheric deposition: East of Austria 1997/98: 0,88 g/ha.a		
	´mg/kg	´range	´g for 100 kg N	´range	´g for 100 kg P	´range
Ammonium nitrate limes	0,34	0,12 - 0,65	0,13	0,03 - 0,24		
K-salts	0,42	0,09 - 5,31				
Magnesium sulfates/nitrates	0,47	0,01 - 12,5				
Ammonium sulfate	0,49	< 0,01 - 5,68				
N-P-K + sulfate	0,91	0,14 - 3,38	0,67	0,11 - 2,60	2,06	0,53 - 12,3
P-K fertilizers	0,96	0,40 - 3,05			1,66	0,79 - 3,85
N-P-K fertilizers	0,99	0,04 - 4,13	0,65	0,02 - 3,09	2,09	0,06 - 12,6
Commercial organics	1,08	< 0,01 - 21,8	1,07	< 0,02 - 139	7,1	< 0,02 - 430
Super phosphates	1,28	0,94 - 3,51			1,58	1,02 - 2,76
NP 24/12	1,28	0,35 - 3,15	0,54	0,43 - 0,76	1,84	1,18 - 2,69
Limes, dolomites	1,99	0,48 - 21,1				
Ca-Mg-phosphates	2,14	1,20 - 5,51				
Basic slags	2,22	1,34 - 7,20			3,31	1,86 - 11,2
Rock phosphates	2,32	1,55 - 3,60			1,90	1,32 - 3,59
Basic slag potash	2,88	2,02 - 6,01				
Di-ammonium phosphates	3,00	1,29 - 4,55	1,65	0,62 - 2,53	1,47	0,61 - 2,51
Triple phosphates	3,00	1,85 - 7,48			1,32	0,89 - 4,76
Manures and dungs	3,71	0,18 - 10,9	3,69	0,38 - 44	26,9	2,89 - 115
Garden moulds	5,36	0,22 - 12,0			341	33,8 - 1293
Composts	14,1	10,2 - 25,8	70	24 - 350	353	65 - 1129

mg/kg indicates occasional additions at this level. Composts, garden moulds and even manures contained Mn within a surprisingly narrow range.

Extraction of manganese along with crops may vary widely, but is usually more than the annual atmospheric deposition. A fertilization of 100 kg N per hectare from mineral fertilizers was likely not to cover plant needs, whereas supply from manures and composts was sufficient. 100 kg N equivalent loads from organic fertilizers were highly variable. NPKs containing sulfate supplied significantly less Mn for 100 kg N than others.

Together with atmospheric deposition, an input of 100 kg P covered plant needs from most sources, whereas organics and basic slags supplied clearly more. Too much Mn is no problem for soils, because it is precipitated, but may be of concern for hydroponic cultures (table 11).

Table 11. Data for manganese

Mn: Mean crust = 950 mg/kg mean soil = 760 mg/kg		Crop uptake: wheat 164+ 68 g potatoes 37 g		Atmospheric deposition: East of Austria 1997/98: 25 g/ha.a		
	´mg/kg	range	´g for 100 kg N	range	´g for 100 kg P	range
K-salts	3,0	0,32 - 9,1				
Magnesium sulfates/nitrates	7,5	0,29 - 228				
N-P-K + sulfate	11,1	3,3 - 797	9,3	2,1 - 634	28,4	7,10 - 2792
P-K fertilizers	11,4	3,5 - 212			17,6	6,77 - 250
Ammonium nitrate limes	11,5	0,5 - 395	4,1	0,92 - 98,3		
Ammonium sulfates	11,6	0,07 - 161				
NP 24/12	14,5	2,3 - 91,5	6,1	3,80 - 48	21,5	8,3 - 104
Super phosphates	15,2	6,6 - 175			19,9	4,68 - 144
Triple phosphates	38,4	17 - 283			20,1	7,6 - 102
Commercial organics	57,3	3,2 - 733	83,7	2,4 - 6017	373	29 - 24515
N-P-K fertilizers	58,0	4,3 - 1737	46,4	2,3 - 1654	132	7,67 - 7605
Di-ammonium phosphates	70,2	22 - 407	42,5	11,6 - 213	43,2	10,9 - 200
Limes, dolomites	86,7	6,7 - 965				
Ca-Mg-phosphates	96,9	8,6 - 352				
Rock phosphates	113,2	10,8 – 346			96	8,8 - 316
Garden moulds	219	75 - 688			15899	5134 - 47033
Manures and dungs	255	43 - 710	386	46 - 6171	1926	563 - 4983
Composts	474	320 - 831	2380	764 - 8406	10719	2147 - 134316
Basic slags	1399	124 - 16526			401	193 - 18494
Basic slag potash	2871	2431 - 6712				

3.13. Molybdenum

Molybdenum mean crust level is just 1,5 mg/kg, but as a trace element it is known to be enriched in the biosphere. Among fertilizers, its concentration correlated strongly with P, except for commercial organics and garden moulds, which may contain excess molybdenum. Thus, highest median concentrations were found in triple phosphates, rock phosphates and super phosphates, at about 3-fold above the mean crust occurrence of 1,5 mg/kg. They often exceeded the level of concern for contaminated sites, which has been set to 5 mg/kg. Mineral NPK fertilizers with or without sulfate frequently contained added molydenum up to 200 mg/kg from intentional additions to cover special plant needs. This should marked on the label. Although plant tissues contain more molybdenum than animal tissues, manures had more molybdenum than composts.

Ammonium nitrate limes, limes and dolomites, K-salts and Mg-salts were at or below the detection limit of the ICP-OES.

Though molybdenum concentrations in fertilizers were largely above mean crust levels, for a fertilization rate of 100 kg N per hectare, the Mo plant uptake was hardly covered from mineral fertilizers, but with high probability from organics, manures and composts. Similarly, an input of 100 kg P per hectare from mineral fertilizers might balance crop extraction, but not in any case. Organics of any kind and basic slags contained more Mo/P, but Mo was accasionally added to NPKs to meet crop needs (table 12).

Table 12. Data for molybdenum

Mo: Mean crust = 1,5 mg/kg Soil medians (Austria) 0,19-0,67 mg/kg Mean soil = 1,9 mg/kg		Crop uptake: wheat 3,1 + 1,1 g potatoes 2,1 g maize 1,3 g		Atmospheric deposition: Below detection limit		
	´mg/kg	´range	´g for 100 kg N	´range	´g for 100 kg P	´range
Ammonium nitrate limes	< 0,5	< 0,5 - 0,8	< 0,2	< 0,2 - 0,28		
Limes, dolomites	< 0,5	< 0,5 - 2,4				
Magnesium sulfates/nitrates	< 0,5	< 0,5 - 0,8				
K-salts	< 0,5	< 0,5 - 1,0				
Ammonium sulfates	0,59	< 0,5 - 14,6				
N-P-K + sulfate	0,87	< 0,5 - 156	0,60	< 0,5 - 92	2,21	< 0,5 - 420
NP 24/12	0,95	< 0,5 - 1,9	0,52	< 0,5 - 0,94	1,63	0,8 - 3,9
Composts	1,06	< 0,5 - 3,1	4,68	1,23 - 18	19,4	4,8 - 339
N-P-K fertilizers	1,41	< 0,5 - 195	0,89	< 0,5 - 122	2,64	< 0,5 - 383
Commercial organics	1,53	< 0,5 - 10,5	2,33	< 0,5 - 83	11,0	< 0,5 - 342
Ca-Mg-phosphates	1,54	< 0,5 - 6,1				
Garden moulds	1,89	< 0,5 - 15,6			201	8,0 - 2246
P-K fertilizers	2,02	< 0,5 - 8,2			3,42	0,55 - 10,1
Basic slags	2,14	1,04 - 8,2			3,14	1,61 - 10,0
Manures and dungs	2,78	< 0,5 - 12,3	4,24	0,95 - 29	20,6	4,8 - 90,2
Basic slag potash	3,96	< 0,5 - 6,1				
Di-ammonium phosphates	4,00	< 0,5 - 9,2	2,16	< 0,5 - 5,41	2,11	< 0,5 - 5,10
Triple phosphates	5,13	3,33 - 21,0			2,58	1,42 - 11,4
Rock phosphates	5,14	1,01 - 11,2			4,16	0,79 - 9,9
Super phosphates	5,62	0,86 - 14,1			5,90	2,11 - 12,2

3.14. Sodium

Sodium occurs in the earth crust at a mean of 2,3%, but an appreciable part is dissolved in the oceans. Large parts of Na in soils (mean soil = 1,1%) is bond to silicates, but probably all Na is soluble in fertilizers. Salinization of soils, i.e. increase of available Na from irrigation or fertilization, is an unwanted process, however.

Among the fertilizer samples, most Na was contained in basic slag potashes (median 3,7% Na), manures (median 0,97% Na) and K-salts (median 0,63%). Na concentrations in K-salts, manures, commercial organics, and Mg-sulfates varied over more than 2 orders of magnitude. K- salts contained top values like up to 5,6% Na. Manures contained about 20 times more Na than composts. PK fertilizers contained more than double the amount of Na per weight than DAP or NPKs. At the low end of the scale there were the ammonium nitrate limes (median 78 mg/kg), limes and dolomites (median 353 mg/kg), and garden moulds (median 334 mg/kg).

Table 13. Data for sodium

Na: Mean crust = 23 g/kg mean soil = 11 g/kg				Atmospheric deposition: Highly variable		
	´g/kg	´range	´kg for 100 kg N	´range	kg for 100 kg P	´range
Ammonium nitrate limes	0,078	0,034 - 0,134	0,029	0,013 - 0,05		
Limes, dolomites	0,253	0,056 - 7,45				
Garden moulds	0,334	0,082 - 1,25			24,3	7,44 - 69,5
Composts	0,75	0,43 - 3,14	4,54	1,87 - 13,39	14,9	4,65 – 139
Ammonium sulfates	1,49	0,03 - 3,74				
Magnesium sulfates/nitrates	1,73	0,08 - 12,8				
NP 24/12	2,30	1,29 - 5,49	1,01	0,67 - 1,61	3,5	1,47 - 4,7
Di-ammonium phosphates	2,33	0,87 - 4,25	1,12	0,45 - 2,34	1,06	0,37 - 2,3
N-P-K + sulfate	2,62	0,91 - 8,50	1,82	0,60 - 9,87	6,9	2,83 - 28,0
N-P-K fertilizers	2,96	0,54 - 9,38	1,94	0,25 - 11,0	6,2	1,61 - 25,1
Triple phosphates	3,30	2,66 - 8,89			1,53	1,09 - 4,9
Commercial organics	3,43	0,21 - 21,1	6,44	0,33 - 33,1	21,2	2,26 – 246
Basic slags	3,47	1,87 - 24,8			5,4	2,09 - 52,0
Ca-Mg-phosphates	4,92	2,36 - 7,33				
Super phosphates	5,07	1,99 - 8,10			4,8	3,32 - 7,9
Rock phosphates	5,57	4,69 - 8,09			4,6	3,47 - 6,3
P-K fertilizers	5,78	2,09 - 11,4			9,2	4,76 - 21,5
K-salts	6,27	0,65 - 56,7				
Manures and dungs	9,41	0,62 - 68,5	10,82	2,61 - 102	52,3	10,1 – 515
Basic slag potash	37,3	25,8 - 47,7				

If salinization gets a problem, it should be considered that 100 kg N might bring up to 100 kg Na from manures, 17 kg from composts, and 33 kg from other organics. Inputs from mineral fertilizers equivalent to 100 kg N were ten times lower, and inputs from ammonium nitrate lime were almost negligible.

The atmospheric deposition of Na varies with the distance to the sea, but even a 100 kg P per hectare load yielded less Na than atmospheric deposition. Inputs from other P-mineral sources were slighly higher than atmospheric deposition. Appreciable amounts came from composts (median 15 kg Na / 100 kg P), organics (median 21 kg Na / 100 kg P), and manures (median 52 kg Na / 100 kg P) (table 13).

3.15. Nickel

Due to its cancerogenic action, the level of concern for nickel in soils has been set to 60 mg/kg, which is below the mean crust occurrence of 80 mg/kg. Among the fertilizers, highest median concentrations were found in triple phosphates (27,7 mg/kg), composts(24,5 mg/kg) and super phosphates (22,3 mg/kg), which are close to the soil background of 21 mg/kg. Di-ammonium phosphates, however, contained till 90 mg/kg, apart from outliers, and 1/5 of the samples had Ni concentrations larger than the levels of concern. On the other end, very low Ni (detection limit to 10 mg/kg) was found in minerals without P as main component.

100 kg N given via mineral fertilizers approximately balanced plant uptake, whereas ammonium nitrate lime supplied less, and di-ammonium phosphates, organics, manures and above all composts supplied much more. The level of concern would be an input of 375 g/ha in 2 years, which was not reached by any 100 kg N fertilization (except for some composts).

Though data about atmospheric deposition of nickel differ widely, the median input 100 kg P per hectare was higher from any source and exceeded crop outputs about 10-fold. Ni/P from composts was 20 times higher than from mineral phosphates, other organics were in-between. The level of concern of 375g/ha in 2 years was not reached from 100 kg P, however (table 14).

Table 14. Data for nickel

Ni: Mean crust = 80 mg/kg Soil medians (Austria) 19-29 mg/kg Mean soil = 34 mg/kg		Crop uptake: wheat 2,7 + 1,1 g potatoes 6,5 g maize 5,7 g		Atmospheric deposition: Marchfeld 1985: 15,2 g/ha.a East of Austria 1997/98: 5,0 g/ha.a		
	´mg/kg	´range	´g for 100 kg N	´range	´g for 100 kg P	´range
Ammonium nitrate limes	0,43	< 0,5 - 3,7	< 0,3	< 0,3 - 1,40		
K-salts	1,08	< 0,5 - 21,5				
Magnesium sulfates/nitrates	1,30	< 0,5 - 21,7				
Limes, dolomites	2,11	< 0,5 - 30,5				
Ammonium sulfates	3,57	< 0,5 - 10,9				
Commercial organics	5,51	< 0,5 - 47,1	8,1	< 0,5 - 241	34,2	< 0,5 – 1225

Ni: Mean crust = 80 mg/kg Soil medians (Austria) 19-29 mg/kg Mean soil = 34 mg/kg		Crop uptake: wheat 2,7 + 1,1 g potatoes 6,5 g maize 5,7 g		Atmospheric deposition: Marchfeld 1985: 15,2 g/ha.a East of Austria 1997/98: 5,0 g/ha.a		
	´mg/kg	´range	´g for 100 kg N	´range	´g for 100 kg P	´range
N-P-K fertilizers	6,51	< 0,5 - 30,5	4,0	< 0,5 - 22,2	16,0	< 0,5 – 582
N-P-K + sulfate	7,31	< 0,5 - 18,7	5,2	< 0,5 - 19,0	16,8	< 0,5 – 101
Garden moulds	7,63	< 0,5 - 32,6			740	59 – 2297
Manures and dungs	10,1	1,2 - 32,7	12,2	2,9 - 133	66,0	20 – 332
Ca-Mg-phosphates	10,7	2,4 - 17,8				
Basic slags	10,8	2,8 - 29,0			14,3	3,2 - 45,3
P-K fertilizers	11,4	2,2 - 23,4			16,3	4,2 - 44,2
Basic slag potash	16,3	0,8 - 33,7				
Di-ammonium phosphates	16,4	4,0 - 89,5	10,3	1,7 - 52,3	10,0	1,6 - 46,8
Rock phosphates	16,5	10,0 - 23,1			13,4	7,7 - 21,9
NP 24/12	17,5	4,1 - 31,7	7,4	4,3 - 10,2	26,9	9,3 - 39,6
Super phosphates	22,3	5,2 - 28,8			20,3	6,9 - 39,6
Composts	24,5	10,0 - 37,4	124	14 - 523	436	127 – 2824
Triple phosphates	27,7	10,3 - 50,7			14,0	3,7 - 31,0

3.16. Lead

Most kinds of mineral fertilizers and commercial organic fertilizers contained Pb at median concentrations below 2 mg/kg, just some of them reached the soil background levels of 15 mg/kg. For rock phosphates, super phosphates, garden moulds and manures, the medians of Pb concentrations were slightly higher, but not the top values. Composts had significantly more Pb, 37 mg/kg as the median (range 14-66 mg/kg), but no sample reached the level of concern, which has been set at 100 mg/kg.

Though annual atmospheric deposition of lead has been strongly declining within the last 20 years and is currently below 10 g/ha.a, inputs of 100 kg N from any source were lower from the atmosphere, except from composts. Similarly, inputs of 100 kg P per hectare were below annual atmospheric deposition, except from basic slags and organics. The level of 375 g/ha might have been reached by 100 kg P per hectare added via composts (table 15).

Table 15. Data for lead

Pb: Mean crust = 14 mg/kg Soil medians (Austria) 5,7 - 22,2 mg/kg Mean soil = 29 mg/kg		Crop uptake: wheat variable potatoes < 0,1 g		Atmospheric deposition: Marchfeld 1985: 86 g/ha.a East of Austria 1997/98: 16,5 g/ha.a		
	´mg/kg	´range	´g for 100 kg N	´range	´g for 100 kg P	´range
K-salts	< 0,5	< 0,5 - 1,10				
Ammonium sulfates	< 0,5	< 0,5 - 3,55				
Magnesium sulfates/nitrates	< 0,5	< 0,5 - 24,0				
Ammonium nitrate limes	1,12	< 0,5 - 10,0	0,35	< 0,3 - 3,72		
N-P-K + sulfate	1,21	< 0,5 - 7,04	0,89	< 0,5 - 9,20	3,08	< 0,5 - 27,4
N-P-K fertilizers	1,25	< 0,5 - 18,4	0,83	< 0,5 - 9,13	2,74	< 0,5 - 75,1
Di-ammonium phosphates	1,30	< 0,5 - 3,24	0,70	< 0,5 - 1,97	0,65	< 0,5 - 2,22
Commercial organics	1,45	< 0,5 - 19,3	2,16	< 0,5 - 135	7,46	< 0,5 - 602
P-K fertilizers	1,95	< 0,5 - 12,3			3,33	0,55 - 12,6
NP 24/12	2,14	0,57 - 5,05	1,06	< 0,5 - 2,75	3,56	1,18 - 4,70
Limes, dolomites	2,16	< 0,5 - 18,0				
Triple phosphates	2,81	1,53 - 5,52			1,44	0,79 - 4,39
Manures and dungs	4,17	< 0,5 - 27,5	4,43	0,83 - 237	25,0	4,9 - 301
Ca-Mg-phosphates	5,31	1,83 - 11,7				
Basic slag potash	6,59	2,43 - 22,0				
Super phosphates	6,60	1,23 - 16,8			5,46	1,97 - 17,7
Rock phosphates	7,54	2,55 - 16,1			6,60	2,08 - 14,1
Basic slags	8,63	5,32 - 24,2			12,4	7,19 - 27,1
Garden moulds	10,4	3,15 - 23,9			724	195 - 2211
Composts	36,9	14,2 - 65,7	182	38 - 877	570	154 - 2561

3.17. Strontium

Strontium mean crust level is 370 mg/kg, but it is known to be strongly enriched in certain carbonates (calcite, aragonite). Among fertilizers, most of the strontium was found in rock phosphates (median 1420 mg/kg), in super phosphates (median 1037 mg/kg), and in Ca-Mg-phosphates (median 948 mg/kg). Strontium concentrations in limes, NPKs, and PKs were at about mean crust levels (370 mg/kg), but some of them also exceeded 1000 mg/kg. Lowest levels were met in ammonium salts, K-salts and garden moulds.

Green plants cannot discriminate between calcium and strontium. Therefore, plant uptake and atmospheric deposition may vary widely, due to the geological formation and Ca-occurrence, but were less than loads from fertilizers (except from ammonium nitrate lime).

For an equivalent of 100 kg N, inputs from NPKs containing sulfate was largely more than from those without. An input of 100 kg P from all sources was higher than the presumable annual precipitation, except for di-ammonium phosphates. Sr/P in composts were highest, but might be exceeded by some NPKs, which covered a very broad range (0,005 - 12,6 kg Sr for 100 kg P) (table 16).

Table 16. Data for strontium

Sr: Mean crust = 370 mg/kg mean soil = 278 mg/kg		Crop uptake: wheat variable maize 61 g		Atmospheric deposition: Highly variable		
	´mg/kg	´range	´g for 100 kg N	´range	´g for 100 kg P	´range
Ammonium nitrate limes	17,1	8,9 - 86,8	6,2	2,9 - 32,9		
Magnesium sulfates/nitrates	19,2	0,2 - 2117				
Ammonium sulfates	44,1	< 0,2 - 178				
Commercial organics	49,5	0,9 - 485	79	1,0 - 1265	354	6,2 - 4201
Garden moulds	52,2	19,1 - 115,52			3811	1280 - 9944
Manures and dungs	53,4	13,1 - 119	80	18,6 - 857	398	138 - 1038
K-salts	58,3	0,99 - 86,2				
Di-ammonium phosphates	96	48 - 317	52	23,5 - 141	48,0	22,2 - 349
Limes, dolomites	110	35 - 1115				
N-P-K fertilizers	260	0,5 - 6789	77	0,4 - 935	261	0,5 - 12578
Composts	146	51 - 238	720	179 - 2945	2749	967 - 7521
N-P-K + sulfate	173	12,1 - 1473	124	11,0 - 1086	416	53 - 6293
Basic slag potash	220	150 - 293				
NP 24/12	246	166 - 445	104	72 - 255	370	217 - 436
P-K fertilizers	486	43 - 1137			867	259 - 1415
Basic slag	592	251 - 823			880	527 - 1180
Triple phosphates	823	512 - 1081			379	228 - 532
Ca-Mg-phosphates	948	486 - 1419				
Super phosphates	1037	724 - 1429			1108	867 - 1301
Rock phosphates	1420	795 - 1562			1140	676 - 1270

3.18. Vanadium

In mineral fertilizers without P as main component, as well as in manures and garden moulds, vanadium was found to be largely below 10 mg/kg. Vanadium concentrations correlated with phosphate, except for basic slags and garden moulds. Composts had significantly more vanadium than manures and garden moulds, but still far below mean soil of 108 mg/kg (median 23 mg/kg, range 11-34 mg/kg). Highest concentrations were found in basic slag, basic slag potash, triple P and diammonium phosphates, which largely exceeded the mean soil value.

In case of vanadium, both export to crops as well as mobility in soils are very low. Like for lead, atmospheric vanadium deposition has been declining within the last 20 years. For 100 kg N, the input of vanadium from ammonium phosphates and composts significantly exceeded atmospheric deposition. Similarly, 100 kg P per hectare from all mineral fertilizers and also many organics supplied more vanadium than annual atmospheric deposition. Top V/P values were found in basic slags, but median V/P were by far highest for composts (0,53 kg V for 100 kg P) (table 17).

Table 17. Data for vanadium

V: Mean crust = 160 mg/kg Mean soil = 108 mg/kg	Crop uptake: wheat 0,03 + 0,34 g potatoes 0,04 g	Atmospheric deposition: Marchfeld 1985: 29,2 g/ha.a East of Austria 1997/98: 1,5 g/ha.a

	´mg/kg	´range	´g for 100 kg N	´range	´g for 100 kg P	´range
K-salts	0,55	< 0,5 - 1,87				
Ammonium nitrate limes	1,33	0,80 - 3,13	0,5	0,2 - 1,2		
Magnesium sulfates/nitrates	1,47	< 0,5 - 19,4				
Ammonium sulfates	2,04	< 0,5 - 19,6				
Limes, dolomites	2,67	< 0,5 - 63,2				
Commercial organics	3,18	< 0,5 - 35,2	6,3	< 0,5 - 218	23	< 0,5 - 1009
Manures and dungs	4,31	< 0,5 - 11,2	5,3	< 0,5 - 351	28	1,8 - 115
Garden moulds	13,3	1,79 - 22,8			728	229 - 2451
Composts	22,8	10,7 - 34,1	113	34 - 513	531	88 - 1949
N-P-K fertilizers	29,1	< 0,5 - 252	18,5	< 0,5 - 107	62	< 0,5 - 503
N-P-K + sulfate	34,0	0,88 - 116	26,9	1,0 - 77	84	2,4 - 213
P-K fertilizers	44,8	13,9 - 90,3			70	35 - 110
Ca-Mg-phosphates	53,6	33,8 - 83,8				
NP 24/12	80,6	1,81 - 197	40	32 - 58	107	81 - 169
Super phosphates	83,5	54,8 - 129			79	60 - 207
Basic slags	103	48 - 4974			147	75 - 5566
Rock phosphates	104	73 - 129			85	65 - 109
Di-ammonium phosphates	121	62 - 411	66	34 - 225	64	28 - 203
Triple phosphates	138	64 - 234			65	28 - 136
Basic slag potash	277	130 - 2857				

3.18. Zinc

Zinc concentrations in K-, ammonium and Mg-salts without P as main component were below the mean crust level of 75 mg/kg. They covered a quite narrow concentration range (less than a factor 10) in garden moulds, PKs, Ca-Mg- phosphates, rock phosphates, triple phosphates and basic slags. To the contrary, zinc concentrations in NPKs as well as in commercial organics had a bimodal frequency distribution, one at about ambient level (soil background = 71 mg/kg), and a second peak at 300 mg/kg for organics, and at 400 for NPKs, due to occasional additions, with some samples well above 700 mg/kg. More than half of the manures, triple P and NP 24/12 samples exceeded the level of concern for soils, which has been set to 300 mg/kg.

Table 18. Data for zinc

Zn Mean crust = 75 mg/kg
Soil medians (Austria) 56 - 89 mg/kg
Mean soil = 60 mg/kg

Crop uptake:
wheat 104 + 42 g
potatoes 72 g
maize 375 g

Atmospheric deposition:
Marchfeld 1985: 289 g/ha.a
East of Austria 1997/98: 160 g/ha.a

	´mg/kg	´range	´g for 100 kg N	´range	´g for 100 kg P	range
K-salts	2,4	< 0,6 - 7,8				
Ammonium nitrate lime,	4,2	1,3 - 63,2	1,4	0,4 - 23,5		
Magnesium sulfates/nitrates	4,3	< 0,6 - 101				
Ammonium sulfate,	4,6	< 0,6 - 39,1				
Lime,, dolomite,	7,5	0,8 - 96,4				
Garden moulds	50,9	16,4 - 119			4689	1446 - 9433
N-P-K + sulfate	75,8	8,2 - 753	56,8	5,3 - 438	168	35 - 2637
Commercial organics	76,1	5,3 - 1407	110	8,5 - 2622	390	34 - 15339
N-P-K fertilizers	85,5	2,0 - 715	59,0	1,4 - 501	172	5,2 - 2590
P-K fertilizers	94,4	42,2 - 157			152	77 - 255
Basic slag potash	100	66,4 - 511				
Ca-Mg-phosphates	103	68,5 - 171				
Basic slags	156	61,6 - 463			198	95 - 592
Rock phosphates	165	79,6 - 236			142	73 - 190
Composts	191	106 - 529	1066	284 - 3139	3706	2369 - 12645
Super phosphates	197	110 - 322			180	142 - 2291
Di-ammonium phosphates	206	8,4 - 605	110	4,4 - 339	99	4,3 - 306
Manures and dungs	316	37,1 - 2327	433	105 - 3595	2355	684 - 11301
Triple phosphates	319	205 - 611			159	72 - 332
NP 24/12	907	16,6 - 1725	562	82 - 730	468	61 - 2291

The input of zinc from an equivalent of 100 kg N as well as of 100 kg P plus atmospheric deposition easily covered plant uptake of wheat and potatoes. For 100 kg N, inputs from composts and manures may exceed the level of concern which is a load of 2500 g/ha within 2 years.

100 kg P from manures and organics add much more to the soil than the plant needs and may exceed the level of concern for contaminations also (table 18).

4. Discussion

4.1. Analytical Procedure

ICP multi- element spectroscopy permits to determine a lot of elements in the sample digest simultaneously, provided suitable dilutions have been done to meet the optimum calibration ranges and to minimize matrix effects. The number of available analytes is limited by volatilizations, incomplete dissolution, and interactions with the usual glass vessels, as well as insufficient detection limits and blanks. Thus, low B, Ba and Na levels are doubtfull. Some samples contained B up to 1000 mg/kg, which caused memory effects. Na and Ba levels were sufficiently high to overcome variations of the blanks from the glasses.

In terms of an environmental screening presented within this context, the detection limits of the ICP-OES were regarded as sufficient. Further elements, like thallium, uranium, and antimony would be accessible by additional measurement on the ICP-MS. Analysis for sulfur, arsenic and selenium in organic matrices requires drying and ashing with magnesium nitrate in order to keep oxidizing conditions; manures lose more of these elements even during drying than during ashing. Arsenic and selenium data have been obtained by hydride AAS. Mercury is largely lost, and has not been treated in these batches.

Due to homogenization problems, dry ashing is still preferred, particularly for composts, manures and organic fertilizers. Thus, up to 100 ml of wet manure can be used for one digest, but this contains just 2-5 g of dry weight. After oven-drying, most of the pathogens have been presumably killed. Respective reference materials have been freeze dried and milled, but this was not possible to be done in our routine process, last not least for hygienic reasons.

From the digests obtained with magnesium nitrate, however, due to matrix interferences, just the lines for Ca-Cd-Cr-Cu-Fe-K-Mn-P-S-Zn could be read, others were spectrally interfered. In these cases, calibrants contained the same amount of Mg as the sample solutions

4.2. Nutrient inputs and sustainability

For all over Austria, the annual input of nutrient elements per hectare arable area has been estimated for 1995 as 51 kg N + 34 kg P_2O_5 + 77 kg K_2O from organic fertilizers produced on site, and 47 kg N + 22 kg P_2O_5 + 25 kg K_2O from mineral fertilizers (Dachler, Kernmayer 1997). Of course there are variations between different regions, crops and soil types. P_2O_5 and K_2O input peaked in the sixties, and N- input peaked in the eighties, but since then there has been a slight and steady decrease. The figures in tables 2-18 refer to straight numbers, to a load of 100 kg N and 100 kg P (=229 kg P_2O_5) per hectare, which represent rather a worst case scenario.

In Austria, 58% of nitrogen load came from ammonium nitrate lime, 36% from various NPK mixtures, 4% from urea, and only 2% from diammonium phosphate. The P- load came mainly from NPK mixtures (54 %), 21% from PK fertilizers, and 16% from super phosphate, triple phosphate and diammonium phosphates (data from 1994; Dachler, Kernmayer 1997).

There are efforts to recyle biological waste from households, gardens and parks as organic fertilizers via composting. This is advantageous to sustain humics, utilize plant nutrient N and P, and substitute peat. In terms of thermal utilization, one kg yields more than 11 M Joule as caloric respective electric energy. In biogas facilities, biogas is extracted for heating and/or electricity production, and the so-called fermentation residue can be taken as organic fetilizer as well (Kehres 2008).

If energy-crops are permanently used for combustion, and supplied by mineral fertilizers, this will result in a reduction of soil humics and decrease soil fertility on a long term (Kehres 2008).

4.3. Elemental Budgets

At background level, major parts of element contents can be found in silicates, sulfides, refractory oxides, as well as bound tightly to humic material (Sager et al. 1990, Sager 1992), whereas it is expectable to encounter element contents in fertilizers in a rather soluble form.

Within decades, depletion or enrichment of trace and nutrient elements in arable soils are governed by differences between input and output. Addition of fertilizers, but also dust and wet precipitation as well as irrigation contribute to the input. Components of output are leaching to the groundwater, washout of soils, and export as crops; beneath the kind of cropped plant, export loads depend on the parts of crops left on the field.

Plant uptakes for wheat were calculated for full extraction of grains and straw. Current debates touch the question if some of the straw should remain on the field fo maintain the level of humics, or all the straw should be used elsewhere, e.g. thermally. In the latter case, plant uptakes would be less than half of the tabulated value. Other crops might have other needs, especially for essential trace elements, which cannot be discussed here in detail, but should be known among agriculturalists. Apart from geology, cycling of elements of low physiological significance have been rarely discussed in detail and this paper should give new insights.

Sources of input to arable soils are fertilization and atmospheric deposition, beneath local additions from domestic waste, and traffic.

Intense mineral fertilization with mineral fertilizers may lead to enrichments of Be, Cd, Cr, and Ni in the soil. In addition, di-ammonium phosphates may enrich soils with Co and V.

Kádár (2005) gives a similar compilation of fertilizer data for neighbouring Hungary, which has about the same climate as the East of Austria, but a different structure of agriculture. The phosphates there contained much more arsenic (100 mg/kg), but less nickel (1 mg/kg) and zinc (20 mg/kg), and also more arsenic but less Ni, Zn and Cu in their manures. Atmospheric precipitation was higher for Cd, but lower for As and Cu. Some KCl had even up to 10 % Na. Microelement balances resulted in enrichment of Zn, Pb, Cu, Ni, As, Cd, and Se at the agricultural area.

Changing from mineral fertilizers to manures and composts means less Cd, but more As, Co, Cu, Pb and Zn. For highly sensitive crops, high Na-inputs from manures should be considered. In Maryland (USA), manure from intense poultry production was applied to fields in close proximity to the production houses. This resulted in significant increase of P, As, Cu, Mn and Zn especially in the surface layers, and in changes of soil characteristics (Codling, Chaney, Mulchi 2008). Similary in the Netherlands, fertilizer and manure application as well as atmospheric deposition accumulated metals in soils in rural areas for decades, and the removal with crops is still less than the total input (Bonten, Römkens, Brus 2008).

High salt concentrations may damage plant roots and their mycorrhizas via plasmolytic and osmotic effects. Too much uptake of Na and Cl interfere with the transport of Ca and essential trace elementes. Salt accumulations preferably occur at high evaporation rates and are more effective, if roots cannot penetrate into deeper layers. Soils of higher cation exchange capacity are more buffered against salt effects (Leifert 2008).

The entire dataset shows that it is not enough to control fertilizers for their nutrient content N-P-K and Cd, but for a broad range of elements to predict soil composition changes.

4.4. Merits of Compost and Manures

In organic farming, addition of nutrient elements from easily soluble and thus readily available mineral fertilizers, and addition of synthetic pesticides and fungicides is not permitted (Bodenschutz in Österreich). Tables 2-18 show that the load of concomitant elements per N or per P is largely higher from composts and manures (except for Cd) than from mineral fertilizers. Most of these composts and manures have been obtained from conventionally operating farms, however.

Besides dangers of enrichment of concomitant elements from fertilization processes, the selection of the suitable fertilizer has to be governed by other criteria also. With respect to P and also to N, composts carry the biggest load of accessory elements, but organics, above all composts, increase water retention, stimulate microbial soil life, and introduce the metals rather as organic complexes instead of readily available ions. Humics is a general term for dead organic residual substances in soils, which serve as nutrient basis for all soil organisms which rely on organic compounds (bacteria, fungi and hiogher trophic levels).

Compost addition seems to be the most efficient measure for recultivation of deteriorated soils (Kehres 2008). It should substitute additions of N, P, K, and Mg from mineral fertilizers. Accessory elements, however, may be widely variable, and inputs of additional elements should be regularly and thoroughly controlled, which is one of the goals of this article.

Peat is usually mined from ecologically intact wetlands and largely imported from Eastern Europe. About 80% of the peat are used for growing vegetables and flowers, or parks and roadside areas. In order to preserve ecologically valuable wetlands, it should be substituted by composts and soil aids. Contrary to fertilizers, soil aids do not contain significant amounts of plant - available nutrients, but they improve soil respiration and the soil water regime.

Composts should contain at least 15% of organic substances within their dry mass.

Organo-mineral fertilizers may be mixed from wood residues, crude phosphates, K-salts, K-Mg salts, limestone, magnesia, gypsum or other ground rocks of natural origin; no industrial by-products are permitted.

Domestic animals excrement up to 10% of their body weight per day. Additions of Cu and Zn to commercial animal feeds are known to raise their levels in the respective manures.

But if manures or dungs are not adequately recycled to soils, they cause nutrient and hygienic problems in surface and ground waters. Within a series of 146 manure and 75 dung samples from Upper Austria, heavy metal contents in manure and dung samples from the same kind of domestic animals were within the same range, if referred to dry mass. Excrements from pigs and poultry carried higher metal loads, particularly zinc and copper. Pig manures frequently exceeded the acceptable limits for sewage sludge with respect to copper and zinc. Lead, nickel, chromium, cadmium and mercury, however, were significantly lower than in sewage sludges (Aichberger 1995).

4.5. Source and Mobility of Accessory Elements

Apatite $Ca_{10}(PO_4)_6F_2$ is by far the most important mineral source for phosphates. The Ca^{2+} can be easily substituted by Cd^{2+}, which is just marginally smaller (ionic radius 0,10 nm). Sr^{2+} (0,11 nm), Eu^{2+} (0,11 nm), Pb^{2+} (0,12 nm), and Ba^{2+} (0,135 nm) are also possible. Phosphate can be substituted by arsenate and vanadate, and very important for biological apatites, 3 phosphates by 4 carbonates. The position of fluoride can be also occupied by hydroxide or chloride (Mc Connell 1973). Substitution leads to various minerals of apatite structure at the atomic level.

These accessory elements, or call them impurities, cannot be removed from the bulk by physical methods (e.g. flotation), but only via complete dissolution and recrystallisation, Phosphates are usually mined open air at large scale at dry locations, and recrystallisation on site is too expensive.

Nitrogenous components, however, get synthesized largely via the gas phase, with ammonia as the first step, thus they are usually very clean.

Contrary to soils, mineral fertilizers are largely soluble, because they should be available to the crops within one season of growth. When high barium loads were applied to Brazilian soils cropped with maize, major parts were recovered as exchangeables in sequential leaching (Souza et al. 2007).

Fertilization with ammonium sulfate, diammonium phosphate, or urea may shift the soil pH towards more acidity, whereas hyperphosphate and ammonium nitrate lime tend to increase soil pH.

Apart from being extracted with the crops, microelements might be washed down to the groundwater, depending widely upon soil properties. When ten times the annual precipitation was added to model chernozem columns without plant cover during one growing season, V was largely retained within the first 5cm, Mo-Se-As moved about 15 cm, and B moved about 30 cm, but there were variations due to soil pH, carbonates and clay contents (Sager 2002 and 2004). Addition of K led to peaks of Na, Li, Mg and Sr in the eluates along with the moving water because of ion exchange (Sager 2002). In a field experiment in Hungary (Németh and Kádár 2005), a load of even 810 kg/ha treatment on a loamy textured calcareous chernozem soil displayed no significant movement for As, Hg, Ni, Cu, Pb, Ba, and Sr, added as soluble salts. Moderate leaching down occurred in case of Zn and Cd, whereas chromate, selenate and molybdate were rather mobile. As a conclusion, most of the added wanted and unwanted microelements is transferred to the crops again on a long term.

However, in the Netherlands, model calculations showed that leaching of arable soil is an important source of surface water contamination. 48% of Ni, 38% Pb, 36% Zn, 26% Cu and

22% Cd of the load to surface waters derive from subsurface runoff and loading of soils (Bonten, Römkens, Brus 2008).

4.6. Long-Term Trends

In Austria, more intense control of fertilizer composition started in 1986. The current dataset can be compared with data compilations 1986-93 (before Austria joined the EC; Sager 1997), and 1996/97 (Sager, Scholger 2002).

In mineral fertilizers, most important, cadmium levels have been about constant, they increased just in diammonium phosphates, and in basic slags; the latter were presumably synthesized from rock phosphates, and did not come from steel production any more. In mineral fertilizers, Pb and Co have been constantly low. In NPKs, which are the most frequently used, Cr and Ni decreased.

In organic fertilizers and composts, Pb levels steadily decrease, whereas Cd-Cr-Ni remain about constant. In manures and dungs, Ni and maybe also Cr are increasing because of the use of steel cages.

4.7. Atmospheric Deposition

Dust and wet precipitation upon the entire agriculturally used area may also contain non-wanted trace elements, which cannot be avoided by current farming practices. However, usual fertilization rates easily exceed atmospheric depostion, except for N-only products and K-salts.

Within the period 1985 - 1998, significant decrease of atmospheric deposition of As, Cu, Ni, Pb, V, and Zn was noted in the East of Austria. This can be explained by the installation of cleaning facilities at atmospheric emission sources, less use of coal, and less use of leaded gasoline. Cadmium and chromium levels remained about constant. Barium, lithium and manganese atmospheric depositions have not been determined in former series.

In Vorarlberg, the most western country of Austria, the annual atmospheric deposition of lead decreased from 300 g/ha in 1990 to 20 g/ha in 2006 (15 Jahre Vorarlberger Bodenschutzkonzept). Likewise, within the same period, phosphorus loads of Lake Constance from arable areas decreased to half, and from domestic effluents decreased to one third (15 Jahre Vorarlberger Bodenschutzkonzept):

Similarly, atmospheric deposition strongly decreased for Pb and Zn in the Netherlands within 1990 - 2004 (Bonten, Römkens, Brus 2008)

4.8. Outlook

From the above dataset, the user might conclude excess and deficiencies of micro-elements with respect to his special crop needs. Control of as many elements as possible gives new insights into the cycling of elements. This dataset shall be extended to include sulfur, selenium, and possibly uranium.

REFERENCES

Aichberger K.: Die Ergebnisse einer Untersuchung von Wirtschaftsdüngern in Oberösterreich, Bundesamt für Agrarbiologie, Linz 1995

Bonten L.T.C., Römkens P.F.A.M., Brus D.J.: Contribution of Heavy Metal Leaching from Agricultural Soils to Surface Water Loads. Environ. Forensics 9, 252-257 (2008)

Bowen H.J.M.: Environmental Chemistry of the Elements. Academic Press, London New York, Toronto, 1979

Codling E.E., Chaney R.L., Mulchi C.L.: Effects of broiler litter management practices of P, Cu, Zn, Mn and As concentration in Maryland Coastal Plain Soils. Comm. Soil Sci. Plant Anal. 39, 1193-1205 (2008)

Dachler M., Kernmayer I.: Düngemittelaufwand in Österreich, in: Bodenschutz in Österreich, Bundesamt und Forschungszentrum für Landwirtschaft, Wien 1997

Danneberg O.H.: Hintergrundwerte von Spurenelementen in den landwirtschaftlich genutzten Böden Ostösterreichs, Mitt. Österr. Bodenk. Ges. 57, 7-24, 1999

Danneberg O.H. et al.: Richtlinien für die sachgerechte Düngung, BMLF, 5. Auflage 1999

Houba V.J.G., Uittenbogaard J.: Chemical Composition of Various Plant Species, Wageningen Agricultural University 1994

Kádár I.: A talaj és a tápláléklánc szennyezödése (Contamination of soils and food chain), Talajvédelem Különszám 129-137. Talajvédelem Alapítvány, Gödöllö 2005

Kehres B.: Bioabfälle vom Nawaro-Bonus ausgenommen. Humuswirtschaft und Kompost aktuell 01/2008

Kehres B.: Perspektiven der Kompostierung in der Biomassewirtschaft. Humuswirtschaft und Kompost aktuell 07/2007

Leifert I.: Verbrennt Kompost Pflanzen?. HuMuss 19/3 (2008)

Mc Connell D.: Apatite, Its Crystal Chemistry, Mineralogy, Utilization, and Geological and Biologic Occurrences. Springer 1973

Németh T., Kádár I.: Leaching of microelement contaminants in a long-term field study. Z. Naturfoschung 60, 260-264 (2005)

Sager M.: Chemical speciation and environmental mobility of heavy metals in sediments and soils; in: Hazardous metals in the environment, M. Stoeppler ed., Elsevier Science Publishers 1992

Sager M., Belocky R., Pucsko R.: Zur Ermittlung der Bindungsformen von Haupt- und Spurenelementen in Sedimenten durch sequentielle Löseverfahren, Acta hydrochim. hydrobiol. 18, 123-139 (1990)

Sager M.: Possible trace metal load from fertilizers, Die Bodenkultur 48(4), 217-223 (1997)

Sager M., Scholger R.: Magnetic Measurements as a Rapid Screening Method for Fertilizers, Journal of the Korean Institute of Mineral and Energy Resources Engineers 39,(5), 307-313 (2002)

Sager M.: Vertical mobility of Selenium, Arsenic and Sulfur in Model Soil Columns. Die Bodenkultur 53(2), 83-103 (2002)

Sager M.: Effects of a fertilization pulse on migration of nutrient and trace elements in chernozem soil columns within a vegetation period. Die Bodenkultur 55 (4), 165-182 (2004)

Sager M.: Multi- Elementbestimmung in Fleisch, Leber und Nieren. Ernährung/ Nutrition 29(4), 151-156 (2005)

Sager M.: Micro- and macro-element composition of animal feedstuffs sold in Austria. Ernährung/Nutrition 30 (11), 455 – 473 (2006)

Sager M.: Trace and nutrient elements in manure, dung and compost samples in Austria. Soil Biology and Biochemistry 39, 1383-1390 (2007)

Sager M.: Ökologie des Aluminiums, Schweiz. Z. Hydr. 48/1, 71-103 (1986)

Scherer J., Buhmann D., Kopf M.: 15 Jahre Vorarlberger Bodenschutzkonzept, eine Bilanz. Schriftenreihe Lebensraum Vorarlberg, Band 58, Bregenz 2008

Souza L.C., Campos H.M., Oliveira L.R., Mouta E.R., Melo G.M.P., Melo V.P., Melo W.J.: Barium in an Oxisol treated for nine consecutive years with sewage sludge and cropped with maize, Geophys. Res. Abstr. 9, 10107, 2007

Spiegel H., Sager M.: Element contents of wheat and potato varieties in different growing zones in Austria (in German). Ernährung/Nutrition 32, 7/8, 297-308 (2008)

Ure A.M., Berrow M.L.: The chemical constituents of soils. In: Bowen H.M. (ed) Environmental Chemistry. R. Soc. Chem, Burlington House, London 2: 94-202 (1982)

In: Organic Farming: Methods, Economics and Structure
Editors: Michael Nelson and Ivan Artamova
ISBN 978-1-60692-864-6

Chapter 8

DO THE AGRICULTURAL PRACTICES IN THE ORGANIC FARMING INCREASE SOIL ORGANIC MATTER AND MICROBIAL BIOMASS CONTENT?

Ademir Sérgio Ferreira de Araújo[1,1]
and Luiz Fernando Carvalho Leite[2,2]
[1]Universidade Federal do Piauí – Centro de Ciências Agrárias,
Campus da Socopo, Teresina, PI, Brazil.
[2]Embrapa Meio-Norte. Av. Duque de Caxias, Teresina-PI, Brazil.

ABSTRACT

Agricultural production systems have to combine management practices in order to sustain soil quality and profitability. Organic farming is gaining worldwide acceptance and has been expanding at an annual rate of 20% in the last decade, accounting for over 24 million hectares worldwide. Organic practices avoid applications of synthetic fertilizers and pesticides, rely on organic inputs and recycling for nutrient supply, and emphasize cropping system design and biological processes for pest management, as defined by organic farming regulation in the world. In comparison with conventional farming, organic farming has potential benefits in improving food quality and safety. Plant production in organic farming mainly depends on nutrient release as a function of mineralization processes in soils. The build-up of a large and active soil microbial biomass is an important pool of accessible nutrients, and therefore is an important priority in organic farming. Nutrient mineralization and immobilization are soil microbial processes governed by C availability and are linked closely to soil organic matter (SOM) content. Furthermore, soil microbial biomass respond rapidly to changes resulting from agronomic practices and other disturbances and have been used to assess early changes due to different soil management practices. Therefore, agricultural practices in organic farming can cause changes in soil organic matter and microbial biomass content.

Keywords: soil microorganisms, microbial activity, organic C, soil enzymes.

[1] E-mail: asfaruaj@yahoo.com.br
[2] E-mail: luizf@cpamn.embrapa.br

INTRODUCTION

Organic farming can be defined as an approach to agriculture where the aim is to create integrated, humane, environmentally and economically sustainable agricultural production systems. Organic agriculture is distinct from conventional agriculture through alternative agricultural practices, world view and values (Watson et al. 2006). Maximum reliance is placed on locally or farm-derived renewable resources and the management of self-regulating ecological and biological processes and interactions. This in necessary to provide acceptable levels of crop, livestock and human nutrition, protection from pests and diseases, and an appropriate return to the human and other resources employed.

In the world, organic farming is regulated legally by the International Federation of Organic Agriculture Movements (IFOAM). In the last decade, the organic farming system is gaining worldwide acceptance and has been expanding at an annual rate of 20%, accounting for over 24 million hectares worldwide (Lotter, 2003). A global survey has recently shown that there are more than 24 million hectares of land under organic management. Australia has the largest land area under organic management (about 10 million hectares), followed by Argentina (almost 3 million hectares), Italy (almost 1.2 million hectares), and the USA (almost 1 million hectares) (Willer and Yussefi, 2004). European countries have the highest proportion of land under organic management. For instance, organic land accounts for about 12% of the total agricultural area in Austria, 10% in Switzerland, 8% in Italy, and 7% in Finland (Willer and Yussefi, 2004). It is estimated that by 2010 organically farmed land will occupy 10–20% of total agricultural land areas in many European countries (Greene and Kremen, 2003). In the USA, the organic market was predicted to grow by about 20% annually between 2002 and 2007 (Willer and Yussefi, 2004). Brazil counts as one of the leading countries worldwide in organic farming (about 100,000 ha) and occupies the 34th position in the world (Araújo et al., 2008).

The practices in organic farming exclude the use of synthetic fertilizers and pesticides, plant growth regulators, livestock feed additives, and genetically modified organisms. The term 'organic' is best thought of as referring to the concept of the farm as an organism, in which all the component parts — the soil minerals, organic matter, micro-organisms, insects, plants, animals and humans — interact to create a coherent and stable whole.

The main characteristics of organic farming include: a) Protecting the long term soil fertility by maintaining organic matter levels, encouraging soil biological activity and careful mechanical intervention; b) Providing crop nutrients indirectly using relatively insoluble nutrient sources which are made available to the plant by the action of soil microorganisms c) Nitrogen self-sufficiency through the use of legumes and biological nitrogen fixation, as well as effective recycling of organic materials, including crop residues and livestock manures; d) Weed, disease and pest control relying primarily on crop rotations, natural predators, diversity, organic manuring, resistant varieties and limited (preferably minimal) thermal, biological and chemical intervention; e) The extensive management of livestock, paying full regard to their evolutionary adaptations, behavioral needs and animal welfare issues with respect to nutrition, housing, health, breeding and rearing; f) Careful attention to the impact of the farming system on the wider environment and the conservation of wildlife and natural habitats.

In the last decades, intensive use of agrochemicals has increased the toxicity in soils and also has degraded the status of soil. Organic farming is becoming a major tool for sustaining soil quality degraded by synthetic chemicals for increasing crop production, and therefore, use of bio-agents as biofertilizers or biopesticides is an integral part of organic farming, especially in vegetable cultivation (Truu et al., 2008).

Plant production depends almost exclusively on nutrient transformations in soils since only limited amounts of permitted fertilizers are used by organic farmers, Because nutrient transformations are primarily controlled by microorganisms, an active soil microflora and a considerable pool of accessible nutrients are very important to the functioning of organic farming systems. Therefore, the maintenance of adequate status of soil microbial biomass and organic matter is crucial for success in organic farming system.

Soil Microbial Biomass and Organic Matter

Soil is a complex environment, where microorganisms play a crucial role in nutrient cycling and the degradation of different pollutants (e.g. pesticides and industrial wastes), contributing in this way to the maintenance of soil quality (Dick, 1997; Haney et al., 2002; Watanabe et al., 2003; Araújo et al., 2003; Araújo and Monteiro, 2006). Additionally, the soil microorganisms perform other important functions, such as to form symbiotic associations with roots, to act as antagonist of pathogens, to influence the weathering and solubilization of minerals and to contribute to soil structure and aggregation (Sparling, 1997). The role of the microbial fractions in mediating the soil process and their relatively high rate in turnover suggest that the microbial fraction could be a sensitive indicator and early predictor of the changing soil organic matter processes (Polsow et al., 1987), as well as changes in soil status affected by agricultural management, such as conventional and organic farming systems (Sampaio et al., 2008; Araújo et al., 2008).

The microbial biomass is defined as the living component of soil organic matter (Jenkinson and Ladd, 1981) but excludes macrofauna and plant roots. Typical methods of determination are by fumigation-incubation, fumigation-extraction or by substrate-induced respiration methods (Sparling and Ross, 1993). Recently, other methods have been used to estimate microbial biomass, such as Biolog™ system (Biolog, Inc., California), phospholipid ester-linked fatty acid analysis (PLFA) (Zelles et al., 1992) and denaturing gradient gel electrophoresis (DGGE) (Muller et al., 2002).

Generally, the amount of microbial biomass in the soil reflects the total soil organic matter (SOM) content, and the proportion present of living microbial cells (microbial biomass C, mg kg^{-1} soil) typically comprises 1-5% of soil organic C (Jenkinson and Ladd, 1981). Microbial biomass is considered an integrative, biologically meaningful, management sensitive and measurable signal of soil (Truu et al., 2008). Its turnover rate is much faster than that of total SOM, and based on the dynamics of soil microbial biomass content, longer-term trends in SOM content can be predicted (Sparling, 1997). Lately, microbial biomass has been considered the active pool in some SOM dynamic simulation models, such as Century (Parton et al., 1987) and RothC (Coleman and Jenkinson, 1996).

SOM is a critical component of the soil resource base, which affects physical, chemical and biological processes of soil and through the effect on these processes, fulfills a very wide range of functions. The role of SOM in the formation of stable soil aggregates has major implications for soil structure and therefore on water infiltration, water holding capacity, aeration, resistance to root growth and surface crusting (Mirsky et al., 2008). The physical protection of SOM within aggregates is an important factor controlling dynamics and decomposition of organic C and the disruption of aggregates is one of the mechanisms proposed for lower SOC in tilled rather than in no-till systems (Six et al., 2000; Leite et al., 2003; Jiao et al., 2006; Murage et al., 2007). SOM is considered the source of soil cation exchange capacity (CEC) by which the cations are retained for plant use in an available form and are readily exchanged for other cations. The exchange quite often is made as a hydrogen ion, from plant roots, replacing one of the cations in the exchange complex. By reducing fixation and leaching losses of the cations, CEC helps maintain a more constant supply of the nutrients, thereby potentially increasing crop yields (Craswell and Lefroy, 2001; Bot and Benites, 2005). Finally, organic matter and SOM are considered essential to support a biological population in the soil. Soil organisms, consisting of microflora and microfauna, do not just live passively in the soil performing their own functions but are affected by one another, either through competition or symbiotic relationship. Often they compete with one another for nutrients or energy, much of which is derived from organic matter, but frequently organisms, having been nourished by organic matter, become an energy and nutrient source for other organisms. (Wolf and Snyder, 2003).

More recently, SOM has attracted great interest because of the phenomenon of global warming and the prospect of using soil as a reservoir of carbon released to the atmosphere (CO_2) by activity human. The best strategies to built-up carbon stocks in the soil are basically those that increase crop residue addition to the soil or decrease the soil organic matter decomposition rate (Lal, 2004). In fact, soil organic carbon (SOC) storage is a balance between C additions from non-harvested portions of crops and organic amendments, and C losses, primarily through organic matter decomposition and release of respired CO_2 to the atmosphere and at small amounts, by erosion and leaching of dissolved organic carbon (Kemmit et al., 2008). Organic matter returned to the soil, directly from crop residues or indirectly as manure, consists of many different organic compounds. The result is a quick formation of microbial compounds and body structures, essential in holding particles together to provide soil structure and the release of carbon dioxide back to the atmosphere through microbial respiration (Kadivko, 2001; Liu et al., 2006). The strategies or practices which can increase SOC stocks include, but are not limited to reduced tillage intensity, decreasing or ceasing the fallow period, using a winter cover crop, changing from monoculture to rotation or altering soil inputs to increase primary cropping (e.g., fertilization and irrigation) (West et al., 2002; Huchinston et al., 2007). No-tillage system was introduced to combat soil erosion by water and this soil management led to the build-up SOM with simultaneous high crop yields (Machado, 2005). Also, manuring increases the C input to soil and consequently enhances SOC concentration (Jarecki and Lal, 2003). According to Paustian et al. (1997) and Fliebach et al. (2007), having passed the digestive tract, manure is enriched in more refractory compounds that can persist as stable SOM associated with clay and silt particles. While farmyard manure return to soils has a positive impact on SOC, the application of composted manure has further benefits that are carried along with the anaerobic decomposition, where less CH_4 develops than in stacked manure (Davis, 2002; Fliebach et al., 2007).

EFFECT OF AGRICULTURAL PRACTICES IN THE ORGANIC FARMING ON SOIL MICROBIAL BIOMASS

Organic farming systems avoid applications of synthetic fertilizers and pesticides, rely on organic inputs and recycling for nutrient supply, and emphasize cropping system design and biological processes for pest management (Rigby and Caceres, 2001). These practices have a potential benefit to improve biological soil quality, such as, soil microbial biomass and activity. In recent years, several studies were conducted around the world in order to evaluate the effect of organic practices on soil microbial biomass (Fliesbach and Mader, 2000; Glover et al., 2000; Oehl et al., 2004; Tu et al., 2005; Melero et al., 2005; Tu et al., 2006; Araújo et al., 2008; Sampaio et al., 2008).

In a long-term field experiment established in the USA, Tu et al. (2005) evaluated the effect of transitional practices from conventional to organic farming on the size of the soil microbial communities. The results showed that the microbial biomass C (MBC) was highest in the organic plots compared to conventional plots (Figure 1). According to the authors, the significant differences in microbial biomass C between the organic and conventional farming likely reflect the accumulative impact of organic C inputs (5,660 to 10,322 kg ha^{-1}) during 2000 and 2002, in organic farming, on the size of microbial biomass. In another field study, carried out in Spain, Melero et al. (2005) evaluating the microbial biomass in soil under organic and conventional farming. The results showed that the microbial biomass was significantly higher under organic than conventional management, in all sampling periods (Figure 2), due to greater supply of available C.

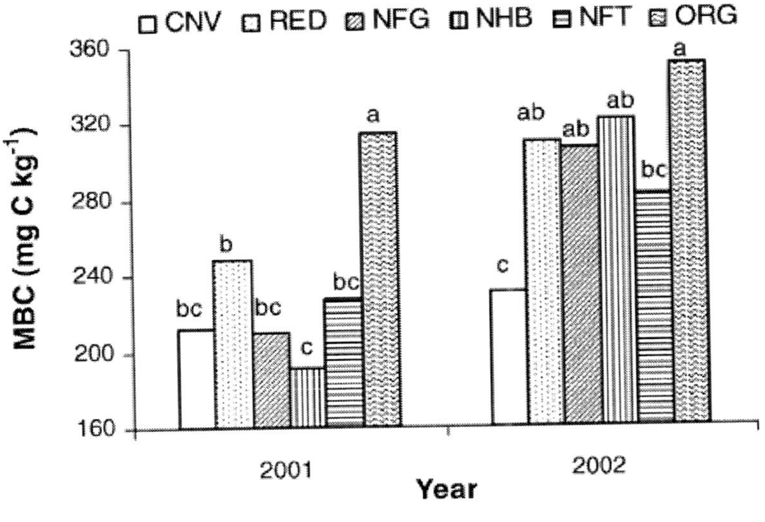

Figure 1. Effects of organic transition strategies on soil microbial biomass C (MBC) in 2001 and 2002 (CNV, conventional practice; RED, gradually reduced synthetic inputs; NFG, no fungicides/insecticides added; NHB, no herbicides added; NFT, no synthetic fertilizer added; ORG, organic practice). Bars with the same letters within a year are not significantly different at $P < 0.05$ (LSD). (Adapted of Tu et al., 2005)

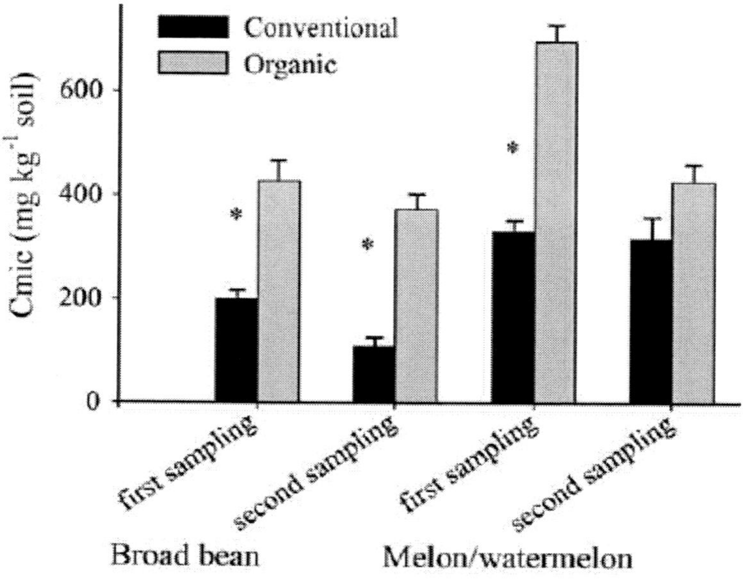

Figure 2. Soil microbial biomas C (Cmic) as affected by conventional and organic management during four sampling periods. Vertical bars are standard deviation. Significant difference between treatments is indicated with (*). (Adapted of Melero et al., 2005)

In organic farming, the quantity and quality of organic inputs are the most important factors affecting microbial biomass and community structure. High content of easily decomposable organic C can lead to fast growth of soil microbes, likely resulting in higher microbial biomass and activity. Chowdhury et al. (2000) observed that a manure compost with high easily decomposable C was more effective than saw-dust and rice husk composts in enhancing soil microbial biomass C. Recently, Tu et al. (2005) examined how different regimes of organic management can impact microbial biomass and activity and determined how the resulting changes in microbial activities influence nutrient (N) availability for the plants. The organic substrates used included composted cotton gin trash (CGT), animal manure (AM) and rye/vetch green manure (RV). The results observed by authors showed that microbial biomass and microbial activity were generally higher in organically than conventionally managed soils with CGT being most effective. Straw mulching further enhanced microbial biomass, activity, and potential N availability by 42, 64, and 30%, respectively, relative to non-mulched soils, likely via improving C and water availability for soil microbes.

Previously, Fließbach and Mader (2000), in the long term experiment (18 years in organic farming) with permanent organic inputs, observed a higher microbial biomass (45–64% higher in organic farming systems than the respective conventional farming). This affirmation that the organic input increased the soil microbial biomass was confirmed recently by Araújo et al. (2008), which evaluated the soil organic C and microbial biomass in different plots planted with "acerola" orchard (*Malphigia glaba*) under six, twelve, eighteen and twenty-four months of soil organic management in Brazil. From the twelve months of organic management, the highest C_{mic} values were found and significant differences between conventional and organic management were observed, while the highest organic C values were observed after twenty-four months (Table 1). These authors attributed the increase in soil microbial biomass to long-term C inputs in organic farming systems. Another study in the same areas was conducted by Araújo et al. (2009), evaluating the soil respiration and physic attributes in a soil under conventional and organic farming systems. Soil respiration in laboratory and field assays was significantly greater in organic farming systems after 1 year of adoption, showing the positive effect of agricultural practices in organic farming on soil microbial activity.

Table 1. Soil organic carbon (C_{org}), microbial biomass C (C_{mic}) and C_{mic}-to-C_{org} ratio of conventional and organic system, at 0-10 cm depth (Araújo et al., 2008).

	C_{org} ($g\ kg^{-1}$)	C_{mic} ($ug\ g^{-1}$ soil)	C_{mic}-to-C_{org} (%)
CNV	11.6 ± 2.1 b	64.0 ± 21 c	0.5 ± 0.09 b
ORG6	12.5 ± 2.9 b	97.1 ± 29 b	0.8 ± 0.10 a
ORG12	11.2 ± 1.9 b	121.2 ± 27 a	1.1 ± 0.13 a
ORG18	13.1 ± 2.0 b	129.4 ± 2.6 a	0.9 + 0.10 a
ORG24	17.8 ± 2.6 a	142.5 ± 22 a	0.8 ± 0.11 a

[a] Mean ± standard error. Means followed by the same letter within each column are not significantly different at 5% level by Duncan's test. [b]The farming systems evaluated were: a) under twelve months of soil conventional management (CNV); b) under six months of soil organic management (ORG6); c) under twelve months of soil organic management (ORG12); d) under eighteen months of soil organic management (ORG18); e) under twenty-four months of soil organic management (ORG24).

EFFECT OF AGRICULTURAL PRACTICES IN THE ORGANIC FARMING ON SOIL ORGANIC MATTER.

Farm comparison and long-term field trials show that organically managed soils have significantly higher organic matter content. Leite et al (2007) determined the changes on total soil carbon and nitrogen stocks and on soil carbon pools due to long term addition of mineral fertilizer (MF1 (10 kg N ha^{-1}, 15 kg P ha^{-1} and 17 kg K ha^{-1}) and MF2 (20 kg N ha^{-1}, 30 kg P ha^{-1} and 34 kg K ha^{-1}) and compost (OC; 40 m^3 ha^{-}) to a Ferric Acrisol under maize/common bean intercrop in Southeastern Brazil. The authors observed that SOC values were higher ($p<0.05$) in MF2 + OC (25.02 and 23.86 Mg C ha^{-1}) than in MF2 (23.07 and 20.17 Mg ha^{-1}) at 0-10 and 10-20 cm soil depth, respectively (Table 2). Stocks of soil nitrogen (TN) followed a similar pattern only for 0-10 cm soil layer (MF2+OC: 2.15 Mg ha^{-1}; MF2: 1.73 Mg ha^{-1}). At 10-20 cm soil depth, with the application of compost, all soils showed higher carbon and nitrogen stocks ($p<0.05$) than those soils, which received only mineral fertilizer or control soils. The magnitude of the differences at 0-10 cm depth was not large, but the application of organic compost did cause an increase on TOC and TN. This tendency also was observed in light fraction carbon and labile carbon (Table 2).

Table 2. Stocks of carbon and nitrogen, and carbon pools of a Ferric Acrisol at 0-10 and 10-20 cm layers as affected by different fertilizer treatments in an intercrop (Leite et al., 2007).

Organic Compost	MINERAL FERTILIZATION 0	1	2	MEAN	Mineral Fertilization 0	1	2	MEAN
	0-10 cm				10-20 cm			
	Soil Organic Carbon (SOC; Mg C ha^{-1})							
AF		35.00 ± 5.29				27.21 ± 2.00		
0	23.43 a	22.90 a	23.07 b	23.13	22.19 a	21.00 a	20.17 b	21.12
OC	24.40 a	24.39 a	25.02 a	24.60	23.03 a	22.63 a	23.86 a	23.17
Mean	23.91	23.64	24.04	-	22.61	21.81	22.00	
	Total nitrogen (TN; Mg N ha^{-1})							
AF		2.59 ± 0.41				2.19 ± 0.16		
0	1.73 a	1.77 a	1.73 b	1.74	1.58 b	1.53 b	1.47 b	1.53
OC	1.94 a	1.95 a	2.15 a	2.01	1.76 a	1.84 a	2.12 a	1.90
Mean	1.83	1.86	1.94	-	1.67	1.67	1.79	-
	Light fraction carbon (C_{LF}; Mg ha^{-1})							
AF		11.74 ± 0.31				5.02 ± 0.57		
0	1.31 b	1.40 b	1.47 b	1.39	1.14 b	1.17 b	1.31 b	1.21
OC	2.14 a	1.81 a	1.92 a	1.96	1.87 a	1.74 a	1.78 a	1.80
Mean	1.85	1.60	1.69	-	1.50	1.45	1.54	-
	Fulvic acids fraction (FAF Mg ha^{-1})							
AF		3.80 ± 0.56				3.65 ± 0.15		
0	3.63 a	3.58 a	3.59 a	3.60	3.32 a	3.44 a	3.51 a	3.42
OC	3.06 b	2.90 b	3.59 a	3.18	2.96 a	3.00 b	3.17 a	3.04
Mean	3.34	3.24	3.59	-	3.14	3.22	3.34	-
	Humic acids fraction (HAF Mg ha^{-1})							
AF		3.70 ± 0.39				3.15 ± 0.22		
0	2.70 a	2.94 a	2.91 a	2.85	2.26 a	2.66 a	2.30 a	2.41
OC	3.09 a	2.70 a	3.93 a	3.24	3.02 a	2.71 a	2.95 a	2.89
Mean	2.89	2.82	3.42	-	2.64	2.68	2.62	-
	Humin fraction (HUM Mg ha^{-1})							
AF		18.10 ± 2.10				15.62 ± 2.35		
0	12.67 a	13.27 a	11.98 a	12.64	13.20 a	12.80 a	11.60 a	12.53
OC	14.58 a	13.84 a	12.44 a	13.62	13.14 a	12.40 a	13.71 a	13.08
Mean	13.62	13.55	12.21	-	13.17	12.60	12.65	-
	Labile carbon (C_L Mg ha^{-1})							
AF		2.10 ± 0.02				1.18 ± 0.04		
0	0.92 b	1.21 b	1.11 b	1.08	1.01 b	0.99 b	1.06 b	1.02
OC	1.73 a	1.56 a	1.79 a	1.69	1.38 a	1.42 a	1.54 a	1.45
Mean	1.32	1.38	1.45	-	1.19	1.20	1.30	-
	Non-labile Carbon (C_{NL} Mg ha^{-1})							
AF		32.90 ± 2.42				26.08 ± 1.39		
0	22.51 b	21.68 b	21.96 b	22.05	21.18 a	20.01 b	19.10 b	20.09
OC	22.67 a	22.83 a	23.23 a	22.91	21.66 a	21.21 a	22.32 a	21.73
Mean	22.59	22.25	22.59	-	21.42	20.61	20.71	-

Means followed by the same letters, in line, are not different by F test at $P < 0.05$. AF: Atlantic Forest, 0: control, MF1: mineral fertilization, 10 kg N ha^{-1}, 15 kg ha^{-1} and 17 kg K ha^{-1}, MF2: mineral fertilization, 20 kg N ha^{-1}, 30 kg P ha^{-1} and 34 kg K ha^{-1}; OC: organic compost, 40 m^3 ha^{-1}.

In Brazil, the addition of organic fertilizer to acid soils leads to an increase on SOM in which the effect may not be significant with only one application. In a study about the effect of 40 years of farmyard manure, mineral and mixed fertilizations on soil organic properties of a Fluvi-calcaric Cambisol from Northeastern Italy, Nardi et al. (2004) also reported a larger influence of the organic treatment alone on total organic carbon than mineral fertilizer alone or mixed with farmyard manure. In contrast to the control treatment, or the treatment in which only mineral fertilizer was applied and carbon input was only due to plant residues, the treatments with the application of organic compost received compost with diverse biochemical composition. In a study comparing different plant residues, Drinkwater et al. (1998) reported large amounts of soil carbon from compost have larger residence time, probably due to its partial decomposition or recalcitrant material.

Compost application has been evaluated with other practices which can increase SOM stocks such as cover crop, crop rotation or no-tillage. In Brazilian semi arid, Menezes and Silva (2008) evaluated in an Entisol, the effects of application of six years of compost and/or cover crop on the soil chemical characteristics and SOC content. The treatments were: animal manure (15 t ha^{-1}); use of *Crotalaria juncea* as cover crop; fertilization with half (7.5 t ha^{-1}) of the usual doses of animal manure combined with *C juncea* as a cover crop e; and control plots without application. The annual fertilization with annual manure, with or without *C. juncea* increased SOC, total N and total P in the 0-20 layer (Table 3). According to the authors, the results showed that the application of lower doses of manure combined with *C. juncea* maintained soil fertility and increased SOM.

Table 3. Soil organic carbon (SOC), total nitrogen (TN) and total phosphorus (P) contents in two layers of the Entisol after six years of organic fertilization with animal manure or Crotalaria juncea (Menezes et al., 2008)

Treatments	SOC	TN	P
	g kg^{-1}		mg kg^{-1}
		0 – 20 cm	
Control	4.5 c	0.48 b	39.7 c
CE	6.5 ab	0.77 a	63.0 ab
E	7.3 a	0.80 a	72.1 a
C	5.3 bc	0.54 b	50.2 bc
Mean	5.9	0,65	56.3
		20 – 40 cm	
Control	4,5 a	0.42 b	30.0 b
CE	5,5 a	0.61 a	41.4 ab
E	5,4 a	0.56 ab	43.5 a
C	4,3 a	0.49 ab	39.3 ab
Mean	4.9	0.52	38.6

[1] Means followed by the same letter within each column and each depth are not significantly different at 5% level by Tukey's test. [2] CE –crotalária + 7.5 t ha^{-1} of animal manure; E – 15 t ha^{-1} of animal manure; C –crotalária [3] CV – coeficiente de variação

In a long term field experiment (1966-2000), in southeastern Italy, Triberti et al. (2008) compared organic and mineral fertilizations on a 2 year rotation: maize – winter wheat, without irrigation and cattle manure (M), cattle slurry (S) and wheat or corn residues (R) together with an unfertilized control (C), combined factorally with increasing rates of mineral N fertilizers in both crops. The authors observed that in 2000, SOC concentration in the first 0.40m soil layer (6.77g kg^{-1}, on the average of all plots) was 12 % lower than in 1966 (7.72g kg^{-1}) (Table 4). Thirty-four years after the start of the trial, the interaction between mineral and organic fertilizations revealed no significant effects on SOC, while single factors caused marked differences (Table 4). Moreover, on the average of N rates, manure, slurry and crop residues increased SOC concentration compared to the control by 51, 30, and 19%, respectively. SOC content of manured soil was, thus, higher than that of slurry (+16%) and crop residues (+27%). SOC content was also augmented by N fertilization, independently from the type of organic supply (Table 4).

Table 4. Concentrations of soil organic carbon (SOC) and total nitrogen (TN) in the 0–0.40 m soil layer, C/N ratio in 2000, 34 years after the start of the experiment, and wheat and maize grain yields, at 13 and 15.5% humidity respectively, averaged over the period 1966–2000 (Triberti et al., 2008).

Treatment	SOC	TN	C/N	Wheat	Maize
	g kg^{-1}			t ha^{-1}	
		Organic Fertilizers			
Control	5,43 c[a]	0,77 d	7,05	5,44 b	8,60 b
Manure	8,17 a	1,09 a	7,52	6,20 a	9,62 a
Slurry	7,04 b	0,96 b	7,32	5,99 a	9,36 a
Crop residue	6,44 b	0,88 c	7,32	5,55 b	8,64 b
		Inorganic fertilizes (kg N ha^{-1})			
N0 (0)	6,66	0,91	7,30	4,18	7,93
N1 (100)	6,68	0,91	7,33	6,33	9,17
N2 (200)	6,87	0,93	7,40	6,48	9,54
N3 (300)	6,87	0,95	7,19	6,18	9,56
ANOVA			Probability		
	SOC	TN	C/N	Wheat	Maize
Organic fertilization	0,0004	<0,0001	ns	0,0001	0,0034
Inorganic fertilization					
Linear component	0,0012	<0,0001	ns	<0,0001	<0,0001
Quadratic component	<0,0001	<0,0001	ns	<0,0001	<0,0001

Means of the principal factors of the split plot design.[a] Organic fertilization means within a column followed by different letters are significantly different at $P \leq 0.05$ by SNK test. ns means not significant.

Also, in Italy, Monaco et al. (2008) analyzed changes in soil quality properties caused by continuous cultivation of irrigated maize and represented different quantities and nature of C and N inputs supplied to the soil for 11 years through the addition of slurry, farmyard manure or plant residue. Soils that received farmyard manure application (MsF) or maize straw (MgU) had statistically higher amounts of both C_{org} and N_{tot} compared with the unfertilized soil (Ms0) and the N fertilized soil (MsU) (Table 5). On the other hand, C_{org} and N_{tot} measured on cattle slurry (MsS) and MsU were not significantly different from the control (Ms0). The authors reported also that, considering only the treatments with additional inputs of organic C (MgU, MsS and MsF), C_{org} was higher after the application of farmyard manure (MsF) than slurry (MsS) or maize straw (MgU), although the latter treatment actually received the highest input of total C. The results that showed an increase in the SOM stocks by farmyard manure have coincided with the positive effect of manure applications on soil health indicators in management systems as related by others authors (Nardi et al., 2004).

Table 5. Soil organic C, total N and C:N in the 0–15 and 15–30 cm layers, sampled in February 2004 after the 11 y different management. (Monaco et al., 2008)

Depth cm	Treatment*	C_{org}	N_{tot}	C:N
		g kg^{-1}		
0-15	Ms0	10,7 cd	1,03 d	10,4 ab
	MsU	10,4 cd	1,06 d	9,8 ab
	MgU	12,8 b	1,41 ab	9,1 ab
	MsS	12,6 bc	1,14 cd	11,0 a
	MsF	15,7 a	1,48 a	10,6 ab
15-30	Ms0	10,1 cd	1,16 bcd	8,7 b
	MsU	9,4 d	0,96 d	9,8 ab
	MgU	11,2 cd	1,15 cd	9,7 ab
	MsS	11,0 cd	1,10 cd	10,0 ab
	MsF	12,5 bc	1,31 abc	9,5 ab
0-30	Ms0	10,4	1,10	9,6
	MsU	9,9	1,01	9,8
	MgU	12,0	1,28	9,4
	MsS	11,8	1,12	10,5
	MsF	14,1	1,39	10,1
0-15	Average	12,4	1,22	10,2
15-30	Average	10,8	1,14	9,5
Source of variation				
Treatment		< 0,001	< 0,001	0,072
Depth		< 0,001	0,008	0,019
Treatment x depth		0,003	0,006	0,032

Concluding Remarks

The different agricultural practices can cause positive and negative effects on soil microbial and organic matter content. In the case of organic farming, several studies have shown that the agricultural practices in this agricultural system, such as addition of compost, straw and natural amendments, promotes positive changes in the soil microbiological and biochemical processes, resulting in the increase of soil microbial biomass and organic matter. In this way, the organic system is extremely important for the increase of the soil fertility and the maintenance of the environmental sustainability.

Acknowledgments

The authors are grateful to "Conselho Nacional de Desenvolvimento Científico e Tecnológico" (CNPq-Brazil) and "Fundação de Amparo a Pesquisa do Estado do Piauí" (FAPEPI) for financial support, and the research of the "Soil Microbial Ecology in Mid-North

Region" group (UFPI-CNPq). The authors are supported by personal fellowships from CNPq-Brazil.

REFERENCES

Araújo, ASF, Monteiro, RTR Microbial biomass and activity in a Brazilian soil amended with untreated and composted textile sludge. *Chemosphere*, 2006, 64, 1028-1032.

Araújo, ASF, Monteiro, RTR, Abarkeli, RB Effect of glyphosate on the microbial activity of two Brazilian soils. *Chemosphere,* 2003, 52, 799-804.

Araújo, ASF, Santos, VB, Leite, LFC, Carneiro, RFV Soil respiration and physic attributes in conventional and organic farming system of a Typic Quartzipsamment from Piauí State. *Acta Scientiarum Agronomy*, 2009 (in press).

Araújo, ASF, Santos, VB, Monteiro, RTR Responses of soil microbial biomass and activity for practices of organic and conventional farming systems in Piauí state, Brazil. *European Journal of Soil Biology*, 2008, 44, 225-230.

Bot. A, Benites J The importance of soil organic matter: Key to drought-resistant soil and sustained food and production. FAO Soils Bulletim. FAO. Roma. 2007. 80p.

Colleman, K, Jenkinson, DS, Roth, C 26.3- a model for the turnover of carbon in soil. In: Powlson, TS, Smith, P, Smith, JU. editors. Evaluation of soil organic matter models using existing long term datasets. NATO ASI, Series I, 1996, pp.237-246.

Craswell, ET, Lefroy, RDB The role and function of organic matter in tropical soils. *Nutrient Cycling and Agroecosystem*, 2001, 61, 7-18.

Davis, J, Livestock manure impacts on C cycling. http:// *www.colostate.edu/depts/soil/crop/extension/newsletter/2002/may/mayweb/may09.html*

Dick, RP. Soil enzymes activities as integrative indicator of soil health. In: Pankhurst, C, Doube, BM, Gupta, VVSR. editors. Biological indicators of soil Health. Cambridge:CAB International, 1997. p.121-156.

Drinkwater, LE, Wagoner, P, Sarrantonio, M Legume-based cropping systems have reduced carbon and nitrogen losses. *Nature*, 1998, 396, 262–265.

Fließbach, A, Mader, P. Microbial biomass and size-density fractions differ between soils of organic and conventional agricultural systems. *Soil Biology and Biochemistry*, 2000, 32, 757–768.

Glover, JD, Reganold, JP, Andrews, PK. Systematic method for rating soil quality of conventional, organic, and integrated apple orchards in Washington State. *Agriculture, Ecosystem and Environment*, 2000, 80, 29–45.

Greene, C, Kremen, A, U.S. Organic Farming in 2000–2001: Adoption of Certified Systems. Agriculture Information Bulletin No. 780. Resource Economics Division, Economic Research Service, USDA, Washington, DC, 2003. 55 pp.

Haney, RL, Senseman, SA, Krutz, LJ, Hons, FM. Soil carbon and nitrogen mineralization as affected by atrazine and glyphosate. *Biology and Fertility of Soils*, 2003, 35, 35-40.

Hutchinson, J, Campbell, CA, Desjardins, RL Some perspectives on carbon sequestration in agriculture. *Agriculture and Forest Meteorology*, 2007, 142, 288-302.

Jareccky, M, Lal, R Crop Management for Soil Carbon Sequestration. *Critical Review in Plant Science*, 2003, 22, 471-502.

Jenkinson, DS, Ladd, JN. Microbial biomass in soil: measurement and turnover. In: Paul. EA, Ladd, JN. Editors. Soil biochemisty. New York: Marcel Dekker, 1981, p.415-471.

Jiao, Y, Whalen, JK, Hendershot, WH No-tillage and manure applications increase aggregation and improve nutrient retention in a sandy-loam soil. *Geoderma*, 2006, 134, 24-33.

Kemmit, SJ, Lanyon, CV, Waite, IS, Wen, Q, Addiscott, TM, Bird, NRA, O'Donnell, AG, Brookes, PC Mineralization of native soil organic matter is not regulated by the size, activity or composition of the soil microbial biomass – a new perspective. *Soil Biology & Biochemistry*, 2008, 40, 61-73.

Kladivko, EJ Tillage system and soil ecology. *Soil Tillage Research*, 2000, 6, 61-76.

Lal, R Soil carbon sequestration to mitigate climate change. *Geoderma*, 2004, 123, 1-22.

Leite, LFC, Mendonça, ES, Machado, PLOA Influence of organic and mineral fertilisation on organic matter fractions of a Brazilian Acrisol under maize/common bean intercrop. *Australian Journal Soil Research*, 2007, 45, 25-32.

Leite, LFC, Mendonça, ES, Machado, PLOA, Matos, ES Total C and N storage and organic C pools of a Red-Yellow Podzolic under conventional and no tillage at the Atlantic Forest Zone, Southeastern Brazil. *Australian Journal Soil Research*, 2003, 41, 717-730.

Liu, X, Herbert, SJ, Hashemi, AM, Zhang, X, Ding G Effects of agricultural management on soil organic matter and carbon sequestration – a review. *Plant Soil Enviroment*, 2006, 52, 531-543.

Lotter, D. Organic Agriculture. *Journal of Sustainable Agriculture*, 2003, 21, 1-4.

Machado, PLOA Carbono do solo e a mitigação da mudança climática global. *Química Nova*, 2005, 28, 329-334.

Melero, S, Porras, JCR, Herencia, JF, Madejon, E. Chemical and biochemical properties in a silty loam soil under conventional and organic management. *Soil Tillage Research*, 2005, 90, 162-170.

Menezes, RSC, Silva, TO Mudanças na fertilidade de um Neossolo Regolítico após seis anos de adubação orgânica. *Revista Brasileira Engenharia Agrícola e Ambiental*, 2007, 12, 251-257

Mirsky, SB, Lanyon, E, Needelman, AB Evaluating Soil Management Using Particulate and Chemically Labile Soil Organic Matter Fractions. *Soil Science Society of American Journal*, 2008, 72, 180-185.

Monaco, S, Hatch, DJ, Sacco, D, Bertora, C, Grignani, C Changes in chemical and biochemical soil properties induced by 11-yr repeated additions of different organic materials in maize-based forage systems. *Soil Biology and Biochemistry*, 2008, 40, 608-615.

Muller, AK, Westergaard, K, Christensen, S, Sorensen, SJ. The diversity and function of soil microbial communities exposed to different disturbances. *Microbial ecology*, 2002, 44, 49-58.

Murage, EW, Voroney, PR, Kay, BD, Deen, B, Beyaert, RP, Dynamics and turnover of soil organic matter as affected by tillage. *Soil Science Society of America Journal*, 2007, 71, 1363-1370.

Nardi, S, Morari, F, Berti, A, Tosoni, M, Giardini, L Soil organic matter properties after 40 years of different use of organic and mineral fertilisers. *European Journal of Agronomy*, 2004, 21, 357-367

Parton, WJ, Schimel, DS, Cole, CV, Ojima, DS Analysis of factors controlling soil organic matter levels in great plains grasslands. *Soil Science Society of American Journal*, 1987, 51, 1173-1179.

Powlson, DS, Brookes, PC, Christensen, BT. Measurement of soil microbial biomass provides an early indication of changes in total organic matter due to straw incorporation. *Soil Biology and Biochemistry*, 1987, 19, 159-164.

Sampaio, DB, Araújo, ASF, Santos, VB. Avaliação de indicadores biológicos de qualidade do solo sob sistemas de cultivo convencional e orgânico de frutas. *Ciência e Agrotecnologia*, 2008, 32, 353-359.

Six, J, Ellliot, EJ, Paustian, K Soil macroaggregate turnover and microaggregate formation: a mechanism for C sequestration under no-tillage conditions. *Soil Biology Biochemistry*, 2000, 32, 2099-2103

Sparling, GP, Ross, DJ. Biochemical methods to estimate soil microbial biomass: Current developments and applications. In: Mulangoy, K, Merckx, R. editors. *Soil organic matter dynamics and sustainable of tropical agriculture*. Chichester: Wiley, 1993, p.21-37.

Sparling, GP. Soil microbial biomass, activity and nutrient cycling as indicators of soil health. In: Pankhurst, C, Doube, BM, Gupta, VVSR. editors. *Biological indicators of soil Health*. Cambridge: CAB International, 1997. p.97-120.

Triberti, L, Nastri, A, Giordani, G, Comellini, F, Baldoni, G, Toderi, G Can mineral and organic fertilization help sequestrate carbon dioxide in cropland. *European Journal of Agronomy*, 2008, 29, 13-20.

Truu, M, Truu, J, Ivask, M. Soil microbiological and biochemical properties for assessing the effect of agricultural management practices in Estonian cultivated soils. *European Journal of Soil Biology*, 2008. (In press, available on line).

Tu, C, Ristaino, JB, Hu, S. Soil microbial biomass and activity in organic tomato farming systems: Effects of organic inputs and straw mulching. *Soil Biology & Biochemistry*, 2006, 38, 247–255.

Watanabe, K, Hamamura, N. Molecular and physiological approaches to understanding the ecology of pollutant degradation. *Applied Soil Ecology*, 2003, 31, 120-135.

Watson, CA, Kristensen, ES, Alroe, HF. Research to support the development of organic food and farming. In: Kristiansen, P, Taji, A, Reganold, J. editors. *Organic Agriculture: A Global Perspective*. Collingwood: CSIRO Publishing, 2006, p.361–383.

West, TO, Post, WM Soil organic carbon sequestration rates by tillage and crop rotation: A global data analysis. *Soil Science Society of American Journal*, 1946, 66, 1930-1946.

Willer, H, Yussefi, M. The World of Organic Agriculture: Statistics and Emerging Trends. *Bonn: IFOAM*, Germany, 2004. p. 167.

Wolf, B, Snyder, GH Sustainable soils. The place of organic matter in sustaining soils and their productivity. Ed. Food Products Press, New York, 2003, 352p.

Zelles, L, Bai, QY, Beck, T, Breese, F. Signature fatty acids in phospholipids and lipopolysaccharides as indicator of microbial biomass and community structure in agricultural soils. *Soil Biology and Biochemistry*, 1992, 24, 317-323.

In: Organic Farming: Methods, Economics and Structure ISBN 978-1-60692-864-6
Editors: Michael Nelson and Ivan Artamova

Chapter 9

U.S. ORGANIC AGRICULTURE

Catherine Greene

U.S. farmland under organic management has grown steadily for the last decade as farmers strive to meet consumer demand. By 2003, the United States had over 2 million acres of certified organic crops and pasture. USDA implemented an organic regulatory program in 2002, and Federal research and education activities have also emerged.

INTRODUCTION

Farmers have been developing organic farming systems in the United States since the 1940s, and organic markets have emerged and expanded greatly since then. USDA implemented national standards for organic production and processing in October 2002, following more than a decade of development, and the new uniform standards have facilitated further growth in the organic farm sector. USDA's organic standards incorporate cultural, biological, and mechanical practices that foster cycling of resources, ecological balance, and protection of biodiversity—practices that have evolved over the last half-century.

An increasing number of U.S. farmers are adopting these systems in order to lower input costs, conserve nonrenewable resources, capture high-value markets, and boost farm income. Despite the time, costs, and effort required to meet these stringent requirements, USDA estimates that farmers and ranchers added more than a million acres of certified organic land for major crops and pasture between 1995 and 2003, doubling organic pasture and more than doubling organic cropland for major crops. Total certified organic cropland and pasture now encompasses 2.2 million acres in 49 States. Organic livestock, which require access to organic pasture, have grown more numerous since USDA lifted restrictions on organic meat labeling in late 1999. Food crops and other animal foods (eggs and dairy) are regulated by the Food and Drug Administration and were allowed to carry an organic label throughout the 1990s.

Consumer demand for organically produced goods has risen for over a decade, providing market incentives for U.S. farmers across a broad range of products. Organic products are now available in nearly 20,000 natural food stores and nearly 3 of 4 conventional grocery stores. Organic sales account for approximately 2 percent of total U.S. food sales, according to recent industry statistics. Farmers'markets and other direct-market venues have also grown in number over the last decade, and are especially popular among organic producers. Organic farmers are also finding ways to capture a larger segment of the consumer food dollar through on-farm processing, producer marketing cooperatives, and new forms of direct marketing, including agricultural subscription services.

U.S. Organic Production Standards

Organic farming systems rely on ecologically based practices, such as biological pest management and composting; virtually exclude the use of synthetic chemicals, antibiotics, and hormones in crop production; and prohibit the use of antibiotics and hormones in livestock production. Under organic farming systems, the fundamental components and natural processes of ecosystems—such as soil organism activities, nutrient cycling, and species distribution and competition—are used as farm management tools. For example, crops are rotated, food and shelter are provided for the predators and parasites of crop pests, animal manure and crop residues are cycled, and planting/harvesting dates are carefully timed.

Organic livestock production systems attempt to accommodate an animal's natural nutritional and behavioral requirements, ensuring that dairy cows and other ruminants, for example, have access to pasture. The 2002 USDA livestock standards incorporate requirements for living conditions, pasture and access to the outdoors, feed ration, and health care practices suitable to the needs of the particular species.

The national organic standards address the methods, practices, and substances used in producing and handling crops, livestock, and processed agricultural products. Although specific practices and materials used by organic operations may vary, the standards require every aspect of organic production and handling to comply with the provisions of the Organic Foods Production Act of 1990. Organically produced food cannot be produced using genetic engineering, sewage sludge, or irradiation. These standards include a national list of approved synthetic substances (such as insecticidal soaps and horticultural oils), and prohibited nonsynthetic substances (such as arsenic, strychnine, and tobacco dust) for use in organic production and handling. (See "National Organic Program" on USDA's Agricultural Marketing Service website.)

Adoption of Organic Farming Systems

U.S. farmland under organic management has grown steadily for the last decade as farmers strive to meet consumer demand in both local and national markets. U.S. certified organic crop acreage more than doubled between 1992 and 1997, and doubled again between 1997 and 2003 for many crops. Organic fruit and vegetable crop acreage, along with acreage used for hay and silage crops, expanded steadily between 1997 and 2003. However, most of the acreage increase for organic grain and oilseed crops took place early in this period (Figure 1), and organic soybean acreage has declined substantially since 2001.

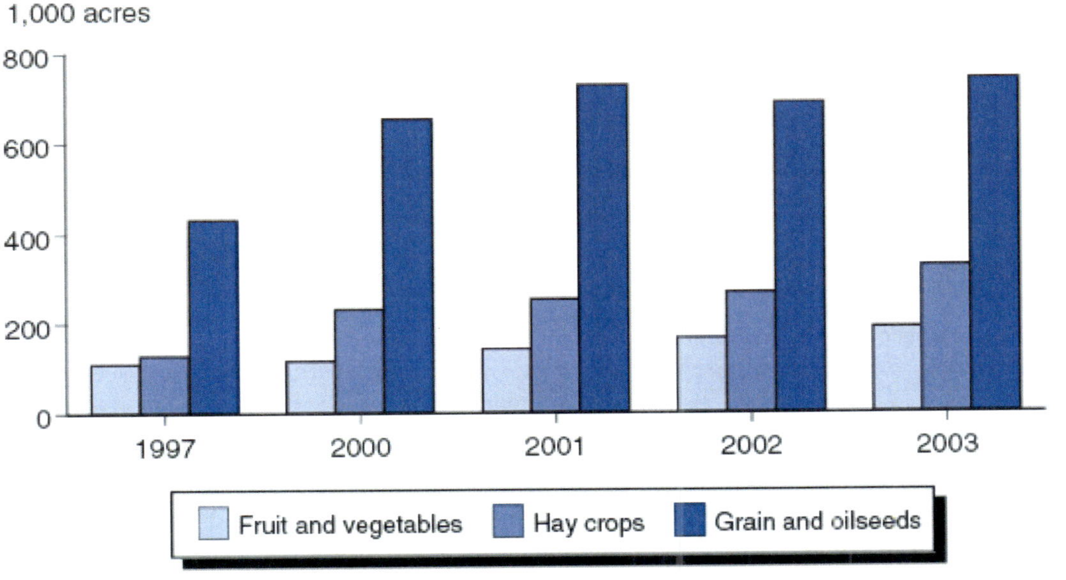

Source: Economic Research Service, USDA.

Figure 1. U.S. certified organic acreage, selected crops, 1997-2003.

Certified organic pasture (including ranchland) declined between 1992 and 1997, but increased 50 percent between 1997 and 2003 after USDA lifted restrictions on organic labeling for meat and poultry. Overall, U.S. farmers and ranchers in 49 States dedicated 2.2 million acres of cropland and pasture to organic production systems in 2003 (Table 31). Many crop/livestock sectors and most States showed strong growth between 2002 and 2003. While certified organic cropland accounted for only 0.4 percent of U.S. cropland in 2001, the share is much higher in some crops, such as fruits (over 2 percent), and vegetables (over 4 percent).

California was the leading State in certified organic cropland in 2003, with nearly 180,000 acres, mostly used for fruit and vegetable production. North Dakota followed with nearly 130,000 acres of cropland, mostly for wheat, soybeans, and other field crops. Minnesota, Montana, Colorado, and Iowa were other top States.

Nearly 40 States had certified pasture and rangeland in 2003, most with under 20,000 acres, although several States had over 100,000 acres and Texas had over 250,000 acres. The number of certified organic beef cows, milk cows, hogs, pigs, sheep, and lambs was up more than five-fold since 1997, and up 15 percent between 2002 and 2003. Dairy has been one of the fastest growing segments of the organic foods industry during this period, and milk cows accounted for over half of the certified livestock animals. Poultry animals raised under certified organic management—including layer hens, broilers, and turkeys—showed even higher levels of growth during 1997-2003.

California had more certified operations than any other State, with just over 1,900 operations in 2003, up 28 percent from the previous year (Figure 2). Wisconsin, Washington, Iowa, Minnesota, New York, Ohio, Vermont, Oregon, and Maine rounded out the top 10.

Many of these States have a high proportion of farms with fruits and vegetables and other specialty crops. Also, some of these States, particularly in the Northeast, have relatively little cropland but a large concentration of market gardeners.

Table 31. U.S. certified organic farmland acreage, livestock numbers, and farm operations, 1997 and 2003

Item	1992	1997	2003	Change 1992-1997	Change 1997-2003
				Percent	
U.S. certified farmland:					
Total	935,450	1,346,558	2,196,874	44	63
Pasture/rangeland	532,050	496,385	745,273	-7	50
Cropland	403,400	850,173	1,451,601	111	71
		Number			
U.S. certified animals:					
Livestock--					
Beef cows	6,796	4,429	27,285	-35	516
Milk cows	2,265	12,897	74,435	469	477
Hogs and pigs	1,365	482	6,564	-65	1,262
Sheep and lambs	1,221	705	4,561	-42	547
Total livestock[1]	11,647	18,513	124,346	59	572
Poultry--					
Layer hens	43,981	537,826	1,591,181	1,123	196
Broilers	17,382	38,285	6,301,014	120	16,358
Turkeys	--	750	217,353	--	28,880
Total poultry[2]	61,363	798,250	8,780,152	1,201	1,000
Total certified operations*	3,587	5,021	8,035	40	60

[1]Total livestock includes other and unclassified animals.

[2]Total poultry includes other and unclassified animals.

*Number does not include subcontracted organic operations. Numbers may not add due to rounding.

Source: Economic Research Service, USDA.

Although consumer demand for organic foods is expected to continue growing rapidly in the United States and other major markets, the competition for these markets is likely to increase considerably. Since 2002, USDA has accredited over 40 organizations in foreign countries, as well as approximately 50 groups in the United States, to certify producers and handlers. USDA's Foreign Agricultural Service estimates that the value of U.S. organic exports in 2002 was between $125 million and $250 million, while the value of U.S. organic imports was between $1.0 and $1.5 billion (U.S. Department of Agriculture, 2005). Cotton and soybeans are among the U.S. organic crops that have declined since 2001, despite the growth in retail sales of cotton and soy-based products. Import competition has likely played a role in the decline.

ECONOMIC CHARACTERISTICS OF ORGANIC SYSTEMS

The rapid increase in organic crop and livestock production reflects the increase in consumer demand for organically produced food—20 percent or more per year throughout the 1990s. According to industry data, retail sales of organic food reached $10.4 billion in 2003, up 20.4 percent from the previous year, and accounting for nearly 2 percent of U.S. food sales.

Farmgate and retail price data, collected by private groups, have indicated substantial organic premiums for fruits, vegetables, and milk over the last decade, and recent government data show similar premiums at the wholesale level. Organic grain and soybean crops also enjoyed substantial price premiums during the 1990s, exceeding 50 percent for corn, soybeans, wheat, and oats during 1993-99, and continue to carry a substantial premium (Streff and Dobbs, 2004).

A number of studies have been conducted on the motivations of consumers who purchase organic foods, such as perceived health attributes and concern about pesticide residues and the environment (Dimitri and Greene, 2002). Potential benefits from organic farming systems include improved soil tilth and productivity, lower energy use, and reduced use of pesticides (USDA, 1980; Smolik et al., 1993), and researchers are beginning to compare differences in the nutritive value of the foods produced from these systems as well (Gold, 2000).

A growing number of studies in the United States have examined the yields, input costs, profitability, managerial requirements, and other economic characteristics of organic farming. A 1990 review of the U.S. literature at Cornell concluded that the "variation within organic and conventional farming systems is likely as large as the differences between the two systems," and found mixed results in the comparisons for most characteristics (see Part 4 for more information on conventional systems).

Several USDA and university studies during the 1990s in California, Ohio, and Texas indicated that organic price premiums are necessary to give organic farming systems comparable or higher whole-farm profits than conventional systems, particularly for crops like processed tomatoes and cotton. A Henry A. Wallace Institute of Alternative Agriculture review of university-based comparative studies in the 1980s and early 1990s on Midwestern organic grain and soybean production found organic systems needed price premiums to be more profitable than conventional systems (Welsh, 1999). Several of these studies, however, found that organic grain and soybean production could be as profitable even without price premiums due to higher yields in drier areas or periods, lower input costs, or higher revenue from the mix of crops used in the system. Other recent studies have also found that some organic systems may be more profitable than conventional systems, even without price premiums (Swezey et al., 1994, Reganold et al., 2001).

Net returns to both conventional and organic production systems vary with factors such as soil type, climate, and proximity to markets, and help explain the wide variation in economic performance within each system. Factors not captured in standard profit calculations—such as convenience, longer-term planning horizons, and environmental ethics—can motivate rational adoption of a particular practice or farming system. Our understanding of the factors influencing net returns to organic farming systems remains imperfect.

Economic research on organic farming has tended to focus narrowly on profitability (Fox et al., 1991), but land-grant universities and others are increasingly examining the long-term economics of organic systems through replicated field trial research and a multidisciplinary systems approach (Table 32).

According to the Organic Farming Research Foundation, 18 States had land-grant institutions with research acres under certified organic management in 2003, up from 6 States in 2001. Organic farming systems trials—in experiment stations and on-farm settings—seek to answer basic research questions about yields, profitability, and environmental impacts, as well as to address farmer-defined management and production obstacles to adoption of organic production systems.

USDA's national standards do not restrict additional eco-labeling of organic products, and some organic certifiers are developing standards on social aspects of agricultural production and food distribution—such as fair trade and local sourcing. The Florida Certified Organic Growers and Consumers organization, for example, recently developed a partnership program for food retailers and restaurants in North Florida to certify their level of commitment to local food sourcing (FOG/QCS, 2003). Most coffee sold in the United States that is certified as fair trade (indicating that farmers receive a fair price) also has a separate organic certification. Some certification groups are trying to improve the efficiency of their efforts by integrating these programs.

Federal Policy Initiatives

Government research and policy initiatives often play a key role in the adoption of new farming technologies and systems. Worldwide, adoption levels for organic farming systems are currently the highest in European Union countries. Governments there have been developing consumer education initiatives and providing direct financial support to producers for conversion since the late 1980s to capture environmental benefits and support rural development.

Table 32.—Examples of U.S. multidisciplinary, long-term research projects with organic trials

Project	Date established	Farming system/Commodity focus
University of Nebraska-Lincoln *Long-Term Experiment Trials*	1975	Compare conventional and organic systems (Rotations include corn, wheat, and soybean)
Rodale Institute—Kutztown, PA *Farming Systems Trial TM*	1981	Examine the transition process from conventional to organic farming (corn and soybeans)
University of California-Davis *Sustainable Agriculture Farming Systems Project*	1988	Compare conventional, low-input and organic systems; evaluate conservation tillage in these systems (tomato, safflower, bean, corn)
Iowa State Univ.—Leopold Center *Neely-Kinyon Long-Term Agroecological Research*	1988	Compare conventional and organic systems (corn, soybeans and alfalfa)
University of Minnesota-Lamberton Experiment Station *Elwell Agroecology Farm*	1989	Compare conventional and organic systems (corn, soybeans, alfalfa and oats)

Table 32. (Continued)

Project	Date established	Farming system/Commodity focus
Michigan Agricultural Experiment Station *Living Field Laboratory*	1993	Compare conventional, organic and other systems (corn, soybeans, and wheat)
USDA Agricultural Research Center-Beltsville, MD *Farming Systems Project*	1993	Compare organic systems typical in the mid- Atlantic region (corn and soybeans)
West Virginia University (WVU) *Horticulture Farm Project*	1999	Evaluate organic systems on the entire Horticulture Farm (market garden and field crop/livestock systems)
North Carolina State University *Farming Systems Trial*	2001	Compare conventional and transitional organic systems (grains, livestock and woodlots)
Ohio State University *John Hirzel Sustainable Agriculture Research and Education Site*	2001	Compare conventional, no-till and high-and lowinput organic systems (soybeans, corn, wheat, and vegetables)

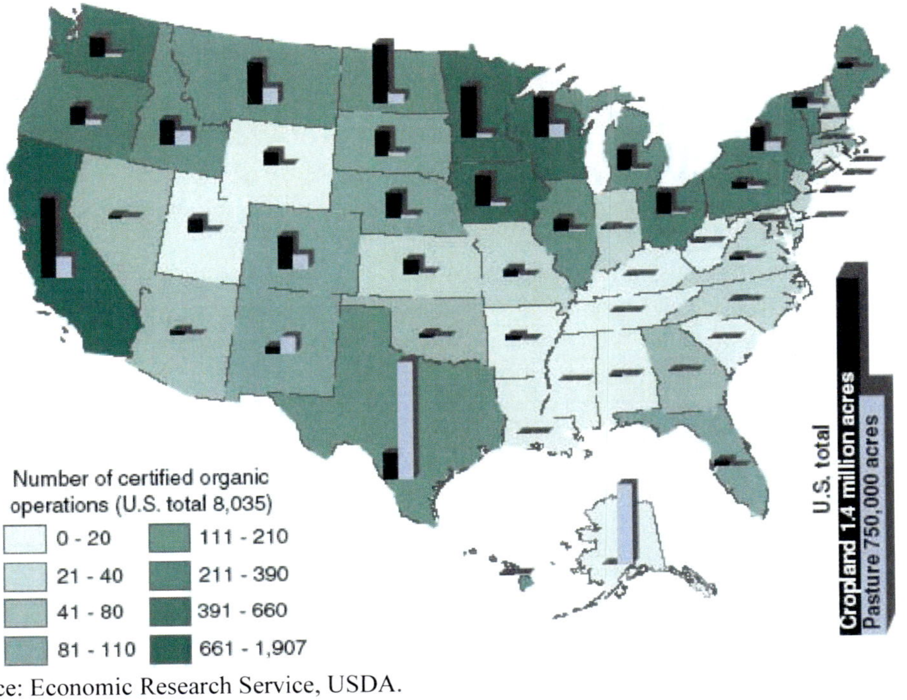

Source: Economic Research Service, USDA.

Figure 2. Certified organic acreage and operations, 2003.

State and Federal support for organic farmers and handlers is also beginning to emerge in the United States. Minnesota and Iowa, for example, began subsidizing conversion to organic farming systems in the late 1990s as a way to capture the environmental benefits of these systems. In 2003, the National Association of State Departments of Agriculture released a policy statement on organic agriculture expressing support for a wide range of activities that would expand public-sector organic research and education and provide technical assistance to organic and transitional farmers.

USDA agencies have started or expanded programs on organic agriculture during the last several years, and the Farm Security and Rural Investment Act of 2002 (the 2002 Farm Act) contains several first-time research and technical assistance provisions to directly assist organic producers. Recent programs and initiatives include:

- *Certification Cost-Share Support.* In 2001, USDA established a certification cost-share program to help farmers defray certification costs in 15 States. The 2002 Farm Act allocated $5 million in cost-share assistance funds for this program and expanded eligibility to growers and handlers in all States.
- *Research and Technical Assistance.* The 2002 Farm Act contains an Organic Agriculture Research and Extension Initiative that authorizes $3 million per year in new mandatory appropriations in fiscal years 2003-07. These funds are being used to administer competitive research grants focused on organic agriculture production, breeding, and processing methods, as well as the marketing and policy constraints.
- *Conservation Initiatives.* The 2002 Farm Act provided funding for the Conservation Security Program, which provides payments to producers for adopting or maintaining land management and conservation practices to address resource concerns. This new program may interest organic farmers who commonly adopt these types of practices as part of their organic farming systems. (See Part 5 for a general discussion of Federal conservation initiatives and for a discussion of working-land programs.)
- *Exemptions From Marketing Assessments.* Another provision in the 2002 Farm Act specifies that certified organic producers who produce and market only organic products and do not produce any conventional or nonorganic products will now be exempt from paying an assessment under any commodity promotion law. In December 2003, USDA published a proposed rule to exempt producers from paying assessments associated with its marketing order programs, and in January 2005 a similar proposed rule was published to cover exemptions for its research and promotion programs.
- *Export Promotion, Crop Insurance, and Other Initiatives.* USDA's Risk Management Agency has provided insurance coverage for organic farming practices as good farming practices by written agreement since 2001, and made organic farmers eligible for a wider range of coverage options in 2004. USDA's Foreign Agricultural Service (FAS) is helping design protocols for working with foreign nations to keep organic trade moving as more countries develop organic standards.

REFERENCES

[1] Dimitri, Carolyn, and Catherine Greene (2002). *Recent Growth Patterns in the U.S. Organic Foods Market*, AIB-777, U.S. Dept. Agr., Econ. Res. Serv., Sept.

[2] FOG/QCS (2003). *Certification Manual of North Florida Local Food Partnership, 2003-2004*, Florida Organic Growers & Consumers/Quality Certification Services. Gainesville, FL.

[3] Fox, Glenn, Alfons Weersink, Ghulam Sarwar, Scott Duff, and Bill Deen (1991). "Comparative Economics of Alternative Agricultural Production Systems: A Review," *Northeastern Journal of Agricultural and Resource Economics,* Vol. 20.

[4] Gold, Mary (2000). *"Organically Produced Foods: Nutritive Content,"* Special Reference Briefs Series SRB 2000-03, Alternative Farming Systems Information Center, National Agricultural Library, U.S. Dept. Agr., Agricultural Research Service.

[5] Greene, Catherine, and Amy Kremen (2003). *U.S. Organic Farming in 2000-2001: Adoption of Certified Systems*, AIB-780, U.S. Dept., Agr., Econ. Res. Serv., June.

[6] National Association of State Departments of Agriculture. 2003. *Organic Agriculture*, NASDA Policy Statement. Sept. 11.

[7] Reganold, J.P., J.D. Glover, P.K. Andrews, and H.R. Hinman (2001). "Sustainability of three apple production systems," *Nature*, 410: 926-930.

[8] Smolik, James D., Thomas L. Dobbs, Diane H. Rickerl, Leon J. Wrage, George W. Buchenau, and Thomas A. Machacek (1993). *Agronomic, Economic, and Ecological Relationships in Alternative (Organic), Conventional, and Reduced-Till Farming Systems*, B718, Agricultural Experiment Station, South Dakota State University, Sept.

[9] Streff, Nicholas, and Thomas L. Dobbs (2004). *'Organic' and 'Conventional' Grain and Soybean Prices in the Northern Great Plains and Upper Midwest: 1995 through 2003,* South Dakota State University Economics Pamphlet 2004-1, Brookings, SD, June.

[10] Swezey, Sean, Jim Rider, Matthew Werner, Marc Buchanan, Jan Allison, and Stephen Gliessman (1994). "Granny Smith conversions to organic show early success," *California Agriculture*, Vol. 48, No. 6.

[11] U.S. Department of Agriculture (2000). "National Organic Program, Final Rule," *Federal Register* 7 CFR Part 205, Dec. 21.

[12] U.S. Department of Agriculture (2005). "U.S. Market Profile for Organic Food Products," Commodity and Marketing Programs—Processed Products Division, International Strategic Marketing Group, Feb. 22.

[13] U.S. Department of Agriculture, USDA Study Team on Organic Farming (1980). *Report and Recommendations on Organic Farming*, U.S. GPO No. 620-220/3641, July.

[14] Welsh, Rick (1999). *The Economics of Organic Grain and Soybean Production in the Midwestern United States*, Policy Studies Report No. 13, Henry A. Wallace Institute for Alternative Agriculture, May.

In: Organic Farming: Methods, Economics and Structure
Editors: Michael Nelson and Ivan Artamova
ISBN 978-1-60692-864-6

Chapter 10

ORGANIC AGRICULTURE IN THE U.S.: PROGRAM AND POLICY ISSUES*

Jean M. Rawson

SUMMARY

Congress passed the Organic Foods Production Act (OFPA) in 1990 as part of a larger law governing U.S. Department of Agriculture (USDA) programs from 1990 through 1996 (P.L. 101-624, the Food, Agriculture, Conservation, and Trade Act of 1990). The act authorized the creation of a National Organic Program within USDA to establish standards for producers and processors of organic foods, and permit such operations to label their products with a "USDA Organic" seal after being officially certified by USDA-accredited agents. The purpose of the program, which was implemented in October 2002, is to give consumers confidence in the legitimacy of products sold as organic, permit legal action against those who use the term fraudulently, increase the supply and variety of available organic products, and facilitate international trade in organic products.

Due to the newness of the National Organic Program, it is difficult to gauge its overall impact on the organic industry. However, a recent USDA report states that the number of people seeking USDA accreditation to become certification agents increased by more than 130% between 2002 and 2005, suggesting significant growth in the number of farm and processing operations going into organic production. USDA estimates that sales of organic foods rose from $6 billion in 2000 to $10.4 billion in 2003. The annual rate of market growth since 1990 has remained steady at about 20%.

Policy issues affecting the National Organic Program since implementation largely reflect the differences in interpretation among stakeholders of the language and intent of OFPA and the actual operation of the program under the final rule. The most recent issues concern, first, USDA's efforts to write a new regulation governing livestock access to pasture, and second, a lawsuit brought by a plaintiff who argued that USDA's administration of the NOP has been too lenient concerning the use of certain synthetic ingredients in foods bearing the "USDA Organic" label, and too lenient concerning the process for converting a dairy herd from conventional to organic production. The final

* This is an excerpted and augmented edition of a Congressional Researach Service Publication RL31595, Updated November 3, 2005

judgment in the case, which was issued in June 2005, requires the Department to develop new rules based on the original statutory language. The revised rules would become effective two years after the date of the judgement, up until which time products made under the old rules would be allowed into commerce.

Subsequently, however, conferees on the FY2006 USDA appropriations bill attached a provision that amends the OFPA in such a way as to largely maintain the NOP regulations as they were before the court decision. This action is likely to worsen the rift between those who welcomed the court ruling because they felt it would force USDA to write stricter regulations that would more closely reflect the statute, and those (a sizeable consortium of major food companies) who predicted that new rules written to satisfy the court decision might severely limit current and future lines of organic products and economically harm all sectors of the industry.

Background

Organic farming, as defined in the final rule establishing the USDA National Organic Program (NOP), is "a production system that is managed in accordance with the [Organic Foods Production] Act and regulations ... to respond to site-specific conditions by integrating cultural, biological, and mechanical practices that foster cycling of resources, promote ecological balance, and conserve biodiversity."[1] This definition indicates that organic agriculture is both an approach to food production based on biological methods that avoid the use of synthetic crop or livestock production inputs (spelled out in detail in the December 2000 rule), and a broadly defined philosophical approach to farming that puts value on resource efficiency and ecological harmony.

Interest in organic farming migrated from Europe to the United States in the early 1900s. Beginning in the 1950s, as the U.S. public became more concerned about the potential adverse environmental and public health effects of agricultural chemicals and so-called "factory farming" methods, private research organizations began to conduct scientific investigations into non-chemical and non-intensive farming techniques, and a small but slowly increasing number of farmers began to adopt organic production practices. Except for a brief period from about 1978 to 1981, USDA did not conduct any activities in support of organic agriculture until OFPA required the Department to begin rulemaking to establish the National Organic Program in 1990.

Organic Sector Statistics [2]

In 2001 (the most recent year for which data are available), there were 2.3 million acres of cropland and pasture under organic management in 48 states, with half of that amount having been added between 1997 and 2001. Nationwide there were about 7,000 certified organic operations in 2001, according to USDA. Fresh produce is the largest sector of the organic industry, with California, Washington, and Colorado having the greatest number of acres devoted to organic fruit and vegetable production. Colorado, Texas, and Montana have the most acreage of organic pasture for livestock. In the Northeast, Southeast, and Upper Midwest, small-scale growers of organic fruits, vegetables, herbs, and flowers are a significant component of individual states' agriculture industries.[3]

Since 1990, sales of organic products have been growing 20% per year on average. Estimated growth rates from 2005 through 2010 range from 9% to 16% annually. The estimated value of sales of organic foods was $6 billion in 2000 and $10.4 billion in 2003. In 2003, 47% of organic foods were sold through conventional retailers, 44% through natural food stores, and 9% through farmers' markets, restaurants, exports, and other marketing channels.[4] Despite sustained market growth, organic food sales constitute about 1.8% of the total U.S. retail sales of food. Various sources of export data estimated U.S. exports of organic foods at between $125 million and $300 million in the 2000-2002 period.[5] The biggest export market is Canada; other major markets are Japan, the European Union, and other countries in Asia.

The high growth rate in sales of organic products can be attributed in part to the higher prices that organic producers and processors receive for their products, according to USDA economists. They speculate that part of the price premium may be due to higher production costs and to higher demand relative to supply. For the 2000-2004 period, the annual average farmgate price premiums for fresh organic broccoli and carrots fluctuated from 75% to 133% above the prices of conventionally grown broccoli and carrots, according to ERS. Price premiums for the two vegetables at the wholesale level never went below 125% in the same period.[6] The ERS study goes on to say that:

> Laws of supply and demand, however, make it unlikely that price premiums contributing to higher profits and market growth can coexist over the long run: as long as higher profits exist, new suppliers will enter the market, and once market supply increases faster than demand, price premiums and the commensurate level of higher profits are likely to decline.... Many organic industry participants and observers believe that the price premiums ... need to decrease if organic foods are to penetrate much beyond the 2- to 3-percent level into the mainstream.[7]

The consumer studies that ERS reviewed did not permit any clear estimate of future demand for organic products. The studies showed that price, size, packaging, appearance, and concerns about health and nutrition, taste, food safety, and the environment all play varying roles in consumer decisions to buy organic food. Surveys on race, ethnicity, and income levels showed significant diversity.[8]

THE ORGANIC FOODS PRODUCTION ACT OF 1990

Congress passed the Organic Foods Production Act (OFPA) of 1990 (Title 21 of P.L. 101-624, the Food, Agriculture, Conservation, and Trade Act of 1990; the 1990 farm bill) with widespread support from organic industry groups, the National Association of State Departments of Agriculture, and other farm and consumer groups. The organic industry petitioned Congress to draft the act in the late 1980s, after it had been frustrated in its attempts to come to an internal consensus on production and certification standards. The industry maintained that federal standards would reduce consumer confusion over the many different state and private standards then in use, and would promote confidence in the integrity of organic products over the long term. Manufacturers of multi-ingredient organic food products stated that uniform standards would facilitate labeling. Others held that regulations would help the organic industry expand product lines and increase marketing opportunities. Industry analysts asserted that a consistent U.S. organic standard would facilitate access to a potentially lucrative international organic market.

The Organic Foods Production Act of 1990 authorized a National Organic Program to be administered by USDA's Agricultural Marketing Service (AMS). The act established a 15-member National Organic Standards Board (NOSB) to "assist in the development of standards for substances to be used in organic production" (referred to as the "National List") and to "provide recommendations to the Secretary regarding implementation."

Under the program, producers, processors and handlers who wish to market their products as organic are required to follow production practices as spelled out in detail in regulations (7 CFR 205). USDA accredits private and state certification agents, who visit producers, processors, and handlers to certify that their operations abide by the standards; they conduct annual reviews to verify continued compliance. It is illegal for anyone to use the word "organic" on a product if it does not meet the standards set in the law and regulations.[9] The presence of the "USDA Organic" seal on a product means it is 95% or more organic. Labels on products having 70% to 95% organic content can say "made with organic (specified ingredients or food groups)," but cannot carry the seal. Foreign organic producers and handlers wishing to export products to the United States may be certified by a USDA-accredited certification agent in their own country, if there is one; or USDA may accept certification by agents accredited by a foreign government; or, USDA may negotiate an equivalency agreement with another nation's organic program.[10]

The regulations under the OFPA are intended to set uniform *minimum* standards for organic production. States may adopt additional requirements after review and approval by USDA. Furthermore, private organic organizations are permitted to affix their own labels in addition to the USDA label, indicating that the product meets their standards as well as the national ones. The private label may indicate only that the organization's standards are in addition to (but not superior to) the national standards. AMS reviews certification agents for re-accreditation every five years. AMS enforces the regulations by revoking or suspending a producer's certification or an agent's accreditation if a satisfactory solution to a program violation cannot be found.

The NOP final rule became effective on February 21, 2001; the program itself became fully operational on October 21, 2002. In the first step toward implementation, USDA accredited private and state certification agents, who in turn began to certify organic producers and handlers according to the standards found in 7 CFR 205. After October 21, 2002, all products sold as organic had to be in compliance with the regulations and carry the "USDA Organic" seal.

Congressional Action on Related Legislation

Cost-Sharing Start-Up Costs. Although the OFPA requires the cost of the National Organic Program to be fully supported by user fees collected for USDA accreditation and certification services, Congress has appropriated funds on several occasions to help the program in its initial stages. The FY2001 USDA appropriations act (P.L. 106-387) contained $639,000 to cover accreditation costs. In FY2002, under the Agricultural Management

Assistance Program authorized by the Federal Crop Insurance Act (P.L. 106-224), Congress made $1 million available to state agriculture departments in 15 designated states to help defray the costs of certification for small-scale producers and processors.

The 2002 farm bill (P.L. 107-171, the Farm Security and Rural Investment Act), which was enacted in May 2002, also provided additional funds to support program start-up. Title X of the farm act gave USDA authority to continue to defray the costs of producers and handlers seeking organic certification through FY2007, and authorized a one-time, mandatory transfer of $5 million from the Commodity Credit Corporation (CCC) to establish a national organic certification cost-share program under the NOP. Federal funds may not cover more than 75% ($500 maximum) of a producer's or handler's costs for becoming certified. The transfer occurred in FY2002 and remains available until fully expended, which is expected to occur in 2006.[11]

Value-Added Producer Grants. The rural development title of the 2002 farm bill established a competitive grant program to promote research on the development and marketing of value-added agricultural products, to be funded through an annual transfer of $40 million from the CCC to USDA through FY2006. Projects on organically produced commodities are eligible for these grants. Congress authorized $40 million in mandatory CCC funds to be made available each year through FY2007, but the actual funding levels so far have been volatile. The program received $50 million in mandatory funds in FY2003, consisting of $40 million in unspent funds from the first CCC transfer in FY2002 and $10 million of the $40 million CCC transfer for FY2003. In FY2004 and FY2005, appropriators prohibited the spending of the mandatory funds and instead appropriated about $15 million annually.

Research. The research title of the 2002 farm bill renewed expiring authority for a competitive grant program to support research and extension activities on organic production, processing, and international marketing. The conference report also added language calling for an emphasis on classical and advanced research on genetics to improve organic crops; research to identify the marketing and policy constraints on the organic industry; and expanded on-farm research. The act authorized $3 million to be transferred annually from the U.S. Treasury to USDA, beginning in FY2003, to support this research.[12] Other provisions in the research title: (1) require USDA's Economic Research Service (ERS) to gather and maintain segregated data on the production and marketing of organic agriculture; and (2) require ERS and the National Agricultural Library to make it easier for U.S. organic producers, researchers and extension professionals to obtain the results of organic research conducted in foreign countries.[13]

Concerning the availability of production and marketing data on organic agriculture, an official of the Organic Trade Association (OTA) states that many gaps exist, and that the Economic Research Service still is relying extensively on external data sources for much of its information. As of September 2005, a comprehensive USDA survey of the entire organic sector has not been conducted. USDA's Agricultural Marketing Service has not yet begun to provide market news on the organic sector, which would permit up-to-date price discovery. Separate export and import data for organic products are not being collected at the borders. These deficiencies hamper business planning and expansion, complicate crop insurance premium-setting and loss payments, among other issues, according to OTA.

Exemption from Check-Off Programs. The 2002 farm bill contained a provision concerning issues related to the organic industry and USDA commodity research and promotion programs. These are programs that support generic advertising to promote an agricultural product (e.g., the milk mustache ads). They are funded by assessments that producers, processors, other handlers, and frequently importers, are required to deduct from revenue at the time of sale (thus they usually are called "check-off" programs). Congress has passed many laws authorizing national check-off programs for various farm commodities; there currently are 15 in operation. (For more information on check-off programs, see CRS Report 95-353, *Federal Farm Promotion ("Check-off") Programs*, by Geoffrey S. Becker.)

The Senate version of the 2002 farm bill (S. 1731) would have established a check-off program for organic commodities, but the provision was not adopted in conference. The enacted bill instead included language exempting producers and handlers who have a certified 100% organic operation from having to pay assessments under any existing commodity check-off programs in which they currently participate. For example, a producer who grows peaches on an entirely organic farm that has been certified under the NOP would be eligible to be exempted from paying an assessment under the marketing order for peaches and nectarines, when he sells his crop to a wholesaler or other handler. The proposed rule on the exemption for producers appeared in the *Federal Register* in April 2004 (69 FR 22690). Final rules for both producers and handlers of organic commodities were published in January 2005 and became effective in February 2005 (70 FR 2744 and 2763, respectively).

Livestock Feed Regulations Changed, Then Reversed. The conference report accompanying the Consolidated Appropriations Resolution for FY2003 (P.L. 108-7/H.J.Res. 2; H.Rept. 108-10; February 21, 2003), contained a provision that made a controversial change in the National Organic Program. Section 771 of the law effectively deleted the NOP standard that requires livestock and dairy producers to provide 100% organically produced feed for their animals in order to qualify for organic certification. It further provided that, if USDA surveys found that in some geographic areas organic feedstuffs cost more than twice what conventionally produced feed cost, then livestock and dairy producers in that area could be certified as organic without having to meet the NOP standard.

Reaction to this provision from several Members of Congress, the organic industry, and a few large food processors who had already committed resources to complying with the original rule, was swift and negative. Senator Leahy, who championed OFPA and its inclusion in the Food, Agriculture, Conservation, and Trade Act of 1990, circulated a "Dear Colleague" letter on February 12, 2003, urging Senators to remove Section 771 from H.J.Res 2, but without success. On February 26, 2003, Senator Leahy introduced S. 457, the Organic Restoration Act of 2003, and on February 27, Representative Farr introduced H.R. 955, a bill to repeal Section 771. The conference report accompanying emergency wartime supplemental appropriations for FY2003, which the President enacted on April 16, 2003, contained the language from H.R. 955 repealing the earlier action taken in P.L. 108-7 (Section 2104 of H.Rept. 108-76; P.L. 108-11).

USDA released the report on organic feed availability in June 2003. It states that "ample acreage is available to provide more than enough feed grains to meet the needs of organic livestock and broiler producers ... [and] with limited exceptions, prices for organic poultry rations are not more than twice the prices of convention poultry feed rations."[14]

USDA Regulatory Activity

Interpretive Issue Statements. In April 2004, the National Organic Program headquarters in USDA released "issue statements," i.e., guidelines for interpretation of program regulations in four areas. These were livestock feed, livestock health, inert ingredients in approved organic pesticides, and the inclusion of non-food items like lotions and cosmetics within the scope of the NOP. AMS issued the statements administratively (rather than through the *Federal Register*). Many program participants were immediately critical of the Department for not giving the issue statements to the National Organic Standards Board for advance review, since the critics viewed them as revisions of the regulations, not simply as clarifications. USDA rescinded the issue statements in May 2004, and asked the Board to provide feedback on them, which it did in October 2004.

Subsequently, on August 23, 2005, the Department issued a memorandum reversing its original position on the "scope of program" issue.[15] The memorandum states that, if all standards for organic agricultural ingredients are met, then the organic seal can be used on personal care products, supplements, and pet foods. This reversal was prompted by a lawsuit that was brought against USDA after it released the issue statement in April 2004 saying that nonfood organic products could not carry the "USDA Organic" seal.

The sequence of events is at issue here, more so than the issue statements themselves. The NOP is a new program covering a sector of U.S. agriculture that wants to be perceived as decidedly different from conventional agriculture, which has been covered by USDA marketing and regulatory programs for almost a century. Unlike other marketing programs, the NOP regulations are much more prescriptive regarding production and manufacturing practices, and NOP officials understandably might still be learning how to deal with these new factors in a regulatory setting. Thus it might be expected that NOP officials could on occasion take administrative action without thinking it necessary to consult the National Organic Standards Board, or might differ with the Board on the form and substance of a proposed regulation. Situations of this sort were largely responsible for the Department's taking 11 years to publish an NOP final rule after OFPA enactment in 1990. Similarly, the National Organic Standards Board and industry stakeholders might perceive program officials' failure to consult the Board on how certain regulations should be interpreted as dismissive of their input.

An audit report on the NOP released in July 2005 by USDA's Office of Inspector General (OIG) directly addresses these concerns.[16] The report found that "AMS has not established protocols for working with the National Organic Standards Board or resolving conflicts with them," and recommends that the NOP "establish procedures for receiving, reviewing, and implementing recommendations from the Board." The report states that "AMS also needs to improve management controls for administering the NOP," and recommends that AMS "resolve and implement internal operating procedures for such things as the resolution of complaints to govern program operations." The OIG made 10 recommendations in all. AMS officials agreed with them and promised to implement solutions by the end of FY2005, or publish draft procedures for public comment by the end of December 2005.

Peer Review of Accreditation. Section 2117of OFPA (7 CFR 205.509) requires that a three-person panel periodically review NOP's accreditation procedures, site evaluation reports, and decisions, in order to determine their adherence to the regulations. This part of

OFPA has not been implemented to date because the statutory language is vague concerning some basic issues; e.g., how to find a sufficient number of experts in organic accreditation who themselves are not USDA-accredited certification agents.

In lieu of the three-person panels, USDA contracted in 2003 with the American National Standards Institute (ANSI), a private, non-profit organization that administers the U.S. voluntary standardization and conformity assessment system (part of a world-wide system commonly referred to as ISO 9000; ISO stands for International Standards Organization).[17] ANSI conducted a peer evaluation of the NOP accreditation system and delivered its report in December 2004.[18] It found a number of discrepancies between NOP's accreditation policies and procedures and those outlined in ISO 9000. Furthermore, ANSI reported that it had difficulty assessing the discrepancies because USDA has not fully documented the quality system it uses to assess applicants for accreditation.

The question of how to implement the original statutory language remains to be addressed by NOP officials and the National Organic Standards Board.

Controversy over "Access to Pasture"

In January 2005 a newspaper article about a Colorado organic dairy operation raised a major controversy within the industry and among some public interest groups.[19] The article focused on a 5,300-cow organic farm where the animals were fed almost exclusively on grain and allowed outdoors mostly for exercise.

The NOP regulation at the core of the dispute is 7 CFR 205.239(a)(1-2): "The producer of an organic livestock operation must establish and maintain livestock living conditions which accommodate the health and natural behavior of animals, including (1)Access to the outdoors, shade, shelter, exercise areas, fresh air, and direct sunlight suitable to the species, its stage of production, the climate, and the environment; (2) Access to pasture for ruminants[.]"

The news article cites organic dairy producers who argue that the regulation means that organic cattle must get some of their nutrition, as well as fresh air and sunshine, from grazing on pasture. The Colorado operator maintains that his animals have outdoor access, but that in an arid state like Colorado, providing sufficient pasturage for all his cows to graze would be an insurmountable requirement.

At a February 2005 meeting of the National Organic Standards Board, members discussed and recommended new language for 7 CFR 205.239, which it forwarded to National Organic Program officials for approval.[20] The emphasis in the revised language was on allowing cows at the appropriate "stage of life" to graze pasture "during the pasture's normal growing season." At the August 2005 Board meeting, the NOP staff rejected the recommendation saying it lacked a "clear and concise regulatory objective," and asked the Board to rework it.[21] It could be a year before a proposed rule reflecting a revised NOSB recommendation could be published for comment in the *Federal Register*, a delay that some speakers at the Board meeting argued was unacceptable. A guidance document that the Board posted on its website for comment in March 2005 is serving unofficially as an interim interpretation of the current "access to pasture" requirement until a proposed revised standard can be published.[22]

The issue is sensitive because it involves the question of farm size. USDA considers the NOP to be a marketing program that is size-neutral. On the other hand, many in the industry hold that organic farming *practices* are an integral part of the meaning of the term "organic," particularly with respect to standards for the treatment and feeding of livestock. Some observers argue that if the pasturage in certain parts of the nation can support only small dairy or beef cattle herds, or none at all, then such farms should be small or nonexistent in those areas. Others are concerned lest the regulations become so prescriptive that they deprive producers and processors of the opportunity to benefit from the expanding market for organic products.

Court Ruling Controversy

In June 2005, a First Circuit Court sided with an organic farmer who had filed suit against USDA claiming that several provisions of the final NOP rule were more lenient than the underlying statutory language allowed.[23] Specifically, the court determined that certain "natural" substances not commercially available in organic form had to be individually reviewed to determine their status for the National List of Approved and Prohibited Substances before they could be used in organic-labeled products. Second, the court determined that synthetic substances that heretofore had been approved in the regulations and used in commerce could not be used in the processing or handling of organic-labeled products. And third, it ruled that the USDA regulation on converting a dairy herd from conventional to organic production did not in any way reflect the statutory language and was thus invalid. The District Court in this case set a one-year time frame for USDA to develop new regulations and allowed another year beyond that (until June 2007) for the industry to come into compliance.

Stark differences of opinion within the organic industry on the impact of the court judgement became immediately apparent. Representatives of consumer groups and some food retailing groups welcomed the decision. They maintained that consumers consider the organic label to mean the absence of synthetic ingredients and that the new, stricter regulations would reconfirm consumer confidence in the OFPA as enacted. These groups have expressed strong opposition to resolving the court decision issue by amending the OFPA.

Conversely, a large number of organic food manufacturers were apprehensive that the ultimate outcome of the court decision would be to force the discontinuation of hundreds of existing, organic-labeled food and beverage products, and/or end manufacturers' ability to use the "USDA Organic" seal, which commands premium prices in the marketplace. These stakeholders supported the efforts of the Organic Trade Association to have Congress effect a legislative solution to the issues raised by the court decision.

In late September 2005, during floor debate on its version of the FY2006 USDA spending bill (H.R. 2744), the Senate adopted an amendment requiring the Secretary to conduct a survey to determine the impact of the decision and of various approaches to addressing it, including amending the OFPA to effectively undo the court's action.

In late October 2005, conferees on the FY2006 agriculture appropriations bill adopted a second provision in committee. The overall effect of this provision is to change the statutory language to make moot the latter two of the three court holdings. Specifically, the provision amends the OFPA: (1) to allow "natural" but non-organic products to be used as long as they are subject to the National List evaluation process; and additionally, to permit USDA to develop an expedited procedure for approving the addition of such products to the National List for a limited time period; (2) to remove the original language that generally prohibits the inclusion of synthetic substances on the National List (although there are other requirements that must be met); and (3) to provide a statutory basis for a new regulation on converting dairy herds to organic that is related to on-farm production of organic feed and forage.

CONCLUSION

It will be necessary for USDA to write new regulations to address these three areas. Barring further legislative changes, the new rules will be based on revised statutory language. The history of the NOP, however, indicates the difficulty in predicting the outcome of the notice and comment process. This area of public policy may remain contentious, depending upon the ability of the organic community's diverse stakeholders to develop a workable consensus among themselves on interpreting the OFPA. Additional debate on the statute could happen when Congress begins consideration of the next farm bill, expected in 2007. Hearings on various parts of a farm bill could begin in 2006.

NATIONAL ORGANIC PROGRAM FAQS

The questions and answers below are categorized by subject matter and will be updated on a monthly basis.

General Topics

Q: Where can I get a copy of the National Organic Program regulations?
A: You can obtain a copy of the NOP regulations by contacting the NOP through this website.

Definitions

There are no Questions and Answers for Definitions at this time

Applicability

Posted 6/26/03

Q: Can salt and water be certified organic under the National Organic Program?
A: No. Salt, in all of its forms, and water can not be certified as organic under the National Organic Program. Only agricultural products produced and handled in accordance with

the National Organic Program may be certified as organically produced. Salts are minerals not agricultural products. Water is a chemical substance used in the production of agricultural products but is not itself an agricultural product. Section 2106(c) (7 U.S.C. 6505 (c)) of the Organic Foods Production Act of 1990, as amended, and section 205.302 of the National Organic Program require the exclusion of salt and water when calculating the percentage of organic ingredients in a multi-ingredient product. Further, National Organic Program sections 205.303(b)(1) and 205.304(b)(1) provide that "Water or salt included as ingredients cannot be identified as organic."

Q: When must organic producers and handlers be certified to continue to market their product as organic?

A: Beginning on October 21, 2002, producers and handlers must be certified by a USDA-accredited certifying agent to sell, label, or represent their products as "100 percent organic," "organic," or "made with organic (specified ingredients or food group(s)." Please see section 205.101 for exemptions from certification.

Q: I am a small farmer. Will I have to be certified?

A: It depends. If your gross agricultural income from organic sales total $5,000 or less annually, you are exempt from certification (see section 205.101(a)(1) of the NOP regulations). Exempt operations must comply with the applicable requirements of subpart C and the labeling requirements at section 205.310 of the NOP regulations.

Q: I am a small processor. Do I have to be certified?

A: It depends. If your gross agricultural income from organic sales total $5,000 or less annually, you are exempt from certification (see section 205.101(a)(1) of the NOP regulations). Exempt operations must comply with the applicable requirements of subpart C and the labeling requirements at section 205.310 of the NOP regulations. For other possible handler exemptions see section 205.101 of the NOP regulations.

Q: I sell less than $5,000 worth of organic product. Can I be certified?

A: Yes. Any qualified agricultural production or handling operation may be certified as an organic production and handling operation. This includes all qualified production and handling operations eligible for exemption or exclusion under section 205.101 of the NOP regulations.

Q: Can non-certified companies use the word "organic?"

A: Producers and handlers that qualify for exemption or exclusion from certification may use the term "organic" in compliance with the labeling requirements specific to their exemption or exclusion (see section 205.101 of NOP regulations).

Q: What type of records would be acceptable to National Organic Program as proof of exempt organic operations showing compliance with the labeling requirements, production requirements, handling requirements, record keeping, audit trail, and non use of prohibited substances.

A: Examples of records to be kept by exempt operations are listed in the preamble of Subpart B, Applicability. Exempt operations must comply with the applicable organic production and handling requirements of subpart C of the national standards and meet the labeling requirements of section 205.310. Therefore, each exempt operation should maintain records which demonstrate compliance with the Organic Foods Production Act (OFPA) of 1990 and the national standards. It is the decision of the exempt operation as to which records it needs to demonstrate compliance.

Q: Are exempt operations subject to National Organic Program (NOP) audit? If so, what fee would be charged to those operations?
A: Yes. Exempt operations that produce or handle agricultural products to be sold, labeled, or represented as "100 percent organic" or "organic" are subject to NOP compliance audits. Costs associated with compliance audits would be borne by the NOP.

Q: How was the gross receipt threshold of $5,000 established? What steps would be necessary to raise the limit?
A: The $5,000 producer or handler exemption in §205.101(a) of the national standards was mandated by OFPA (7 U.S.C. 6505(d)). Actions necessary to raise the limit of the $5,000 producer/handler exemption would include congressional amendment of OFPA §6505(d).

Q: Please explain who may use the term organic and how the term is to be used.
A: Any production or handling operation certified according to the provisions of subpart E, Certification, may use the term "organic" (§205.100). Production or handling operations that are exempted or excluded under §205.101 may use the term "organic" according to the regulations specified in §205.310, Labeling; provided, they comply with the production and handling requirements of subpart C of the national standards.

Q: What are the penalties for misuse of the term "organic"?
A: Any operation that knowingly sells or labels a product as "organic", except in accordance with the Act (OFPA) and the national standards, may be subject to a civil penalty of not more than $10,000 per violation and the provisions of 18 U.S.C 1001.

Q: Who will be responsible for the enforcement of the National Organic Program and how will a typical prosecution proceed?
A: USDA, accredited certifying agents, and where applicable, approved State Organic Programs will be responsible for enforcement of the national regulations. Compliance procedures for certified organic operations, accredited certifying agents, and State Organic Programs are specified in sections 205.660 through 205.668 of the national standards.

Q: When a retail establishment markets products supplied from a certified producer, in the event the producer is found non-compliant, will the retailer be subject to any legal recourse from the NOP? Do the same rules apply (to the retailer) for products produced by exempt and non-exempt producers?
A: If a provider of product to a retail food establishment is found to be in violation of the national organic standards and the retail food establishment is not a party to that violation, there will be no action by the National Organic Program (NOP) against the

retail food establishment. This holds true for certified producers and handlers as well as those claiming exemption under section 205.101.

Any person, including a retail food establishment, who knowingly sells or labels a product as organic, except in accordance with the OFPA and the national organic standards, shall be subject to a civil penalty of not more than $10,000 per violation.

Products that have entered the channels of commerce before the certified operation's suspension or revocation will not result in a product recall, unless the non-compliance involves a food safety issue. For further information see page 80627 of the national organic standards.

Organic Production and Handling Requirements

Posted 11/29/05

Q: With Avian Influenza becoming an increasingly serious issue in the poultry industry, what can organic poultry producers and their certifying agents do to help recognize and prevent the spread of this disease?

A: USDA's Animal and Plant Health Inspection Service has provided guidance to poultry producers, including free range and organic producers, regarding biosecurity considerations for their operations – this guidance is found at http://www.aphis.usda.gov/vs/birdbiosecurity/

Integral to the National Organic Program is a management plan for each organic production facility – among other things, this plan must address preventative livestock health care practices, including sanitation practices that will be followed to minimize the occurrence and spread of diseases and parasites.

USDA has brought the APHIS biosecurity guidance materials to the attention of its accredited organic certifying agents for their use in working with organic poultry producers.

If it is determined by a producer and certifier that for the health, safety, and welfare of poultry there is need for temporary confinement of birds, provision exists within the National Organic Program for such confinement without a loss of organic certification.

If Federal or State animal health authorities determine that more prescriptive actions are required, organic producers could be required to adhere to those actions.

Posted 1/14/05

Q: If a producer adheres to all aspects of the National Organic Program (NOP), including never utilizing biotech-derived seeds, but a certifying agent tests and detects the presence of biotech-derived material in the crop, is that crop's status determined to be no longer "certified organic?" And, if so, what in the NOP supports this conclusion?

A: It is particularly important to remember that organic standards are process based. Certifying agents attest to the ability of organic operations to follow a set of production standards and practices that meet the requirements of the Act and the regulations. This regulation prohibits the use of excluded methods in organic operations (§205.2—Terms defined, and §205.105—Allowed and prohibited substances, methods, and ingredients in organic production and handling). The presence of a detectable residue of a product of excluded methods alone does not necessarily constitute a violation of this regulation. As long as an organic operation has not used excluded methods and takes reasonable steps to avoid contact with the products of excluded methods as detailed in their approved organic system plan, the unintentional presence of the products of excluded methods will not affect the status of the organic operation. As to the status of the commodity, USDA's position is that this is left to the buyer and seller to resolve in the marketplace through their contractual relationship. (See page 80556 of the preamble, "Applicability—Clarifications; (1) "Genetic drift").

Posted 1/14/05

Q: Do insufficient buffers or barriers that result in unintended contact with a product of genetic modification threaten the farm's certification or use of the field for the production of organic crops? If an organic producer or handler is found to have not implemented measures necessary to prevent commingling of organic and non-organic products, would that threaten the certification of the producer or handler?

A: In order to become a certified organic operation, a producer must submit an Organic System Plan (plan) to a USDA-accredited certifying agent for approval. That plan must include, among other things, evidence that sufficient buffer zones have been incorporated into the operation to ensure the integrity of the organic crop operation. The certifying agent must not approve a plan that does not provide evidence of sound measures taken to ensure the integrity of the organic crop operation, including buffer zones and other steps to prevent commingling with unapproved non-organic materials or conventional crops. If a producer does not adhere to such preventive measures, the certifying agent is expected to denote such failure as a noncompliance and take appropriate measures toward correction by the producer. Inadequate buffer zones should not be approved in the first place and failure to comply with approved buffer zones constitutes a noncompliance with the approved organic system plan. (See the preamble, page 80558, on Subpart C—General Requirements, which describe what must be contained in an organic system plan, and §205.2 under terms defined – Buffer zone.)

However, even when all precautions have been taken, and an approved buffer zone fails to provide the protection that both the operator and the certifying agent reasonably expected, certifying agents must not "retroactively" punish the producer by an enforcement action or "de-certify" the organic crop. The appropriate action to take in this case is to re-evaluate the buffer zone and other preventive measures in the plan to ensure improved integrity and performance in the future. As to the status of the commodity, USDA's position is that this is left to the buyer and seller to resolve in the marketplace through their contractual relationship. (See page 80556 of the preamble, "Applicability—Clarifications; (1) "Genetic drift").

Posted 1/14/05

Q: Has any certified organic operation that refrains from intentional use of biotech seeds ever lost certification for the inadvertent presence of biotech material in its crop?

A: No accredited certifying agent has reported to us that certification has been lost due to adventitious presence of biotech material. In one instance, a producer admitted to deliberately planting GM seed and representing the crop as organic, for which we took enforcement action and he is no longer certified.

Posted 1/14/05

Q: Are food labels stating "GM, GE, or GMO-free" part of the National Organic Standards?

A: They are not. Truthful labeling is embodied in the National Organic Standards, as supported by USDA's Food Safety and Inspection Service (FSIS), the Food and Drug Administration (FDA), and the Federal Trade Commission (FTC) – the agencies with respective jurisdiction over truthful labeling laws. In the preamble of the National Organic Program final regulations, we stated that organic is not synonymous with "GM-free," when we said: "These phrases may…be used as additional, eco-labels, provided they are truthful statements…[but] they are not permitted as replacements for the term 'organic.'" (See page 80586 of the preamble, under "Labeling—Changes Requested But Not Made: (7) Use of Other Terms as Synonymous for "organic").

Posted 1/14/05

Q: Is there a working definition of the word "contamination" within the NOP? Are all products of genetic modification considered "prohibited substances" as defined in the final regulations? And, what actions are authorized or required when organic crops or products are found to contain unintended or inadvertent genetically modified hybrids or other genetically modified substances?

A: There is no definition in the final regulations of the National Organic Standards for the word "contamination." By our count, "contamination" is mentioned nearly 50 times in the regulations.

All genetically modified practices or products are indeed considered prohibited, as cited in 205.105, the paragraph that describes "excluded methods." Please refer back to the above issue when considering the adventitious presence of a genetically modified or genetically-engineered substance. Such adventitious presence does not affect the status of the certified operation and does not necessarily result in loss of organic status for the organic product, provided it was produced in adherence with all of the organic requirements under 7 CFR 205.

Again, the action regarding the final product's status in this case is left to the determination by the buyer and seller of the product. Contamination by a prohibited substance, when mandated by a government body, however, would result in loss of organic status for the product, even when all other regulations had been followed. In the case of an emergency spray program, for example, if the spray is a prohibited substance but is mandated by a State or Federal program, the crop's organic status is lost and that crop must be diverted for sale in the conventional market. Neither the operation nor the land's organic status is altered by an emergency spray program, however. (See §205.672 Emergency pest or disease treatment.)

Posted 3/6/03

Q: Do I have to subtract the use of steam condensate when calculating the percentage of organically produced ingredients? For example; blanching fruits or vegetables, cooking organic ingredients with steam, or other uses for processing with steam.

A: No.

Posted 1/6/03

Q: One of my clients extracts the oil from organic mint leaves. Although he suspected that a recent crop of mint leaves may have come in contact with a prohibited pesticide, residue testing showed that the unprocessed mint leaves had a residue level which does not exceed 5 percent of the EPA tolerance for that particular pesticide.

Is it possible that although the pesticide residue level was okay in the raw mint leaves, the mint oil could contain a residue level above 5 percent of the EPA tolerance level, meaning the mint oil couldn't be sold as organic?

A: It is possible. The Environmental Protection Agency has established pesticide residue tolerance levels for many processed products (mint oil is a processed product), and these tolerance levels are different from the levels established for the raw product from which the processed product is made.

In this case, there is a separate tolerance level established for mint oil, and the mint oil cannot contain pesticide residue greater than 5 percent of the EPA tolerance established for mint oil for that particular pesticide and still be sold as organic.

When there is no EPA tolerance level established for a processed product, the tolerance level established for the raw product from which the processed product is made will be used to determine if the pesticide residue level is acceptable for that processed product.

EPA pesticide residue tolerance levels can be found at www.access.gpo. gov/nara/cfr/waisidx_01/40cfr180_01.html.

Posted 12/9/02

Q: I operate a certified organic farm, and I wish to purchase compost from a commercial compost operation. Does this compost have to be certified organic?

A: No. Although a commercial compost operation may become USDA certified, there is no requirement for it to do so, and there is no requirement for you to use certified organic compost.

However, any compost you use must meet all the requirements of the NOP regulations, section 205.203. The burden of proving that a compost used on a certified organic operation meets all the requirements of section 205.203 rests with the certified organic operation that uses the compost. That operation must show in its organic systems plan, and be able to prove to its certifying agent, that the compost it uses meets the NOP requirements.

Posted 10/18/02

Q: Must seeds for cover crops be organic?

A: Yes, unless conditions meet those outlined in Section 205.204, provided below:

§ 205.204 Seeds and planting stock practice standard.

(a) The producer must use organically grown seeds, annual seedlings, and planting stock: Except, That,

(1) Nonorganically produced, untreated seeds and planting stock may be used to produce an organic crop when an equivalent organically produced variety is not commercially available, Except, That, organically produced seed must be used for the production of edible sprouts;

(2) Nonorganically produced seeds and planting stock that have been treated with a substance included on the National List of synthetic substances allowed for use in organic crop production may be used to produce an organic crop when an equivalent organically produced or untreated variety is not commercially available;

(3) Nonorganically produced annual seedlings may be used to produce an organic crop when a temporary variance has been granted in accordance with § 205.290(a)(2);

(4) Nonorganically produced planting stock to be used to produce a perennial crop may be sold, labeled, or represented as organically produced only after the planting stock has been maintained under a system of organic management for a period of no less than 1 year; and

(5) Seeds, annual seedlings, and planting stock treated with prohibited substances may be used to produce an organic crop when the application of the materials is a requirement of Federal or State phytosanitary regulations.

Posted 10/16/02

Q: Can wood treated with fungicides or other prohibited substances be used for fence posts, trellis systems, etc., in organic production?

A: As provided in Section 205.206(f), the producer must not use lumber treated with arsenate or other prohibited materials for new installations or replacement purposes in contact with soil or livestock. The Preamble (Crop Production--Changes Based on Comments (7)) elaborates on 205.206(f) as follows:

> "This provision prohibits the use of lumber treated with arsenate or other prohibited materials for new installations or replacement purposes in contact with an organic production site. We included this modification to clarify that the prohibition applies to lumber used in direct contact with organically produced and handled crops and livestock and does not include uses, such as lumber for fence posts or building materials, that are isolated from production. The prohibition applies to lumber used in crop production, such as the frames of a planting bed, and for raising livestock, such as the boards used to build a farrowing house."

Q: Do fields have to meet any size requirements in order to produce crops certified as organic?

A: No. There are no field-size requirements relative to whether or not an operation can be certified organic.

Q: Is the feed that is fed to organic livestock and poultry "vegetarian" feed?

A: Not necessarily. Although as stated in the National Organic Standards subpart C, section 205.237(b)(5), "The producer of an organic operation must not feed mammalian or poultry slaughter by-products to mammals or poultry." there is no restriction against organic livestock feed containing appropriate fish products.

Q: I know organic agriculture prohibits the use of GMOs, but do the guidelines allow for a small amount of contamination?

A: As GMO contamination of organic crops relates to genetic drift, the Preamble to the National Organic Program regulations, Applicability, Clarifications (1) Genetic Drift, states:

> "This regulation prohibits the use of excluded methods [which include GMOs] in organic operations. The presence of a detectable residue of a product of excluded methods alone does not necessarily constitute a violation of this regulation. As long as an organic operation has not used excluded methods and takes reasonable steps to avoid contact with the products of excluded methods as detailed in their approved organic system plan, the unintentional presence of the products of excluded methods should not affect the status of an organic product or operation."

However, if a certifying agent has reason to suspect that an organic product has come into contact with prohibited substances or been produced using excluded methods, the certifying agent can call for testing, which under certain conditions could result in that product no longer being considered "organic."

For a complete discussion of this issue, you must read the National Organic Standards, subpart G, Administrative, sections 205.670 – 205.671.

Q: Can reclaimed water be used for irrigation on organic farms?

A: Generally, the National Organic Standards place no further restrictions on reclaimed water used for irrigation beyond those imposed by State Departments of Natural Resources. However, to fully answer this question, we would need to know how and from what source the water is being reclaimed.

Q: We are a meat processing plant interested in processing all natural pork and beef. What kind of general information can you give us?

A: There are now specific USDA standards that must be met by all producers and processors who wish to label their agricultural products as organic. (Please note that "natural" is not synonymous with "organic.") These producers and handlers must be certified by USDA-accredited certifying agents by October 21, 2002. The organic standards for producing and processing livestock can be found in the National Organic Standards, subpart C, Sections 205.236 – 205.272.

A list of USDA accredited certifying agents can be found on the NOP website at www.ams.usda.gov/nop.

Q: Must manure used to fertilize organic crops come from an organic source, that is from an animal that has been raised in accordance with the organic standards?

A: No. There are no restrictions as to the source of manure. For greater detail on soil fertility and crop nutrient management, please read the National Organic Program regulations, subpart C, section 205.203. This topic is further discussed in the Preamble to the National Organic Program regulations, under the section headed Crop Production, Changes Requested but Not Made, (2) No Prohibition on Manure from Nonorganic Operations.

Q: Are there any known health hazards associated with using animal waste as fertilizer for organic crops?

A: Properly managed animal waste should not cause any known health hazards. Animal waste used to fertilize organic crops must be managed in accordance with the National Organic Program regulations' soil fertility and crop nutrient management practice standard, as stated in the National Organic Program regulations, subpart C, section 205.203(c), which states, "The producer must manage plant and animal materials to maintain or improve soil organic matter content in a manner that does not contribute to contamination of crops, soil, or water by plant nutrients, pathogenic organisms, heavy metals, or residues of prohibited substances." See this section for a description of plant and animal materials, including manure and compost.

Q: Is it safe to say that using organic fertilizers and other organic farming practices is better for the soil and less of a threat to ground and surface water than commercial methods?

A: USDA's National Organic Program is a marketing program and makes no claims that organic farming is "better" in any respect than conventional farming.

Q: Has the National Organic Program issued any guidance or sample plans regarding the development of production and handling system plans?

A: The National Organic Program has issued no specific guidance or samples relating to the development of organic production and handling system plans. However the NOP is helping to fund the development of crop and livestock checksheets designed for organic farmers to use in assessing their compliance with the National Organic Standards. The checksheets are being developed by the National Center for Appropriate Technology (NCAT). For more information, visit NCAT's website at www.ncat.org.

Q: Is the use of ionizing radiation allowed at any stage of production or handling of agricultural products labeled organic?

A: No. Irradiation in the production and handling of organic food is prohibited by the National Organic Standards, subpart B, Applicability, section 205.105, Allowed and prohibited substances, methods, and ingredients in organic production and handling, which states, "To be sold or labeled as '100 percent organic,' 'organic,' or 'made with organic (specified ingredients or food group(s)),' the product must be produced and handled without the use of: (f) Ionizing radiation, as described in Food and Drug Administration regulation, 21 CFR 179.26 ...

Q: Under the National Organic Standards, can the certifying agent approve ion exchange for the processing of organic foods?

A: The National Organic Standards do not prohibit the use of ion exchange in organic food processing as a technology. However, the National Organic Standards do prohibit the use of synthetic substances used in or on processed products unless included in National List section 205.605 as an allowed substance. Accordingly, ion exchange may not be used in the processing of organic foods unless the synthetic substances used in the ion exchange process are listed as approved substances in National List section 205.605.

Q: Can I use homeopathic treatments to treat my livestock?

A: The National Organic Standards do not prohibit use of homeopathic treatments in the production of organic livestock. The producer must make certain that such treatment(s) does not contain a prohibited substance. (Section 205.105(a))

Q: Can I remove organic calves from my farm, raise them conventionally for a year, bring them back to the farm, and then manage them organically for one year prior to the production of organic milk and milk products?

A: No. Livestock or edible livestock products that are removed from an organic operation and subsequently managed on a nonorganic operation may not be sold, labeled, or represented as organically produced. (Section 205.236(b)(1))

Q: Can I sell an organic dairy animal as slaughter stock?

A: Dairy animals that have been under continuous organic management since the last third of gestation may be sold, labeled, or represented as organic slaughter stock. Conversely, dairy animals that have not been under continuous organic management since the last third of gestation may not be sold, labeled, or represented as organic slaughter stock. (Section 205.236(b)(2))

Q: Do all dairy animals have to be organic from the last third of gestation to produce organic milk or milk products?

A: No. Milk or milk products must be from animals that have been under continuous organic management beginning no later than 1 year prior to the production of the milk or milk products that are to be sold, labeled, or represented as organic. Except that, in the case of the conversion of an entire distinct herd, the animals must receive a minimum of 80 percent feed that is either organic or raised from land included in the organic system plan and managed in compliance with organic crop requirements during the first 9 months of the 1-year conversion period. (Section 205.236(a)(2))

Q: Does Breeder Seed need to be raised organically in order for its progeny; i.e., Foundation, Registered, and/or Certified1 Seed, to be sold, labeled, or represented as organic seed?

A: No. The seed of any generation planted with conventional, untreated seed and produced under organic conditions can be certified as organic. In other words, untreated seed of Foundation, Registered, or Certified generations may be sown in an organic seed field/cage/greenhouse, and the progeny will qualify as organic seed. Many varieties are only legal for sale as a class of certified seed according to the Plant Variety Protection Act (PVP), Title V. Check with your official state seed certification agency for the specific varieties protected by PVP, Title V. Additionally, it is not necessary to use only genetically1 certified 2 untreated seed to produce organic seed. Any generation of "common," untreated seed may also be used.

1. The use of the term "genetically" in this sentence in no way implies or authorizes the use of genetically modified organisms.
2. Certified – As defined by the Federal Seed Act Regulations (7 C.F.R. 201(2) (ee)).

Q: Does organic seed always need to be used to produce an organic crop?

A: For edible sprouts, yes; for all other organic crops, no. The National Organic Standards (7 C.F.R. 205.204 (a) (1)) provides that nonorganically produced, untreated seeds may be used to produce an organic crop when an equivalent organically produced variety is not commercially available. The National Organic Standards (7 C.F.R. 205.204 (a) (2)) also provides that nonorganically produced, treated seeds may be used to produce an organic crop when an equivalent organically produced or untreated variety is not commercially available. The seed treatment, however, must be with a substance included on the National List of synthetic substances allowed for use in organic crop production. The only time that a seed treated with a prohibited substance may be used to produce an organic crop is when use of the prohibited substance is a requirement of Federal or State phytosanitary regulations.

Q: How does the National Organic Program (NOP) interpret "equivalent variety" in 7 C.F.R. 205.504(a) (1-2)?

A: An equivalent variety means a variety exhibiting the same "type" (such as head lettuce types, leaf lettuce types, etc.) and similar agronomic characteristics such as insect and disease resistance when compared to the original varietal choice.

"Type" is defined by the Federal Seed Act of 1939 (7 U.S.C. 1551-1661.) as either (A) a group of varieties so nearly similar that the individual varieties cannot be clearly differentiated except under special conditions, or (B) when used with a variety name means seed of the variety named which may be mixed with seed of other varieties of the same kind and similar character, the manner of and the circumstances connected with the use of the designation to be governed by the rules and regulations prescribed under section 1592 of the Federal Seed Act.

Variety is defined by the Federal Seed Act as a subdivision of a kind which is characterized by growth, plant, fruit, seed, or other characters by which it can be differentiated from other sorts of the same kind, for example, Marquis wheat, Flat Dutch cabbage, Manchu soybeans, Oxheart carrot, and so forth. Kind means one or more related species or subspecies which singly or collectively is known by one common name, for example, soybean, flax, carrot, radish, cabbage, cauliflower, and so forth.

Q: How many generations removed from the present must a grower research the breeding techniques for a given variety to ensure compliance with the NOP? In many cases it is impossible to research breeding techniques of older varieties as breeding methods are usually highly confidential and in some cases lost to history.

A: Effective October 21, 2002, plant breeders developing new varieties for use in certified organic production must comply with the requirements of the NOP.

All varieties (open pollinated and hybrid) in existence prior to October 21, 2002 may be used by organic producers provided that these varieties have not been produced using Genetically Modified Organisms (GMO).

What constitutes GMO is defined by NOP through the term "excluded methods," (7 C.F.R. 205.2).

Q: How detailed does my Organic Systems Plan (OSP) have to be?

A: The OSP must be sufficient in detail for the certifying agent to determine that your production or handling operation is in compliance with the applicable NOP regulations. The extent of this detail will be determined in collaboration with your certifying agent.

Q: Do I have to follow the Organic System Plan (OSP) I have filed with my certifying agent?

A: Yes. Your OSP is a detailed description of how your operation will achieve, document, and sustain compliance with all applicable provisions of the NOP regulations. Before granting certification to your operation, your certifying agent must concur that your OSP fulfills the requirements of NOP regulations. Your OSP must be annually updated and approved by your certifying agent (see section 205.406 of the NOP regulations). In addition, your OSP may be modified at any time upon request to and written approval from your certifying agent. If you deviate from your previously approved OSP without written approval from your certifying agent, you are no longer in compliance with NOP regulations and could be subject to suspension or revocation of your certification.

Q: Do NOP regulations require the use of organic seed?

A: NOP regulations require the use of organic seed when commercially available. For your options when organic seed is not commercially available see section 205.204 of the NOP regulations or consult your certifying agent.

Q: During the implementation period can a processor (handler) continue to use product ingredients sourced from production operations that have not been certified by a USDA accredited certifier, but have been certified by other certifiers?

A: Yes. During the implementation period, a handler would be allowed to source organic agricultural product ingredients from operations that have not been certified by a USDA accredited certifier. All organic handling operations must discontinue such practices and comply with NOP regulations by October 21, 2002.

Q: How will I find out about changes regarding laws, regulations, policies, and procedures?

A: The NOP will notify accredited certifying agents of changes and pending changes to laws, regulations, policies, and procedures and post such changes and pending changes to its website. Certifying agents should, in turn, notify their clients of such changes and pending changes. Amendments to the NOP regulations will require rulemaking with an opportunity for public comment. This rulemaking is published in the Federal Register. The NOP will issue news releases on Federal Register publications and post such news releases and Federal Register documents to its website.

Q: How will a National Organic Program audit of a producer be conducted?

A: Producer audits/evaluations will be conducted by accredited certifying agents according to the provisions specified in subpart E, Certification, of the national standards.

Q: When are the greenhouse production and apicultural standards expected to be published?

A: Proposed greenhouse and mushroom standards were announced by the National Organic Standards Board's (NOSB) Crops Committee at the June 2001 NOSB meeting. The proposals were posted at the NOSB link on the National Organic Program (NOP) website for public comment through July 31, 2001. The NOSB created an apiculture task force at its June 2001 meeting. Final recommendations for greenhouse, mushroom and apiculture standards will be submitted to the full NOSB for review and approval at its October 15 - 17, 2001, meeting in Washington, D.C. Pending the NOSB decision on the proposed recommendations, the NOP will begin rulemaking for greenhouse, mushroom and apiculture standards in October 2001. NOP's goal will be implementation of these standards by October 21, 2002.

Q: Please provide discussion, examples, and guidance. on buffer zones. Section 205.202 requires distinct boundaries and "adequate" buffer zones, but minimums are not specified, nor is the term adequate defined.

A: Section 202.202(c) requires distinct, defined boundaries and buffer zones to prevent the unintended application of a prohibited substance to land under organic management.

In examining this issue, USDA concluded that imposing a specific size for buffer zones could impose unnecessary burdens on some organic producers without offering greater protection of organic fields and crops from unintended contact with prohibited substances. For example, buffer zones might not be needed for an organic farm if it were completely surrounded by wilderness or areas not in agricultural production. Accordingly, the national standards do not specify specific dimensions for buffer zones, but leaves the determination of their size to the organic producer and the certifying agent on a case-by-case basis.

It has always been the responsibility of organic operations to manage potential contact of organic products with other substances not approved for use in organic production systems. The organic system plan must outline steps that an organic operation will take to avoid drift from neighboring operations, particularly drift of synthetic chemical pesticides.

When considering drift issues, both certifying agents and producers must remember that organic standards are process based. Certifying agents attest to the ability of organic operations to follow a set of production standards and practices that meet the requirements of the Organic Foods Production Act and the national standards. The national standards prohibit the use of genetically modified organisms (defined in the standards as excluded methods) in organic operations. The presence of a detectable residue of a product of excluded methods alone does not necessarily constitute a violation of the regulations. As long as an organic operation has not used excluded methods and takes reasonable steps to avoid contact with the products of excluded methods as detailed in their approved organic system plan, the unintentional presence of the products of excluded methods should not affect the status of an organic product or operation.

Therefore, while the national organic standards provide significant discretion in establishing buffer zone dimensions, buffer zones should not be sized at distances which attempt to achieve a zero tolerance for prohibited substances. The intent of the regulations are to foster a collaborative effort between the certifying agents and their grower clients to determine an appropriate buffer zone with each party being fully cognizant of the process-based nature of the organic label claim.

Q: Does the National Organic Program plan to provide any training or workshops at the producer/grower level?

A: USDA does not envision conducting any training or workshops at the producer or handler level. USDA has entered into a cooperative agreement with the National Center for Appropriate Technology to provide compliance tools for organic agricultural producers and certifying agents.

Q: How does the National Organic Program (NOP) envision that producers/growers can become familiar with the intricacies of this new regulation?

A: Under the NOP, USDA provides oversight to accredited certifying agents. Certifying agents provide oversight to producers and handlers. Section 205.501(a)(8) requires that certifying agents provide sufficient information to persons seeking certification to enable them to comply with OFPA and national organic standards. USDA has entered into a cooperative agreement with the National Center for Appropriate Technology to provide compliance tools for organic agricultural producers and certifying agents.

Q: Do nonagricultural substances included on the National List of Allowed and Prohibited Substances have to be produced without the use of volatile synthetic solvents? My certifying agent says yes because of the prohibition on the use of volatile synthetic solvents found in section 205.270(c)(2).

A: No. Section 205.270(c)(2) prohibits the use of a volatile synthetic solvent unless included on the National List as an allowed substance. However, synthetic solvents do not have to be on the National List to be allowed in the production of an allowed nonagricultural substance found on the National List. The use of volatile synthetic solvents in the production of allowed nonagricultural substances included on the National List is considered approved through the materials review process, unless otherwise stated through an annotation to the approved substance. (Example: § 205.605(a)(9), Flavors -- nonsynthetic sources only and must not be produced using synthetic solvents and carrier systems or any artificial preservative.)

Labels, Labeling, and Market Information

Posted 12/9/02

Q: We are a company that makes bread using certified organic flour and other ingredients. The yeast we use is grown and cultured using certified organic substrates, and the yeast producers follow the NOP regulations for production and processing. Using this yeast, can we label our bread 100 percent organic?

A: You do not say whether or not the yeast has been certified organic by a USDA-accredited certifying agent. If the yeast is certified organic, and all the rest of the ingredients in your bread have been certified organic, you may label your bread as 100 percent organic. If the yeast has not been certified, then your bread does not contain 100 percent organic ingredients. Therefore, you may not label your bread as 100 percent organic.

Posted 11/12/02

Q: I use flavorings in the manufacture of my organic agricultural products. I have proof that these flavorings are from nonsynthetic sources, and are produced without the use of synthetic solvents, carriers, or artificial preservatives as required in section 205.605 (a) 9. However, these flavorings do not come from a certified organic operation. In which categories of organic agricultural products am I allowed to use these flavorings (100 percent organic; organic; made with (at least 70 percent) organic ingredients)?

A: 100 percent organic—No. Products labeled 100 percent organic must contain 100 percent certified organic ingredients.

Organic—Yes. As long as the flavorings constitute 5 percent or less of the total ingredients, and meet the requirements of section 205.605(a)(9).

Made with (at least 70 percent) organic ingredients—Yes. As long as the flavorings meet the requirements of section 205.605(a)(9).

Posted 11/12/02

Q: How are producers whose company names include the term "organic" affected by the NOP regulations? For example, could a company selling organically produced vegetables call itself Blue Sky Organic?

A: The short answer is yes, as long as that company is not trying to mislead by using that name. For a more thorough explanation, I refer you to a pertinent section of the Preamble to the NOP regulations below:

Labeling - Changes Requested But Not Made

(1) "Organic" in Company Names. Many commenters stated that the term, "organic," must not be used as part of a company name if the company does not market organically produced foods. They are concerned that the term in a company name would incorrectly imply that the product, itself, is organically produced.

While we understand commenter concerns, we do not know the extent of the problem. We do not believe those concerns require such a prohibition in the regulations at this time. These regulations may not be the best mechanism to address the issue. Section 6519(b) of the Act provides the Secretary with the authority to take action against misuse of the term, "organic." USDA will monitor use of the term, "organic," in company names and will work with the FTC to take action against such misuse of the term. These determinations must be made on a case-by-case basis.

Q: On what agricultural products can the USDA Organic Seal be displayed?

A: The USDA Organic Seal may appear on organic agricultural products that are certified 100 percent organic or products that are certified as containing at least 95 percent organic ingredients.

Q: Is this seal in use now?

A: The seal may not be used until October 21, 2002.

Q: Are there any size, font, or color requirements placed on the use of the USDA Organic Seal? How can I get a copy?

A: Yes. The required specifications for the USDA Organic Seal can be found in the National Organic Program regulations, subpart D, Label, Labeling, and Market Information, section 205.311, which can be accessed at the NOP website: http://www ams.usda. gov/nop/NOP/standards.html. A camera-ready copy of the seal is also available at http://www.ams.usda.gov/nop/ Consumers/Seal.html.

Q: I know the USDA Organic Seal cannot appear on any products before October 21, 2002. Does this prohibition also apply to advertising materials?

A: The USDA organic seal cannot appear on products or on any advertising materials, including, but not limited to signage or catalogs before October 21, 2002.

Q: We want to label our poultry "natural." Where in the National Organic Standards do we look for the labeling requirements for using this term?

A: Please note that "organic" is not synonymous with "natural." There is nothing in USDA's National Organic Standards defining or regulating the use of the term "natural." USDA's Food Safety and Inspection Service (FSIS) regulates the term "natural" on meat and poultry labels.

Basically, FSIS defines "natural" in the following way: "A product containing no artificial ingredient or added color and is only minimally processed (a process which does not fundamentally alter the raw product) may be labeled natural. The label must explain the use of the term natural (such as--no added colorings or artificial ingredients; minimally processed). You will find this definition and many other labeling terms, including "free range" at the FSIS web site: www.fsis.usda.gov/oa/pubs/lablterm.htm. The FSIS labeling policy book, which goes into much greater technical detail can be accessed at: www.fsis.usda.gov.

Q: As the manager of a retail store, I purchase bulk, certified organic product, then put it in bins where consumers scoop out the amount they want. What signs can I post regarding the organic status of the product?

A: You may provide the same information as provided on the original container or shipping documents, as described in the National Organic Standards, section 205.308. For example, "If the product is prepared in a certified facility, the retail display, labeling, and display containers may use: the USDA seal; and the seal, logo, or other identifying mark of the certifying agent ..."

Q: In the produce area of my retail store, I display bulk product. I don't want to confuse consumers by displaying several different certifier names, seals and logos. What signage options do I have?

A: For non-packaged organic agricultural products, a retail store may make the same claims as those on the shipping documents or container. The USDA Organic Seal, or the certifier's logo, or both may be used for certified products. The only restriction is that the certifier's logo may not be more prominently displayed than the USDA seal (see National Organic Standards, section 205.308). As an alternative, retail stores may simply refer to a product as "organic" and not use any seals or names of certifiers.

Q: I have a deli in my retail store that makes various multi-ingredient products, where 70-95 percent of the ingredients are certified organic. These products are packaged at the customer's request, and the only thing we put on the package is price information. What organic claims can we make on the signs describing the products?

A: Your signs may indicate that the products you package are "made with (specified ingredients or food group(s))," as explained in the National Organic Standards, section 205.309. You may not list more than three ingredients/food groups. For example, you may sell chicken salad, with a sign that reads, "chicken salad made with organic chicken, celery, and grapes."

You may not represent the chicken salad as "certified organic," use the USDA Organic Seal, or the seal or name of any certifying agent (see section 205.310.)

Q: As a manager of a retail store, if I buy organic products from a small-scale organic producer who is exempt from certification, how can I label these products?

A: As explained in the National Organic Standards section 205.310, if you buy an organic product from an exempt operation, and you do not process this product any further, you may identify this product as "organic." For example, you could buy apples from an exempt producer and identify these apples to your customers as "organic." But—you may not identify these apples as being "certified organic," you may not display the USDA Organic Seal in conjunction with these apples, nor in any way represent these apples as certified organic.

If you further process these same apples, to make applesauce, for example, you may not identify the applesauce as being made with organic apples, or call it organic applesauce.

Q: Are there any restrictions on putting "place of origin" information on the label of an agricultural product that is certified organic; for example, "Organically Grown in Montana?"

A: No. The National Organic Standards do not prohibit the placement of truthful information on the labels of certified organic agricultural products.

Q. Section 205.303(b)(2) states that I must identify the certifying agent, preceded by "Certified organic by * * *" or similar phrase, on the information panel below the name of the handler or distributor. Can I meet this requirement by placing the name, acronym, logo, or seal of the certifying agent anywhere on the information panel?

A. To meet the requirements of section 205.303(b)(2), the full name or registered trade name of the certifying agent (which may include acronyms, if the acronym is a registered trade name) must be placed on the information panel below or otherwise near the information identifying the handler or distributor's name. No other printed material or information may be placed between the certifying agent's name and the name of the manufacturer or distributor. Certifying agents' logos or seals and non-registered acronyms may not be used to meet the requirements of section 205.303(b)(2).

Q. Is there a rule for rounding when calculating the percentage of organically produced ingredients?

A. As provided in section 205.302(b) of the National Organic Standards, the percentage of all organically produced ingredients in an agricultural product must be rounded down to the nearest whole number. For example, 69.99 percent would be rounded to 69 percent.

Q. To meet the requirements of section 205.303(b)(2) and 205.304(b)(2), can a handler identify the name of the certifying agent of the finished product using the following statement: "Certified organic by (insert certifying agent's name here) in accordance with the organic standards of the U.S. Department of Agriculture?"

A. Yes.

Q. With respect to application of the USDA seal on packaging material, does the term "transparent" in section 205.311(b)(2) refer to the backgrounds of the upper half of the circle and the word "organic" prior to application on packaging material, or does it refer to the transparency of the packaging material on which the USDA seal has been applied?

A. The term "transparent" in section 205.311(b)(2) refers to the backgrounds of the upper half of the circle and the word "organic" prior to application on packaging material. For example, if a certified organic handler of pasta chooses to apply the black and transparent version of the USDA seal on a brown shipping container, the brown color of the container may show through the transparent portions of the USDA seal.

Q. "As a certified handler can I make specific certification or production claims on the PDP such as Certified Organic," "Grower Certified," "Facility Certified," "Product Certified," "Organically Raised," or "Organically Grown" in addition to or in place of "100 percent organic," "organic," or "made with organic (specified ingredients or food group(s))"?"

A. Certified operations are not prohibited from making truthful PDP label claims. However, alternative labeling claims must not be used to misrepresent a product. For example, alternative PDP label claims must not be constructed so as to mislead a consumer into thinking that a "made with" product was actually an "organic" product.

Further, the NOP may, in the future, engage in rulemaking to limit the use of alternative PDP claims when it has determined that these alternative claims are being used to misrepresent the product.

Finally, exempt or excluded producers and handlers may not represent their operations as certified or use the term "certified" when labeling their product.

Q: How do I label my products until the NOP regulations are fully implemented on October 21, 2002?

A: Until October 21, 2002, producers and handlers of organically produced agricultural products may continue their current labeling practices. All certified organic operations must discontinue the use of their current labeling practices and adhere to the labeling standards provided in subpart D, Labeling, of the Final Rule (October 20, 2002—old labeling practices allowed; October 21, 2002—old labeling practices not allowed).

Organic products that enter the chain of commerce before October 21, 2002 will not be in violation of NOP regulations. However, beginning October 21, 2002, all organic agricultural products that enter the chain of commerce must be labeled according to NOP regulations.

Q: Who should approve my label before I go to print?

A: Your certifying agent will review your label for compliance with NOP labeling requirements. Labels normally approved by a Federal agency such as FDA, FSIS, or BATF must be approved by those regulatory agencies.

Q: Can I make label claims in addition to "organic" on my product?

A: NOP regulations do not prohibit a producer or handler from making additional claims regarding their product as long as they are truthful and not misleading to the consumer. Such label claims may have to be approved by Federal agencies such as FDA, FSIS, or BATF.

Q: Do I have to include the certifier's address on the information panel of my product's package?

A: No. You must identify the certifier's name on the information panel. All other information related to your certifying agent is optional.

Q: When does organic product have to be labeled "For Export Only"?

A: A product must be labeled "For Export Only" when that product has been produced in the United States to an organic standard other than the National Organic Standards (NOS).

Q: Can domestically produced organic product that meets the NOS and an international organic standard make a claim regarding the international standard on the PDP without violating the NOS?

A: Yes. Truthful label claims are allowed under the NOS.

Certification

Posted 11/29/05

Q: With Avian Influenza becoming an increasingly serious issue in the poultry industry, what can organic poultry producers and their certifying agents do to help recognize and prevent the spread of this disease?

A: USDA's Animal and Plant Health Inspection Service has provided guidance to poultry producers, including free range and organic producers, regarding biosecurity considerations for their operations – this guidance is found at http://www.aphis.usda.. gov/vs/birdbiosecurity/

Integral to the National Organic Program is a management plan for each organic production facility – among other things, this plan must address preventative livestock health care practices, including sanitation practices that will be followed to minimize the occurrence and spread of diseases and parasites.

USDA has brought the APHIS biosecurity guidance materials to the attention of its accredited organic certifying agents for their use in working with organic poultry producers.

If it is determined by a producer and certifier that for the health, safety, and welfare of poultry there is need for temporary confinement of birds, provision exists within the National Organic Program for such confinement without a loss of organic certification.

If Federal or State animal health authorities determine that more prescriptive actions are required, organic producers could be required to adhere to those actions.

Posted 3/25/03

Certification Status during the Appeals Process described in 7 C.F.R. §205.680-681

USDA's National Organic Program has received questions from producers, processors, handlers, distributors, and retailers regarding the status of a certified operation during the appeals process described in 7 C.F.R. §205.680-681.

7 C.F.R. §205.680-681 provides that persons subject to the Organic Foods Production Act of 1990 who believe they are adversely affected by a noncompliance decision of the National Organic Program's program manager, State Organic Program, or accredited certifying agent may appeal the decision to the Administrator of the Department of Agriculture's (USDA) Agricultural Marketing Service or a State Program's Governing State Official.

Certification status, including the ability to market products as organic, is not diminished during the appeals process. Certification continues in full force and effect throughout the entire process including any formal administrative proceeding conducted pursuant to USDA's Uniform Rules of Practice, 7 C.F.R. Part 1, Subpart H, or a State Organic Program's Rules of Procedure.

Posted 11/8/02

Q: I am a certifying agent who has an application from a wool producer and processor. How is wool production and processing addressed in the NOP regulations?

A: Wool is considered an inedible fiber, along with cotton, flax, etc. Inedible fibers are addressed in the Preamble to Final Rule as follows:

(6) Nonedible Fibers Products in the NOP. Some commenters asked the NOP to clarify the certification status of fibers such as cotton and flax. The final rule allows for certification of organically produced fibers such as cotton and flax. However, the processing of these fibers is not covered by the final rule. Therefore, goods that utilize organic fibers in their manufacture may only be labeled as a "made with..." product; e.g., a cotton shirt labeled "made with organic cotton."

In other words, the sheep must be certified organic in accordance with the NOP livestock standards, or the cotton, flax, etc., must be certified organic in accordance with the NOP crop standards to be identified as organic in a finished product. Also, since processing is not covered in the final rule, there are no synthetic processing aids used in fiber processing on the National List.

Posted 10/18/02

Q: I am an organic certifying agent in New England. An organic maple syrup producer has applied to me for certification. Should I require that the producer implement best management practices as recommended by the Vermont Cooperative Extension maple specialists; for example, requiring that the tree trunk reach a certain diameter before tapping, prohibiting the use of a vacuum system, or requiring that all health spouts be removed each year for proper healing of the tap-hole?

A: The National Organic Program encourages all producers to employ best management practices as outlined by the authorities in that specific field of agriculture, as long as they are not prohibited by the National Organic Program regulations. There is nothing in the NOP regulations that would prohibit any practice that you've listed in your question.

Posted 10/16/02

Q: A processing facility contracts with two farms for their production. Is it possible to certify those two farms under the name of the processing facility if they are contracted to produce only for that facility?

A: Yes. Production operations that are contracted to supply only a specific processing/ handling facility can be certified as part of that unit. They must be included in the certified operation's organic systems handling plan and be inspected by the certifying agent.

Posted 10/16/02

Q: If a certified operation uses a warehouse or storage facility to store organic product must the warehouse be certified, too?

A: No. However, the warehouse/storage facility must be included in the certified operation's organic systems plan so that the certifying agent is aware of where the product is being stored and can verify that the facility is compliant with any applicable standards; for example, prevention of commingling and contact with prohibited substances, as outlined in Section 205.101 (b)(1).

Posted 10/16/02

Q: I am applying for organic certification of my farm. However, I used a substance during the 2002 growing season that had been on the OMRI list of approved substances, but is not allowed by the National Organic Program. Can I still be certified, assuming I meet all the requirements of the National Organic Program?

A: Yes. Since the substance had previously been accepted as part of good organic farming practices and you used it in good faith, the status of your land and your eligibility for certification is not affected.

Posted 10/16/02

Q: Does a feed mill producing a feed supplement that will be mixed with feed to be fed to cows at an organic dairy need to be certified?

A: No. As discussed in the Preamble, "Livestock Production—Changes Based on Comments (4)" a natural feed additive [supplement] can be from any source, provided it is not classified as a prohibited substance on the National List, and must be in compliance with the Federal Food, Drug and Cosmetic Act.

However, if the feed mill is also producing feed to be used in organic livestock production, the mill must be certified.

Q. Can an accredited certifying agent include an expiration date or renewal date on a certificate of organic operation?

A. No. Section 205.404(c) provides that once certified, a production or handling operation's organic certification continues in effect until surrendered by the organic operation or suspended or revoked by the certifying agent, the State organic program's governing State official, or the Administrator. Accordingly, the NOP prohibits any language on the certificate of organic operation that indicates an expiration or renewal date.

In order to minimize confusion, NOP encourages all accredited certifying agents to prominently display the phrase, "Certification good until surrendered, suspended, or revoked." on the certificate of organic operation.

Q: Will organic producers and handlers apply to the USDA to be certified?

A: No. Producers and handlers will apply directly to the USDA-accredited certifying agent of their choice for certification (see section 205.401 of the NOP regulations).

Q: How long is my Organic Certificate good for?

A: Your organic certificate remains good until you voluntarily surrender your certification or your certification is suspended or revoked by the certifying agent, the State Organic Program's governing State official, or the Administrator for violation of the Act or NOP regulations. The certifying agent will issue an updated certificate of organic operation as needed.

Q: How, and in what ways, do the national organic standards differ from the organic certification requirements for international trade, such as the European Union standards?

A: The USDA national organic standards should be considered in two segments: (1) the verification system that includes certification of organic agricultural products and accreditation of certifying agents and (2) the production, handling and labeling standards under which organic agricultural products are produced and sold. NOP is consistent with the internationally accepted guidelines for certification and accreditation, International Standardization Organization Guides 65 and 61, respectively. However, NOP requirements for production, handling, labeling and allowed and prohibited materials differ significantly from those of other countries, such as the European Union (EU), particularly in livestock production standards.

Q: If U.S. national organic standards differ from foreign organic standards, such as the EU, are there plans to standardize the regulations?

A: No. The NOP, in conjunction with the USDA Foreign Agriculture Service and the U.S. Trade Representative, has begun to establish a process through which equivalency or other trade agreements can be negotiated with governments of foreign countries to which U.S. organic products are exported.

Q: Is it possible for producers/growers to meet the U.S. national organic standards and foreign organic standards?

A: Section 205.300(b) allows for production and export of products produced in the U.S. and certified to foreign national organic standards or foreign contractor buyer requirements.

Such products may be labeled in accordance with the organic labeling of the receiving country or contract buyer. Such products must be exported. The shipping containers and shipping documents must meet the labeling requirements of section 205.307(c), including the requirement that such containers and documents be clearly marked "For Export Only." For further information see section 205.307(c).

Accreditation of Certifying Agents

Posted 1/2/03

The National Organic Program (NOP) has been asked a number of questions regarding how to fulfill the requirement of section 205.501(a)(15)(ii) of the National Organic Standards (NOS) that states certifying agents must submit to the Administrator a copy of a list, on January 2 of each year, including the name, address, and telephone number of each operation granted certification during the preceding year. Based on the questions asked by certifying agents, the NOP provides the following responses to clarify section 205.501(a)(15)(ii):

Q: Does the NOP favor a method of transmission for the lists (i.e. fax, e-mail, US mail, FedEx, or a combination of these)?

A: NOP prefers to receive the lists via e-mail at Arthur.Neal@usda.gov

Q: Is there a particular format preferred by the NOP for the lists? (Word document, Excel spreadsheet, data base listing, etc.)?

A: When submitting lists via email, NOP prefers to receive them in Excel Spreadsheet format.

Q: Does the NOP want the mailing address for the operators, or the physical location of the operation?

A: NOP requires both addresses, if applicable.

Q: The NOS requires that the lists should show the operations "granted certification during the preceding year". Should the lists only contain new operations granted certification, or all operations, including certification renewals?

A: Because this is the initial year for full implementation of the NOS, the NOP requires a complete list of all organic operations certified by the certifying agent as of midnight, December 31, 2002. This would include all operations going through an annual update.

Q: The NOS only asks for the name, address, and telephone number of the operations. Does the NOP also want the categories for which the operations are certified?

A: The NOS does not require identification of the categories for which the operations are certified. However, NOP encourages the submission of this information. Such information will be valuable in identifying the size, scope, and characteristics of the organic production and handling industry.

Q: The NOS requires that the lists be submitted "on" January 2. Can the lists be submitted "on or before" January 2?

A: Yes, the lists may be submitted "on or before" January 2; provided they are complete through the end of the calendar year.

Q: How will a National Organic Program audit of a certifying agent be conducted?

A: The National Organic Program only conducts site evaluations and on-site review audits of certifying agents. Auditors conducting the reviews will follow the procedures specified in the International Organization for Standardization (ISO) 10011 guidelines, "Guidelines for Auditing Quality Systems." Accredited certifying agents will be reviewed based on their ability to comply with the national standards.

December 31, 2002

Q: Please discuss fees associated with audits.

A: Through October 21, 2002, applicants seeking initial accreditation under the National Organic Program (NOP) will only be assessed travel expenses. The NOP will absorb all labor charges for accreditation services through October 21, 2002. After October 21, 2002, applicants for initial accreditation and renewal of accreditation will be assessed fees in accordance to sections 205.640 and 205.641 of subpart G, Fees. Fees for certification are assessed by the certifying agent. Such fees must be in compliance with section 205.642 of the national standards.

Q: Does the National Organic Program (NOP) have plans to produce a national list of accredited certifying agents?

A: Yes. The first announcement of accredited certifying agents will not occur until on or about April 21, 2002. At this time, the NOP will provide a list of accredited certifying agents which will be accessible on the NOP website at http://www.ams.usda.gov/nop/CertifyingAgents/Accredited.html or by request through the NOP office at 1400 Independence Avenue, SW; Room 2510 South Building; Washington, D.C.; 20250.

Q: As a result of the conflict of interest provisions in the national standards, what mechanisms, if any, are there for certifying agents to provide compliance information?

A: Section 205.501(a)(8) requires that certifying agents provide applicants with sufficient information to enable them to comply with the OFPA and regulations. Section 205.501(a)(11)(iv) prohibits certifying agents from giving advice or providing consultancy services to certification applicants or certified operations for the purpose of overcoming barriers to certification. In other words, certifying agents must explain the regulations, but they cannot tell producers or handlers how to correct a noncompliance. Additional discussion of this issue may be found on page 80601 of the national standards.

Q: Based on the conflict of interest provisions in the national standards, how does National Organic Program envision the producer obtaining the correct information to achieve compliance?

A: To address barriers to certification and other organic issues particular to the geographical area, certifying agents may sponsor in-house publications, conferences, workshops, informational meetings, field days, or other educational activities for which participation is voluntary and open to the general public. To overcome barriers to certification, producers and handlers may seek out consultancy services from sources such as educational institutions, State cooperative extension, private consultants, and other producers and handlers.

National List of Allowed and Prohibited Substances

Q: Do nonagricultural substances included on the National List of Allowed and Prohibited Substances have to be produced without the use of volatile synthetic solvents? My certifying agent says yes because of the prohibition on the use of volatile synthetic solvents found in section 205.270(c)(2).

A: No. Section 205.270(c)(2) prohibits the use of a volatile synthetic solvent unless included on the National List as an allowed substance. However, synthetic solvents do not have to be on the National List to be allowed in the production of an allowed nonagricultural substance found on the National List. The use of volatile synthetic solvents in the production of allowed nonagricultural substances included on the National List is considered approved through the materials review process, unless otherwise stated through an annotation to the approved substance. (Example: § 205.605(a)(9), Flavors -- nonsynthetic sources only and must not be produced using synthetic solvents and carrier systems or any artificial preservative.)

Use of Chlorine in Organic Handling Operations - 7 C.F.R 205.601(a)(2), 205.603(a)(3), and 205.605(b)(9) provides for the use of chlorine materials as algicides, disinfectants and sanitizers in organic crop, livestock and handling operations. The annotation on the use of chlorine materials restricts the residual chlorine levels in the water to the maximum residual disinfectant limit under the Safe Drinking Water Act. This limit is currently established by the Environmental Protection Agency (EPA) at 4 mg/L for chlorine. The National Organic Program has received a number of questions regarding the use of chlorine in certified operations and the sampling protocol to be used by accredited certifying agents (ACA) in monitoring the maximum residual disinfectant limit for chlorine materials. The following Q and A's are designed to clarify these issues.

Q. As an ACA, at what point in crop, livestock or handling operations should I monitor for the maximum residual disinfectant limit?

A. ACA's must monitor the discharge or effluent point to ensure that certified operators are meeting the 4 mg/L limit as set forth by the Safe Drinking Water Act.

Q. As a crop, livestock or handling operation, am I restricted to use chlorine at the maximum residual disinfectant limit specified under the Safe Drinking Water Act, currently 4 mg/L, at the beginning of the wash/rinse water cycle?

A. No. Levels of chlorine used to prepare water to be used to disinfect/sanitize tools, equipment, product or food contact surfaces may be higher than 4 mg/L and should be at levels sufficient to control microbial contaminants. Therefore, chlorine use at the beginning of the applicable water cycle in an organic operation is not limited to 4 mg/L.

Q: Can nonagricultural substances not appearing on the National List of Allowed and Prohibited Substances be used as ingredients in or on a product labeled as "made with organic (specified ingredients or food group(s)).?"

A: A "made with organic (specified ingredients or food group(s))" product must, in accordance with section 205.105(c) of the Final Rule, be produced and handled without the use of nonagricultural substances used in or on processed products, except when the nonagricultural substances are included in section 205.605 of the National List of Allowed and Prohibited Substances. Accordingly, the reference to nonorganic ingredients in section 205.301(c) refers to agricultural ingredients only and should not be construed to include nonagricultural ingredients.

To further clarify the Department's intent, a "made with organic (specified ingredients or food group(s))" product must contain at least 70 percent organic agricultural ingredients that have been produced without the use of:

1. Synthetic substances unless the substances and their use are allowed under section 205.601 or section 205.603 of the National List of Allowed and Prohibited Substances.
2. Nonsynthetic substances prohibited under section 205.602 or section 205.604 of the National List of Allowed and Prohibited Substances.
3. Nonagricultural substances unless the substances are allowed under section 205.605 of the National List of Allowed and Prohibited Substances.

Additionally, the remainder of the ingredients in a "made with organic (specified ingredients or food group(s))" product (up to 30 percent) may include:

1. Nonagricultural products listed in section 205.605 of the National List.
2. Nonorganically produced agricultural products, raw or processed, that have been produced using synthetic, nonsynthetic, and nonagricultural substances without regard to sections 205.601 through 205.605 of the National List of Allowed and Prohibited Substances, except that the use of excluded methods, sewage sludge, and ionizing radiation are prohibited. Nonorganically produced agricultural products listed in section 205.606 of the National List of Allowed and Prohibited Substances must comply with the restrictions placed on that product by section 205.606.

Q: Will there be a generic list of allowed naturals?

A: No, there will not be a generic list of allowed natural materials, because all naturals are allowed unless prohibited on the National List.

Q. What is the "maximum residual disinfectant level?"

A. "Maximum residual disinfectant level" is a term defined by the Environmental Protection Agency (EPA) as the highest level of a disinfectant allowed in drinking water. This level is currently established by EPA at 4 mg/L for chlorine. Practically applied under the National Organic Standards, the term "maximum residual disinfectant level" refers to the chlorine level of the waste water at the discharge or effluent point.

State Organic Programs

Q: Can a State request review of a proposed State Organic Program (SOP) before all the statutory and regulatory rules have been changed? For instance, can a State specify its intent to adopt legislation or rules in its request for approval, rather than having those statutes or rules finalized?

A: A State has two options: (1) It can submit a final SOP to USDA for approval; or (2) It may submit to USDA, a draft SOP for review prior to finalization of statutes, regulations, or procedures. When this option is used, USDA will provide feedback to the State regarding any changes necessary to receive approval. A State can then use the feedback to finalize its SOP. The SOP would, then, be submitted to USDA for approval.

Q: Can a State require registration of all certifying agents operating within the State?

A: Yes; provided that the registration program does not discriminate against the certifying agents.

Q: Can a State with an approved State Organic Program (SOP) review the operations of certifying agents operating within the State?

A: Only States with approved SOPs may review and investigate complaints of noncompliance with the Act or regulations concerning accreditation of certifying agents operating in the State (See section 205.668(c)).

Q: Can a State require certifying agents operating within the State to comply with the approved State Organic Program's (SOP's) more restrictive requirements?

A: Accredited certifying agents operating within a State with an approved SOP must comply with the SOP's more restrictive requirements. A violation of the State's approved more restrictive requirements is a violation of the National Organic Standards (NOS). Only the USDA may take enforcement action against an accredited certifying agent for violating the NOS. States with approved SOPs may review and investigate complaints of noncompliance and report their findings to the NOP for enforcement action (See section 205.668(c)).

Q: When is a State considered to have a State Organic Program (SOP)?

A: A State is considered to have an SOP when it receives USDA approval. To receive approval, a proposed SOP must meet the National Organic Standards, receive approval for any identified more restrictive requirements, and contain noncompliance, mediation, and appeals procedures that meet the requirements of USDA. The basis for the recommendation to approve or disapprove will be based on compliance with the requirements of Sections 205.620 through 205.622, 205.661 through 205.663, 205.668, and 205.680 and 205.681 of the NOS (7 CFR 205.620-.622, 205.661-.663, 205.668, and 205.680-.681).

Q: Who is responsible for handling a U.S. District Court appeal of a State's final decision?

A: The appeal is on a State action, and the State is responsible for defending its action.

Q: What discretion, if any, does a State Organic Program (SOP) have in handling an appeal of an enforcement action initiated by an accredited certifying agent?
A: None. Approved SOPs are required to handle all appeals of enforcement actions initiated by certifying agents against certified operations operating within the State.

Q: Can a State Organic Program (SOP) include additional enforcement provisions, such as authority to issue cease and desist orders, obtain injunctions to stop the sale of noncompliant products, or summarily suspend certifications?
A: A State may request approval of an SOP that includes more restrictive enforcement requirements that do not deny due process.

Q: Will an approved State Organic Program (SOP) be responsible for revoking the certification of a certified operation based in foreign countries?
A: When a certified entity operates in multiple locations and one of those locations is within the jurisdiction of an SOP, the State is responsible for enforcing compliance by that portion of the certified operation.

Q: If a certified producer or handler fails to meet the continuation of certification requirements and contests a proposed revocation of certification, can a State with a State Organic Program (SOP) stop that entity from selling its product as "organic" prior to completion of the appeals process?
A: A stop sale before completion of the appeals process would be a denial of due process.

Q: Is it possible for organic growers to be in compliance with the National Organic Program if their State does not create a State Organic Program.
A: Yes. A State does not have to create a State Organic Program or become an accredited certifying agent for organic growers within the State to be in compliance with the requirements of the National Organic Program.

Q: Please explain the options available to a State considering the possibility of creating a State Organic Program (SOP) under the National Organic Program (NOP).
A: *Option 1*: A State may choose not to establish an SOP or provide certification services under the NOP. Under this option, organic growers may seek organic certification by any certifying agent accredited under the NOP. The State would not be responsibly connected to enforcement of the NOP. Enforcement would be shared jointly by the NOP and the certifying agent.

Option 2: A State may choose to only provide certification services under the NOP. Under this option, the State would have to apply for accreditation. As an accredited certifying agent, the State would be responsible for conducting certifications, enforcing the production and handling standards of the NOP, and maintaining compliance with other applicable NOP regulations. USDA will be responsible for oversight of the State certifying agent. Organic producers and handlers within that State could choose to be certified by the State or any other accredited certifying agent.

Option 3: A State may choose to establish an SOP. Under this option, all organic producers or handlers in the State would have to be certified according to the SOP, which would include the requirements of the NOP and the more restrictive provisions unique to that State and approved by the USDA. The State would assume enforcement responsibility, within its borders, for the requirements in the national standards and its SOP (§205.620(d)). However, the State may not initiate proceedings to suspend or revoke the accreditation of any certifying agent accredited by USDA. Suspension or revocation of an accredited certifying agent may only be pursued by the USDA. Organic producers and handlers may seek and obtain organic certification from any certifying agent accredited under the NOP.

Option 4: A State may choose a combination of Options 2 and 3.

Exporting Organic Products

Q: When does organic product have to be labeled "For Export Only"?

A: A product must be labeled "For Export Only" when that product has been produced in the United States to an organic standard other than the National Organic Standards (NOS).

Q: Can domestically produced organic product that meets the NOS and an international organic standard make a claim regarding the international standard on the PDP without violating the NOS?

A: Yes. Truthful label claims are allowed under the NOS.

References

[1] 7CFR205.2.

[2] USDA, Economic Research Service. *Recent Growth Patterns in the U.S. Organic Foods Market.* Report No. AIB777. September 2002. Available online at [http://www.ers.usda.gov/publications/aib777/].

[3] For a detailed graphic representation of the distribution of organic crop and pasture acreage and number of operations per state, see [http://www.ers.usda.gov/Amberwaves /Feb03/ pdf/indicators.pdf].

[4] USDA. Economic Research Service. Price premiums hold on as U.S. organic produce market expands. Report VGS-308-01. May 2005. Available online at [http://www. ers.usda.gov].

[5] Ibid.

[6] Ibid.

[7] Ibid.

[8] Ibid.

[9] Farms and handling operations that sell less than $5,000 a year in organic agricultural products are exempt from certification; however, these producers and handlers must abide by the national standards for organic products and may label their products as organic.

[10] A July 2005 audit by the USDA Office of Inspector General (OIG) states that USDA had 41 accredited certification agents in foreign countries. The report also found that as of August 2004, AMS had negotiated only one equivalency agreement with a foreign

country (Japan). The OIG recommends implementation of internal operating procedures "to assure that the NOP is achieving its intended objectives to ensure that organic products meet consistent, uniform standards." The audit report is available online by going to [http://www.usda.gov/oig/audits.htm] and following the links from "View Audit Reports and News Releases," to the listing by "Agency Subject of Report." It is under the Agricultural Marketing Service, July 2005.

[11] The CCC is a wholly owned government financing institution for USDA agencies that administer mandatory programs, such as the farm commodity price and income support programs for wheat, cotton, rice, and certain other crops; agricultural export subsidies; and certain conservation and trade programs. CCC funds are considered mandatory funds that must be made available for the purposes authorized. In practice, however, appropriators sometimes prohibit or place restrictions on funding for mandatory programs in the annual appropriations bill. A July 2005 audit of the NOP by USDA's Office of Inspector General (OIG) found that AMS has not been properly documenting its procedures for maintaining and controlling cost-share programs, which makes it difficult if not impossible to monitor the use of funds. AMS officials have promised to implement proper procedures by the end of FY2005.

[12] Information on Integrated Organic Program research grants are available on the Cooperative State Research, Education, and Extension Service website at [http://www.csrees.usda.gov/fo/funding.cfm].

[13] ERS published a paper in August 2005 entitled, "Market-Led Growth vs. Government-Facilitated Growth: Development of the U.S. and EU Organic Agriculture Sectors." Available online at [http://www.ers.usda.gov/ Briefing/Organic/]. Congress has not appropriated funds to date for a National Agricultural Library International Organic Research Collaboration.

[14] USDA, Agricultural Marketing Service. *Organic Feed for Poultry & Livestock: Availability and Prices*. Available online at [http://www.ams.usda.gov/nop/prod handlers/feedstudyjune2003.pdf].

[15] Available online at [http://www.ams.usda.gov/nop/TodaysNews.html]; see August 24, 2005.

[16] USDA, Office of Inspector General. *Audit Report: Agricultural Marketing Service's National Organic Program.* The audit report is available online by going to [http:// www.usda.gov/oig/audits.htm] and following the links from "View Audit Reports and News Releases," to the listing by "Agency Subject of Report." It is under the Agricultural Marketing Service, July 2005.

[17] Notice was published in the *Federal Register* on August 13, 2003 (68 FR 48334).

[18] Available online at [http://www.ams.usda.gov/nop/CertifyingAgents/ ANSIReport.pdf].

[19] "Organic Milk Debate; Dairies dispute 'organic' values; Ex-hippie farmers contest practices of big producers," *Chicago Tribune*, January 10, 2005. See [http://www.organicconsumers.org/organic/milk011105.cfm].

[20] Minutes of February 28-March 3, 2005, NOSB meeting available online at [http://www. ams.usda.gov/nosb/meetings/meetings.html].

[21] Minutes of August 15-17, 2005, NOSB meeting available online at [http://www. ams.usda.gov/nosb/meetings/meetings.html].

[22] "NOSB Livestock Committee Recommendations for Guidance; Pasture Requirements for the National Organic Program," National Organic Standards Board, March 2, 2005. Available online at [http://www.ams.usda.gov/nosb/meetings/meetings.html].

[23] *Harvey v. Veneman*, 396 F.3d 28 (1st Cir. 2005).

In: Organic Farming: Methods, Economics and Structure ISBN 978-1-60692-864-6
Editors: Michael Nelson and Ivan Artamova

Chapter 11

ORGANIC FARMING AND MARKETING: PUBLICATIONS FROM THE UNITED STATES DEPARTMENT OF AGRICULTURE, 1977-2005*

Mary V. Gold†
Alternative Farming Systems Information Center, National Agricultural Library, Agricultural Research Service, U.S. Department of Agriculture, 10301 Baltimore Avenue, Room 132, Beltsville, MD 20705-2351 USA

This compilation cites only documents published directly by the U.S. Department of Agriculture's primary agencies during the years 1977 through August 2005. The full range of published work by USDA scientists, analysts, advisors and other employees at stations throughout the world appears in many other publishing venues including peer-reviewed journals, research reports, newsletters, conference presentations, Web pages, bulletins and factsheets, databases and more. USDA funded projects and organizations also publish documents that appear in many guises and places.

In terms of subject, only publications pertinent to organic production, food and marketing are represented. There are many others that are relevant to organic systems that are not cited. In addition, this listing does not address the large body of research published before 1940 when reliance on synthetic fertilizers and pesticides in agriculture began to become widely adopted. Much of this historic information, which predates the common use of the term "organic agriculture," is directly applicable to current organic systems. For information on other organic-related resources from USDA and elsewhere, or for information about accessing materials from the National Agricultural Library collection, contact the Alternative Farming Systems Information Center (AFSIC).

* Note: This bibliography expands and updates the previously published, Organic Farming and Marketing: Publications from USDA, 1993-2003. Citations are arranged by agency name, and then by publication date, earliest to most recent. Over the years, agency names have changed; publications published under the auspices of older agency entities are placed with their current agency counterpart. Many publications are cited with current Internet access information; each URL was verified as correct September 8, 2005. National Agricultural Library call numbers are included when available

† E-mail address: mgold@nal.usda.gov; 301-504-6559;

Agricultural Marketing Service (AMS)
Agricultural Marketing Service (AMS), National Organic Program (NOP)
Agricultural Research Service (ARS)
Agricultural Research Service (ARS), National Agricultural Library (NAL)
Cooperative State Research, Education, and Extension Service (CSREES)
Cooperative State Research, Education, and Extension Service (CSREES), Sustainable Agriculture Research and Education (SARE)
Economic Research Service (ERS)
Foreign Agricultural Service (FAS)
Natural Resources Conservation Service (NRCS)
Risk Management Agency (RMA)
Rural Development (RD)
World Agricultural Outlook Board (WAOB)
Other USDA Publications

Agricultural Marketing Service (AMS)

Organic Food and Fiber: An Analysis of 1994 Certified Production in the United States, by Julie Anton Dunn. USDA, Agricultural Marketing Service, Transportation and Marketing Division, 1995. 5p. [NAL Call Number: aS441.D86 1995]

Agricultural Marketing Service (AMS), National Organic Program (NOP)

Labeling Alcoholic Beverage Containers. USDA, Agricultural Marketing Service, National Organic Program, 2002.
Full text online: http://www.ams.usda.gov/nop/ProdHandlers/AlcoholLabel.htm

Organic Food Standards and Labels: The Facts. USDA, Agricultural Marketing Service, National Organic Program, 2002.
Full text online: http://www.ams.usda.gov/nop/Consumers/brochure.html

How Retail Food Establishments Can Comply with National Organic Program Regulations. USDA, Agricultural Marketing Service, National Organic Program, 2003.
Full text online: http://www.ams.usda.gov/nop/ProdHandlers/RetailFoodEstablishments.html

Labeling Packaged Products. USDA, Agricultural Marketing Service, National Organic Program, 2003.
Full text online: http://www.ams.usda.gov/nop/ProdHandlers/LabelTable.htm

Organic Feed for Poultry and Livestock: Availability and Prices. USDA, Agricultural Marketing Service, National Organic Program, 2003.
Full text online: http://www.ams.usda.gov/nop/ProdHandlers/FeedStudyJune2003.pdf

Value Through Verification: USDA National Organic Program, by Barbara Robinson. USDA, Agricultural Marketing Service, National Organic Program, 2004. Note: Available in both PowerPoint and HTML formats.
Full text online: http://www.ams.usda.gov/nop/NOPPresentation/home.html

Accredited Certifying Agents. USDA, Agricultural Marketing Service, National Organic Program, 2005.
Full text online: http://www.ams.usda.gov/nop/CertifyingAgents/Accredited.html

Questions and Answers. USDA, Agricultural Marketing Service, National Organic Program, updated monthly.
Full text online: http://www.ams.usda.gov/nop/Q&A.html

AGRICULTURAL RESEARCH SERVICE (ARS)

A Bibliography for Small and Organic Farmers, 1920 to 1978, by James W. Schwartz. USDA, Science and Education Administration, Agricultural Research, 1978. 157p. [NAL Call Number: aZ5074.O7S3]

Description: "The bibliography presented here consists of 1,176 publications of long-term research by scientists of the Science and Education Administration - Agricultural Research's Soil, Water, and Air Sciences staff (SWAS) from 1920-1978 that relate to the needs of both the small and organic farmers. Only those publications that were considered most useful to these groups of farmers were selected. They are listed under 19 subject areas according to the year published. A brief summary of the subject area appears at the beginning of each section." [From the Introduction]

"Postharvest Shelf Life of Organically Grown Produce: Current Perceptions and Need for Further Study in the USA," by T. Schettini. USDA, Agricultural Research Service. In Workshop on Biological Control of Postharvest Diseases of Fruits and Vegetables. (June 1991): 218-225. (ARS, 92) Note: Paper presented at meeting held September 12-14, 1990, Sheperdstown, WV. [NAL Call Number: aS21.R44A7]

Sustainable Production of Fresh-Market Tomatoes and Other Summer Vegetables with Organic Mulches, by Aref A. Abdul-Baki and John R. Teasdale. Rev. edition. (Farmers' Bulletin, 2279) USDA, Agricultural Research Service, Sustainable Agricultural Systems Laboratory (Beltsville MD), August 1997. [NAL Call Number: 1 Ag84F no. 2279]
Full text online: http://www.ars.usda.gov/is/np/tomatoes.html

"Research for MeBr Alternatives May Benefit Organic Growers." USDA, Agricultural Research Service. Methyl Bromide Alternatives 5, no. 4 (October 1999). [NAL Call Number: aSB952.B75M47]
Full text online: http://www.ars.usda.gov/is/np/mba/oct99/organic.htm

"Turning to the Sun—Instead of Methyl Bromide," by Judy McBride. USDA, Agricultural Research Service. Agricultural Research 47, no. 10 (October 1999): 18-19. [NAL Call Number: 1.98 Ag84]
Full text online: http://www.ars.usda.gov/is/AR/archive/oct99/sun1099.htm

Organic Growers Join with ARS to Retain Crop Diversity, by Kathryn Barry Stelljes. (ARS Press Release) USDA, Agricultural Research Service, October 20, 1999.
Full text online: http://www.ars.usda.gov/is/pr/1999/991020.htm

"Fungal Protein Slows Broadleaf Weeds," by Jan Suszkiw. USDA, Agricultural Research Service. Agricultural Research 48, no. 2 (February 2000): 10-11. [NAL Call Number: 1.98 Ag84]
Full text online: http://www.ars.usda.gov/is/AR/archive/feb00/weeds0200.htm

"Soil Fungi Critical to Organic Success," by Don Comis. USDA, Agricultural Research Service. Agricultural Research 49, no. 5 (May 2001): 20-21. [NAL Call Number: 1.98 Ag84]
Full text online: http://www.ars.usda.gov/is/AR/archive/may01/fungi0501.htm

"An Organic Cure for a Turfgrass Disease?" by Ben Hardin. USDA, Agricultural Research Service. Agricultural Research 49, no. 8 (August 2001): 12-13. [NAL Call Number: 1.98 Ag84]
Full text online: http://www.ars.usda.gov/is/AR/archive/aug01/turf0801.htm

The Rodale Institute: USDA ARS - 044 Final Report: August 1997-December 2000. USDA, Agricultural Research Service, 2002? 24p. [NAL Call Number: aS605.5.R63 2002]

"Forum - Adding OOMPH to Organic Farming," by Michael D. Jawson. USDA, Agricultural Research Service. Agricultural Research 50, no. 2 (February 2002). [NAL Call Number: 1.98 Ag84]
Full text online: http://www.ars.usda.gov/is/AR/archive/feb02/form0202.htm

"Organic Grows on America!" by Marcia Wood et al. USDA, Agricultural Research Service. Agricultural Research 50, no. 2 (February 2002): 4-9. [NAL Call Number: 1.98 Ag84]
Full text online: http://www.ars.usda.gov/is/AR/archive/feb02/amer0202.htm

Better Genes, Not Pills, for Healthier Animals, by Don Comis. (ARS Press Release) USDA, Agricultural Research Service, February 22, 2002.
Full text online: http://www.ars.usda.gov/is/pr/2002/020222.htm

Organic Research Matters [periodic newsletter]. USDA, Agricultural Research Service, Organic Research Program (Salinas, CA), 2004 - .
Full text online: http://www.ars.usda.gov/Research/docs.htm?docid=6607

"Soil Fungi Plants' Natural Friend," by David Elstein. USDA, Agricultural Research Service. Agricultural Research 52, no. 5 (May 2004): 7. [NAL Call Number: 1.98 Ag84]
Full text online: http://www.ars.usda.gov/is/AR/archive/may04/fungi0504.htm

Organic Farming is a Winner for Sustainability, by Don Comis. (ARS Press Release) USDA, Agricultural Research Service, November 26, 2004.
Full text online: http://www.ars.usda.gov/is/pr/2004/041126.htm

AGRICULTURAL RESEARCH SERVICE (ARS), NATIONAL AGRICULTURAL LIBRARY (NAL)

Organic Farming and Gardening: 1969-1976 and 1969-1978, by Charles N. Bebee. (Quick Bibliography Series, 77-17; 79-28) USDA, National Agricultural Library, 1977; 1979. [NAL Call Number: aZ5071.N3]

Organic Farming and Gardening: 1970-1988, by Jayne T. MacLean. (Quick Bibliography Series, 80-27; 83-40; 84-42; 85-25; 86-57; 87-10; 88-38) USDA, National Agricultural Library, 1980-1988 . Note: Seven bibliographies in this series, covering 1970-1986. [NAL Call Number: aZ5071.N3]

Educational and Training Opportunities in Organic, Low Input or Sustainable Agriculture. USDA, National Agricultural Library, Alternative Farming Systems Information Center, 1988. 10p. [NAL Call Number: aS605.5.E38]

Tracing the Evolution of Organic/Sustainable Agriculture: A Selected and Annotated Bibliography, by Jane Potter Gates. USDA, National Agricultural Library, Alternative Farming Systems Information Center, 1988. 20p. [NAL Call Number: aZ5076.A1U54 no.72]
Full text online: http://www.nal.usda.gov/afsic/AFSIC_pubs/tracing.htm

Organic, Low-input or Sustainable Agriculture, 1984 -1989, by Jayne T. MacLean. (Quick Bibliography Series, 88-38; 89-15; 89-54) USDA, National Agricultural Library, Alternative Farming Systems Information Center, 1988-1989. Note: Three bibliographies in this series, covering 1984-March 1989. [NAL Call Number: aZ5071.N3]

Oral History Interview with Mr. Robert Rodale, by Jane Potter Gates and Robert Rodale. USDA, National Agricultural Library, Alternative Farming Systems Information Center, 1989. Note: 1 videocassette (69 min., 36 sec.): sd., col. (VHS). [NAL Call Number: Videocassette no. 670]

Educational and Training Opportunities in Sustainable Agriculture [annual directory]. USDA, National Agricultural Library, Alternative Farming Systems Information Center, 1990- . [NAL Call Number: aS605.5.E38]
Full text online: http://www.nal.usda.gov/afsic/AFSIC_pubs/edtr.htm

Oral History Interview with Fred Kirschenmann, by Jane Potter Gates and Fred Kirschenmann. USDA, National Agricultural Library, Alternative Farming Systems Information Center, 1990. Note: 1 videocassette (53 min.): sd., col. (VHS). [NAL Call Number: Videocassette no. 877]

Oral History Interview with James Patrick Madden, by Jane Potter Gates and J. Patrick Madden. USDA, National Agricultural Library, Alternative Farming Systems Information Center, 1990. Note: 1 videocassette (55 min.): sd., col. (VHS). [NAL Call Number: Videocassette no. 1009]

Organic Certification, by Jane Potter Gates. (Special Reference Briefs, 90-04) USDA, National Agricultural Library, Alternative Farming Systems Information Center, January 1990. 9p. Note: Bibliography [NAL Call Number: aS21.D27S64]

Societal Impacts of Adoption of Alternative Agricultural Practices, January 1979-April 1990, by Jayne T. MacLean. (Quick Bibliography Series, 90-57) USDA, National Agricultural Library, Alternative Farming Systems Information Center, June 1990. 16p. Note: Bibliography. [NAL Call Number: aZ5071.N3]

Oral History Interview with Dick Thompson, by Jane Potter Gates and Dick Thompson. USDA, National Agricultural Library, Alternative Farming Systems Information Center, 1991. Note: 1 videocassette (60 min.): sd., col. (VHS). [NAL Call Number: Videocassette no. 1008]

Oral History Interview with I. Garth Youngberg, by Jane Potter Gates and Garth Youngberg. USDA, National Agricultural Library, Alternative Farming Systems Information Center, 1991. Note: 1 videocassette (60 min.): sd., col. (VHS). [NAL Call Number: Videocassette no. 1128]

Organic Certification, by Jane Potter Gates. (Special Reference Briefs, 91-10) USDA, National Agricultural Library, Alternative Farming Systems Information Center, August 1991. 18p. [NAL Call Number: aS21.D27S64]

Societal Impacts of Adoption of Alternative Agricultural Practices, January 1985-August 1992, by Jayne T. MacLean. (Quick Bibliography Series, 93-01) USDA, National Agricultural Library, Alternative Farming Systems Information Center, October 1992. [NAL Call Number: aZ5071.N3 no.93-01]
Full text online: http://www.nal.usda.gov/afsic/AFSIC_pubs/qb93-01.htm

Organic Production: Recent Publications and Current Information Sources, by Mary V. Gold. (Special Reference Briefs, 96-07) USDA, National Agricultural Library, Alternative Farming Systems Information Center, 1996. [NAL Call Number: aS21.D27 S64 no. 96-07]
Full text online: http://www.nal.usda.gov/afsic/AFSIC_pubs/srb96-07.htm

Organic Production: Economic Aspects. Books, Articles, and Videocassettes, 1991-March 1997, by Mary V. Gold. USDA, National Agricultural Library, Alternative Farming Systems Information Center, 1997.
Full text online: http://www.nal.usda.gov/afsic/AFSIC_pubs/organic1.htm

Organically Produced Foods: Nutritive Content, by Mary V. Gold. (Special Reference Briefs, 2000-03) USDA, National Agricultural Library, Alternative Farming Systems Information Center, 2000. 21p. [NAL Call Number: aS21.D27 S64 no. 2000-03]
Full text online: http://www.nal.usda.gov/afsic/AFSIC_pubs/srb0003.htm

Organic Foods: Markets and Marketing - Recent Books, Articles, News Items, Reports, and Videos, by Mary V. Gold. USDA, National Agricultural Library, Alternative Farming Systems Information Center, 2001.
Full text online: http://www.nal.usda.gov/afsic/AFSIC_pubs/orgmrkt.htm

Organic Information Resources: What Are They? Where Are They? How Can I Find Them? by Mary V. Gold. USDA, National Agricultural Library, Alternative Farming Systems Information Center, 2001.
Full text online: http://www.nal.usda.gov/afsic/ofp/orgfind.htm
Organic Gardening: A Guide to Resources 1989-September 2003, by Stephanie M. Boehmer. USDA, National Agricultural Library, Alternative Farming Systems Information Center, September 2003.
Full text online: http://www.nal.usda.gov/afsic/AFSIC_pubs/org_gar.htm

Organic Livestock Production: A Bibliography: Selected Citations from 1974-September 2004, by Mary V. Gold. (Special Reference Briefs, 2004-05) USDA, National Agricultural Library, Alternative Farming Systems Information Center, September 2004.
Full text online: http://www.nal.usda.gov/afsic/AFSIC_pubs/srb0405.htm

Organic Aquaculture, by Stephanie Boehmer et al. (AFSIC Notes, 5) USDA, National Agricultural Library, Alternative Farming Systems Information Center, January 2005.
Full text online: http://www.nal.usda.gov/afsic/AFSIC_pubs/afnotes5.htm

Organic Agricultural Products: Marketing and Trade Resources, by Mary V. Gold. Updated edition. (Special Reference Briefs, 2003-01) USDA, National Agricultural Library, Alternative Farming Systems Information Center, September 2005. 142p.
Full text online: http://www.nal.usda.gov/afsic/AFSIC_pubs/OAP/srb0301.htm

COOPERATIVE STATE RESEARCH, EDUCATION, AND EXTENSION SERVICE (CSREES)

Organic-related Farming Systems Research FY 82: A Directory of USDA and State Projects in CRIS. USDA, Current Research Information System (CRIS), 1985. 459p. [NAL Call Number: aS605.5.O7]

Growth in Organic Markets Helps Launch Florida Research Center. (Selected Results and Impacts) USDA, Cooperative State Research, Education, and Extension Service, July 2004.
Full text online:
http://www.csrees.usda.gov/nea/ag_systems/sri/sustain_ag_sri_sare_growth.html

Second Year for Integrated Organic Program. (In Focus) USDA, Cooperative State Research, Education, and Extension Service, April 2005.
Full text online: http://www.csrees.usda.gov/nea/ag_systems/in_focus/organic_if_iop.html

Cooperative State Research, Education, and Extension Service (CSREES), Sustainable Agriculture Research and Education (SARE)

The Real Dirt: Farmers Tell about Organic and Low-Input Practices in the Northeast, by Miranda Smith, Members of the Northeast Organic Farming Association, and Cooperative Extension. USDA, Cooperative State Research, Education, and Extension Service, Sustainable Agriculture Research and Education, Northeast Region, 1994. 264p. [NAL Call Number: S605.5.R43 1994]

CSA Farms in the United States 1999-2000: Community Supported Agriculture (CSA): A New Partnership Between Farmers and Consumers. USDA, Cooperative State Research, Education, and Extension Service, Sustainable Agriculture Research and Education/Sustainable Agriculture Network, 1999. 51p. [NAL Call Number: aHD1484 .C73 1999]

Explore Organic Production. (Farming for Profit, Stewardship & Community Tip Sheet, 4) USDA, Cooperative State Research, Education, and Extension Service, Sustainable Agriculture Research and Education/Sustainable Agriculture Network, 1999.

Naturalize Your Farming System: A Whole-Farm Approach to Managing Pests. (Information Bulletin) USDA, Cooperative State Research, Education, and Extension Service, Sustainable Agriculture Research and Education/Sustainable Agriculture Network, 2000. 20p. [NAL Call Number: SB950.W463-2000]
Full text online: http://www.sare.org/publications/farmpest.htm

Profitable Poultry: Raising Birds on Pasture. (Information Bulletin) USDA, Cooperative State Research, Education, and Extension Service, Sustainable Agriculture Research and Education/Sustainable Agriculture Network, 2002. 15p. [NAL Call Number: SF487.8.A1 P76 2002]
Full text online: http://www.sare.org/bulletin/poultry/

Exploring Sustainability in Agriculture. USDA, Cooperative State Research, Education, and Extension Service, Sustainable Agriculture Research and Education/Sustainable Agriculture Network, 2003. 13p. [NAL Call Number: aS441.E98 2003]
Full text online: http://www.sare.org/bulletin/explore/

Transitioning to Organic Production. (Opportunities in Agriculture) USDA, Cooperative State Research, Education, and Extension Service, Sustainable Agriculture Research and

Education/Sustainable Agriculture Network, 2003. 32p. [NAL Call Number: S605.5.O673 2003]
Full text online: http://www.sare.org/publications/organic.htm

ECONOMIC RESEARCH SERVICE (ERS)

"Organic Certification Programs," by Charlene C. Price and Judy Brown. USDA, Economic Research Service. National Food Review, no. 15 (Summer 1981): 31-33. [NAL Call Number: aHD9001.N275]

"Organic Farming Grows in Western Europe," by S. Sposato. USDA, Economic Research Service. National Food Review, no. 24 (Fall 1983): 22-23. [NAL Call Number: aHD9001.N275]

"Organically Speaking, Western Europe is Top Seed," by S. Sposato. USDA, Economic Research Service. Farmline 5, no. 4 (April 1984): 16-17. [NAL Call Number: AGE aHD1401.A2U52]

"Natural Foods: What's in a Name," by Charlene C. Price. USDA, Economic Research Service. Farmline 5, no. 11 (November 1984): 14. [NAL Call Number: AGE aHD1401.A2U52]

"Eating 'Natural' Gains Popularity," by Charlene C. Price. USDA, Economic Research Service. National Food Review, no. 28 (1985): 14-18. [NAL Call Number: aHD9001.N275]

Alternative Agriculture: Growing Concerns. USDA, Economic Research Service, 1989. Note: 1 videocassette (19 min.): sd., col. (VHS). [NAL Call Number: Videocassette no. 479]

"The Food Industry Responds to Consumers' Pesticide Fears," by E. Van Ravenswaay. USDA, Economic Research Service, Commodity Economics Division. FoodReview 12, no. 3 (July/Sept 1989): 17-20. [NAL Call Number: aHD9001.N275]

"Success Steady in Organic Produce," by Catherine Greene. USDA, Economic Research Service. Agricultural Outlook, no. 185 (May 1992): 15-17. [NAL Call Number: aHD1751.A42]

"Organic Certification: Standards in the Works," by Julie Anton and Betsy Frazão. USDA, Economic Research Service. Agricultural Outlook, no. 199 (1993): 26-31. [NAL Call Number: aHD1751.A42]

"Some Barriers to Organic Produce at the Wholesale Level," by S. Payson, B. Lin, and J. Wertz. USDA, Economic Research Service, Resource and Technology Division. FoodReview 17, no. 2 (May/Aug 1994): 23-26. [NAL Call Number: aHD9001.N275]

"Organic Foods Find Opportunity in the Natural Food Industry," by Julie Anton Dunn. USDA, Economic Research Service. FoodReview 18, no. 3 (September/December 1995): 7-18. [NAL Call Number: aHD9001.N275]

Organic Vegetable Growers Surveyed in 1994: Updates on Agricultural Resources and Environmental Indicators, by Jorge Fernandez-Cornejo, Renata Penn, and Doris Newton. USDA, Economic Research Service, May 1996. [NAL Call Number: aHD1751.R732]
Full text online: http://www.ers.usda.gov/publications/arei/96upd/upd96-4.pdf

"'Organically Grown' Vegetables: U.S. Acreage and Markets Expand During the 1990's," by Linda Calvin and Catherine Greene. USDA, Economic Research Service. Vegetables and Specialties Situation and Outlook (May 1997). [NAL Call Number: aHD9220.U5S47]
Full text online: http://usda.mannlib.cornell.edu/reports/erssor/specialty/vgs-bb/1997/vgs271f.asc

"Consumers Pay a Premium for Organic Baby Foods," by J. Michael Harris. USDA, Economic Research Service. FoodReview 20, no. 2 (May/Aug 1997): 13-16. [NAL Call Number: aHD9001.N275]
Full text online: http://www.ers.usda.gov/publications/foodreview/aug1997/may97d.pdf

Organic Fruit Growers Survey: Updates on Agricultural Resources and Environmental Indicators, by Jorge Fernandez-Cornejo, Renata Penn, and Doris Newton. USDA, Economic Research Service, June 1997. [NAL Call Number: aHD1751.R732]
Full text online: http://www.ers.usda.gov/publications/arei/97upd/upd97-4.pdf

"Demand for Frozen Vegetables: A Comparison of Organic and Conventional Products," by Lewrene Glaser et al. USDA, Economic Research Service. Vegetables and Specialities (November 1998). [NAL Call Number: aHD9220.U5S47]
Full text online: http://usda.mannlib.cornell.edu/reports/erssor/specialty/vgs-bb/1998/vgs276f.asc

"U.S. Organic Agriculture Gaining Ground," by Catherine Greene. USDA, Economic Research Service. Agricultural Outlook, no. 270 (April 2000): 9-14. [NAL Call Number: aHD1751.A42]
Full text online: http://www.ers.usda.gov/publications/agoutlook/apr2000/ao 270d.pdf

"Organic Foods: Niche Marketers Venture into the Mainstream," by Carolyn Dimitri and Nessa Richman. USDA, Economic Research Service. Agricultural Outlook, no. 272 (June/July 2000): 11-14. [NAL Call Number: aHD1751.A42]
Full text online: http://www.ers.usda.gov/publications/AgOutlook/Jun2000/a o272f.pdf

Economics of Food Labeling, by Elise Golan et al. (Agricultural Economic Report, 793) USDA, Economic Research Service, 2001. vi, 41p. [NAL Call Number: A281.9 Ag8A no. 793]
Full text online: http://www.ers.usda.gov/publications/aer793/

U.S. Organic Farming Emerges in the 1990s: Adoption of Certified Systems, by Catherine Greene. (Agriculture Information Bulletin, 770) USDA, Economic Research Service, 2001. iv, 24p. [NAL Call Number: 1 Ag84Ab no. 770]
Full text online: http://www.ers.usda.gov/publications/aib770/

"Organic Marketing Features Fresh Food and Direct Exchange," by Catherine Greene, Carolyn Dimitri, and Nessa Richman. USDA, Economic Research Service. FoodReview 24, no. 1 (January/April 2001): 31-37. [NAL Call Number: aHD9001.N275]
Full text online: http://www.ers.usda.gov/publications/FoodReview/jan2001/ frv24I1f.pdf

"Organic Wheat Production in the United States: Expanding Markets and Supplies," by Catherine Greene and Thomas Dobbs. USDA, Economic Research Service. Wheat Situation and Outlook Yearbook (March 2001).
Full text online: http://www.ers.usda.gov/publications/so/view.asp?f=field/whs-bby/whs2001.pdf

"Tracking Wholesale Prices for Organic Produce," by Emy Sok and Lewrene Glaser. USDA, Economic Research Service. Agricultural Outlook, no. 285 (October 2001): 7-8. [NAL Call Number: aHD1751.A42]
Full text online: http://www.ers.usda.gov/publications/agoutlook/oct2001/ao285d.pdf

Organic Agriculture Provisions [2002 Farm Bill]. (ERS Analysis) USDA, Economic Research Service, 2002.
Full text online: http://www.ers.usda.gov/Features/farmbill/analysis/organicagriculture.htm

Recent Growth Patterns in the U.S. Organic Foods Market, by Carolyn Dimitri and Catherine Greene. (Agriculture Information Bulletin, 777) USDA, Economic Research Service, September 2002. iii, 39p. [NAL Call Number: 1 Ag84Ab no. 777]
Full text online: http://www.ers.usda.gov/publications/aib777/

"Organic Food Industry Taps Growing American Market," by Carolyn and Catherine Greene Dimitri. USDA, Economic Research Service. Agricultural Outlook, no. 295 (October 2002): 4-7. [NAL Call Number: aHD1751.A42]
Full text online: http://www.ers.usda.gov/publications/agoutlook/oct2002/ao295b.pdf

Organic Production [data set]. USDA, Economic Research Service, October 2002.
Full text online: http://www.ers.usda.gov/Data/Organic/

"U.S. Organic Farming: A Decade of Expansion," by Catherine Greene and Amy Kremen. USDA, Economic Research Service. Agricultural Outlook, no. 296 (November 2002): 31-34. [NAL Call Number: aHD1751.A42]
Full text online: http://www.ers.usda.gov/publications/agoutlook/Nov2002/ao296i.pdf

U.S. Organic Farming in 2000-2001: Adoption of Certified Systems, by Catherine Greene and Amy Kremen. (Agriculture Information Bulletin, 780) USDA, Economic Research Service, 2003. iv, 51p. [NAL Call Number: 1 Ag84Ab no. 780]

Full text online: http://www.ers.usda.gov/publications/aib780/

"Organic Agriculture: Gaining Ground," by Catherine Greene and Carolyn Dimitri. USDA, Economic Research Service. Amber Waves (February 2003). [NAL Call Number: aHD1751.A52]
Full text online: http://www.ers.usda.gov/AmberWaves/Feb03/Findings/Organic Agriculture.htm

Organic Farmgate and Wholesale Prices [data set]. USDA, Economic Research Service, May 2005.
Full text online: http://www.ers.usda.gov/Data/OrganicPrices/

Price Premiums Hold on as U.S. Organic Produce Market Expands, by Lydia Oberholtzer, Carolyn Dimitri, and Catherine Greene. (Outlook Report, VGS30801) USDA, Economic Research Service, May 2005. 22p.
Full text online: http://www.ers.usda.gov/Publications/vgs/may05/VGS30801/

Market-Led Growth vs. Government-Facilitated Growth: Development of the U.S. and EU Organic Agricultural Sectors, by Carolyn Dimitri and Lydia Oberholtzer. (Outlook Report, WRS0505) USDA, Economic Research Service, August 2005. 26p.
Full text online: http://www.ers.usda.gov/Publications/wrs0505/

"Organic Price Premiums Remain High," by Carolyn Dimitri and Lydia Oberholtzer. USDA, Economic Research Service. Amber Waves (September 2005). [NAL Call Number: aHD1751.A52]
Full text online:
http://www.ers.usda.gov/AmberWaves/September05/Findings/OrganicPrice.htm

Foreign Agricultural Service (FAS)

"European Yuppies - The New Niche Market for U.S. Food Exporters." USDA, Foreign Agricultural Service. AgExporter 1, no. 12 (December 1989): 4-8. [NAL Call Number: aHD1401.A7]

"French Organics Market Awaits Development," by R. Gauthier. USDA, Foreign Agricultural Service. AgExporter 8, no. 6 (August/September 1996): 17-19. [NAL Call Number: aHD1401.A7]

"Japan Buys into Organics," by J. K. Jenni. USDA, Foreign Agricultural Service. AgExporter 8, no. 7 (October 1996): 20-22. [NAL Call Number: aHD1401.A7]

"Dutch Organic Food Market Offers All-Natural Potential for U.S. Firms," by Tatiana van der Harst-Collaris and Laura Gabel Scandurra. USDA, Foreign Agricultural Service. AgExporter 9, no. 8 (August 1997): 10-15. [NAL Call Number: aHD1401.A7]

Full text online: http://www.fas.usda.gov/info/agexporter/1997/dutch.html

"Defining Moment: Canadian Organic Regs Are Coming," by George C. Myles. USDA, Foreign Agricultural Service. AgExporter 9, no. 11 (November 1997): 24-26. [NAL Call Number: aHD1401.A7]
Full text online: http://www.fas.usda.gov/info/agexporter/1997/defining.html

Organic Perspectives [quarterly newsletter]. USDA, Foreign Agricultural Service. 1998-.
Full text online: http://www.fas.usda.gov/agx/organics/organics.html

"France's Baby Food Boom," by Mattie R. Sharpless and Roselyne Gauthier. USDA, Foreign Agricultural Service. AgExporter 10, no. 2 (February 1998): 18-19. [NAL Call Number: aHD1401.A7]
Full text online:
http://www.fas.usda.gov/info/agexporter/1998/February%201998/frances.html

"Organic Food and Vegetable Products, a Sino- the Times," by Henry Lin. USDA, Foreign Agricultural Service. AgExporter 10, no. 3 (March 1998): 8-9. [NAL Call Number: aHD1401.A7]
Full text online: http://www.fas.usda.gov/info/agexporter/1998/March%201998/organic.html
"Organic Food Trend Sprouts in Hong Kong," by W. Garth Thorburn II. USDA, Foreign Agricultural Service. AgExporter 10, no. 9 (September 1998): 15. [NAL Call Number: aHD1401.A7]
Full text online: http://www.fas.usda.gov/info/agexporter/1998/Sept%201998/orgHoKoc.html

"Organic Food Trend Wins in Taiwan," by Chiou Mey Perng. USDA, Foreign Agricultural Service. AgExporter 10, no. 9 (September 1998): 13-14. [NAL Call Number: aHD1401.A7]
Full text online: http://www.fas.usda.gov/info/agexporter/1998/Sept%201998/orgtaiw.html

"Singapore: A Growing Market for U.S. Health Foods," by Dale L. Good. USDA, Foreign Agricultural Service. AgExporter 10, no. 9 (September 1998): 10-12. [NAL Call Number: aHD1401.A7]
Full text online: http://www.fas.usda.gov/info/agexporter/1998/Sept%201998/singapor.html

"Organic and Hispanic, These West Coast Chipsters Are Bagging Exports," by Jill Lee. USDA, Foreign Agricultural Service. AgExporter 11, no. 12 (December 1999): 21. [NAL Call Number: aHD1401.A7]
Full text online: http://www.fas.usda.gov/info/agexporter/1999/organic.html

"Five Tips for Organic Trade Success," by Jill Lee. USDA, Foreign Agricultural Service. AgExporter 12, no. 6 (June 2000): 11-12. [NAL Call Number: aHD1401.A7]
Full text online: http://www.fas.usda.gov/info/agexporter/2000/june/fivetips.htm

"France: More Than a Nice Niche?" by Frank J. Piason. USDA, Foreign Agricultural Service. AgExporter 12, no. 6 (June 2000). [NAL Call Number: aHD1401.A7]
Full text online: http://www.fas.usda.gov/info/agexporter/2000/june/france.htm

"Germany: The Customer-Competitor Duality," by Dietmar Achilles. USDA, Foreign Agricultural Service. AgExporter 12, no. 6 (June 2000): 20. [NAL Call Number: aHD1401.A7]
Full text online: http://www.fas.usda.gov/info/agexporter/2000/june/germany.htm

"Lacking a Domestic Source, Hong Kong Must Import Organics," by Carolyn Yuen. USDA, Foreign Agricultural Service. AgExporter 12, no. 6 (June 2000): 22. [NAL Call Number: aHD1401.A7]
Full text online: http://www.fas.usda.gov/info/agexporter/2000/june/lackinga.htm

"Organic Markets Offer U.S. Agriculture Current and Future Sales Opportunities," by Janise Zygmont. USDA, Foreign Agricultural Service. AgExporter 12, no. 6 (June 2000): 4-7. [NAL Call Number: aHD1401.A7]
Full text online: http://www.fas.usda.gov/info/agexporter/2000/june/organic.htm

"Organic Products Finding a Ready Market on Taiwan," by Chiou Mey Perng. USDA, Foreign Agricultural Service. AgExporter 12, no. 6 (June 2000): 17-19. [NAL Call Number: aHD1401.A7]
Full text online: http://www.fas.usda.gov/info/agexporter/2000/june/organic3.htm

"Organic Trade Association Sees Opportunities in Exports," by Janise Zygmont. USDA, Foreign Agricultural Service. AgExporter 12, no. 6 (June 2000): 8-10. [NAL Call Number: aHD1401.A7]
Full text online: http://www.fas.usda.gov/info/agexporter/2000/june/organic2.htm

"Practical Contacts for Organic Producers and Exporters" USDA, Foreign Agricultural Service. AgExporter 12, no. 6 (June 2000). [NAL Call Number: aHD1401.A7]
Full text online: http://www.fas.usda.gov/info/agexporter/2000/june/practica.htm

"Timeline of Organic Milestones," USDA, Foreign Agricultural Service. AgExporter 12, no. 6 (June 2000). [NAL Call Number: aHD1401.A7]
Full text online: http://www.fas.usda.gov/info/agexporter/2000/june/timeline.htm

"The U.K. Organic Sector: Go for It!" by Jennifer Jones. USDA, Foreign Agricultural Service. AgExporter 12, no. 6 (June 2000): 15-16. [NAL Call Number: aHD1401.A7]
Full text online: http://www.fas.usda.gov/info/agexporter/2000/june/uk.htm

"Wizard's Cauldron Conjures International Markets," by Jill Lee. USDA, Foreign Agricultural Service. AgExporter 12, no. 6 (June 2000): 13-14. [NAL Call Number: aHD1401.A7]
Full text online: http://www.fas.usda.gov/info/agexporter/2000/june/wizards.htm

Danish Organic Dairy Production, by Melissa Schmaedick. (FAS Division Reports) USDA, Foreign Agricultural Service, July 2000.
Full text online: http://www.fas.usda.gov/dlp2/circular/2000/00-07Dairy/danish.html

U.S. Organic Fruit: Export Opportunities and Competition in the International Market, by Janise Zygmont. (FAS Division Reports) USDA, Foreign Agricultural Service, 2001.
Full text online: http://www.fas.usda.gov/agx/organics/speech.htm

"Organics East: Postcards from the Pacific Market," USDA, Foreign Agricultural Service. AgExporter 13, no. 7 (July 2001): 15-17. [NAL Call Number: aHD1401.A7]
Full text online: http://www.fas.usda.gov/info/agexporter/2001/july/p15-17.pdf

"Big Changes for the Netherlands' Organic Industry," by Wendalin Kolkman. USDA, Foreign Agricultural Service. AgExporter 13, no. 12 (December 2001): 16. [NAL Call Number: aHD1401.A7]
Full text online: http://www.fas.usda.gov/info/agexporter/2001/dec/page16.pdf

"Benelux Opting for Organics," by Marcel Pinckaers. USDA, Foreign Agricultural Service. AgExporter 14, no. 11 (November 2002): 8-9. [NAL Call Number: aHD1401.A7]
Full text online: http://www.fas.usda.gov/info/agexporter/2002/November/pgs8-9.pdf

European Union: Organic Products: European Action Plan for Organic Food and Farming, by Hilde Brans. (GAIN Report, E23020) USDA, Foreign Agricultural Service, Global Agriculture Information Network, 2003.
Full text online: http://www.fas.usda.gov/gainfiles/200302/145884678.pdf

Exporting Organic: Are You Ready for the Challenge? by Kelly Strzelecki. USDA, Foreign Agricultural Service, 2003. Note: PowerPoint presentation.
Full text online: http://www.fas.usda.gov/agx/organics/exporting.htm

Organic, Natural and Wellness Products at Natural Products Expo West Anaheim, CA - March 6-9, 2003. USDA, Foreign Agricultural Service in partnership with Department of Commerce, U.S. Commercial Service, 2003.
Full text online: http://www.fas.usda.gov/agx/organics/BuyerAlert/BAOrganic.htm

United Kingdom: Organic Products: Market Brief 2003, by Phil Bicknell. (GAIN Report, UK3006) USDA, Foreign Agricultural Service, Global Agriculture Information Network, 2003.
Full text online: http://www.fas.usda.gov/gainfiles/200303/145884996.pdf

"Satisfying the Appetite of the Danes," by Hasse Kristensen. USDA, Foreign Agricultural Service. AgExporter 15, no. 6 (June 2003): 10. [NAL Call Number: aHD1401.A7]
Full text online: http://www.fas.usda.gov/info/agexporter/2003/June/pg%2010.pdf

"Popularity of Organic Foods on the Rise in the United Kingdom," by Philip Bicknell. USDA, Foreign Agricultural Service. AgExporter 15, no. 8 (August 2003): 8-9. [NAL Call Number: aHD1401.A7]
Full text online: http://www.fas.usda.gov/info/agexporter/2003/August/pgs%208-9.pdf

“South Koreans Buying Organic,” by Young Sook Oh and Susan B. Phillips. USDA, Foreign Agricultural Service. AgExporter 15, no. 8 (August 2003): 6-7.
[NAL Call Number: aHD1401.A7]
Full text online: http://www.fas.usda.gov/info/agexporter/2003/August/pgs%206-7.pdf

“Portuguese Looking to Organics,” by Maria do Monte Gomes. USDA, Foreign Agricultural Service. AgExporter 15, no. 10 (October 2003): 14. [NAL Call Number: aHD1401.A7]
Full text online: http://www.fas.usda.gov/info/agexporter/2003/October/pg%2014.pdf

Attaché Reports: Organics [directory]. USDA, Foreign Agricultural Service, 2004.
Full text online: http://www.fas.usda.gov/agx/organics/attache.htm

“Organics in Austria: From Niche to Mainstream,” by Roswitha Krautgartner. USDA, Foreign Agricultural Service. AgExporter 16, no. 2 (February 2004): 6-7.
[NAL Call Number: aHD1401.A7].
Full text online: http://www.fas.usda.gov/info/agexporter/2004/February/pgs%206%20-%207.pdf

“Opportunity Knocks for Organics in Singapore,” USDA, Foreign Agricultural Service. AgExporter 16, no. 6 (June 2004): 14-15. [NAL Call Number: aHD1401.A7]
Full text online: http://www.fas.usda.gov/info/agexporter/2004/June/pgs%2014-15.pdf

“Organics Earning Respect on Taiwan,” by Welly Liu-Chen and Jeff Miller. USDA, Foreign Agricultural Service. AgExporter 16, no. 6 (June 2004): 13.
[NAL Call Number: aHD1401.A7]
Full text online: http://www.fas.usda.gov/info/agexporter/2004/June/pg%2013.pdf

U.S. Market Profile for Organic Food Products, by James M. Tringe. USDA, Foreign Agricultural Service, Commodity and Marketing Programs-Processed Products Division, International Strategic Marketing Group, February 2005. 19p.
Full text online:
http://www.fas.usda.gov/agx/organics/USMarketProfileOrganicFoodFeb2005.pdf

Natural Resources Conservation Service (NRCS)

The Knife Roller (Crimper): An Alternative Kill Method for Cover Crops. (Soil Quality/Agronomy Technical Note, 13) USDA, Natural Resources Conservation Service, Soil Quality Institute, September 2002.
Full text online: http://soils.usda.gov/sqi/files/13.pdf

RISK MANAGEMENT AGENCY (RMA)

2001 Organic Crop Insurance Underwriting Guide. (FCIC-24140 (03-2001)) USDA, Risk Management Agency, Federal Crop Insurance Corporation, Product Development Division, 2001.
Full text online:
http://www.rma.usda.gov/FTP/Publications/directives/24000/pdf/01_24140.pdf

RURAL DEVELOPMENT (RD)

The State of U.S. Organic Producer Marketing Cooperatives in 1990, by Alan D. Borst. (Working Paper, 91-W1) USDA, Agricultural Cooperative Service, 1990. v, 18p. [NAL Call Number: ArU HD1491.U5W67 no.91-W1]

The State of U.S. Organic Producer Marketing Cooperatives in 1991, by Alan D. Borst. (Staff Report, 91-S7) USDA, Agricultural Cooperative Service, 1991. v, 23p. [NAL Call Number: aHD9005.B67 1991]

"New Model, Old Ways: Farmers Find New Life Serving Burgeoning Organic Markets," by Pamela J. Karg. USDA, Rural Development, Rural Business-Cooperative Service. Rural Cooperatives 67, no. 1 (January 2000). [NAL Call Number: aHD1491.U6R87] 23-27.
Full text online: http://www.rurdev.usda.gov/rbs/pub/jan00/models.htm

"A Fresh Advantage: Co-op Helps Small Farms Market Produce to High-end Restaurants," by Pamela J. Karg. USDA, Rural Development, Rural Business-Cooperative Service. Rural Cooperatives (May/June 2005): 24-26. [NAL Call Number: aHD1491.U6R87]
Full text online: http://www.rurdev.usda.gov/rbs/pub/may05/may05.pdf

"Going with the Grain: New Mexico Organic Wheat Cooperative Provides Lift to Farmers, Rural Economy," by Catherine Merlo. USDA, Rural Development, Rural Business-Cooperative Service. Rural Cooperatives (May/June 2005): 10-14. [NAL Call Number: aHD1491.U6R87]
Full text online: http://www.rurdev.usda.gov/rbs/pub/may05/may05.pdf

"Slice of the Market: Penns Corner, Tuscarora Co-ops Target Growing Restaurant Trade," by Stephen A. Thompson. USDA, Rural Development, Rural Business-Cooperative Service. Rural Cooperatives (May/June 2005): 27-29. [NAL Call Number: aHD1491.U6R87]
Full text online: http://www.rurdev.usda.gov/rbs/pub/may05/may05.pdf

"USDA B&I Program Supports Growth of Organic Cooperatives," by Peter J. Thomas. USDA, Rural Development, Rural Business-Cooperative Service. Rural Cooperatives (May/June 2005): 43. [NAL Call Number: aHD1491.U6R87]
Full text online: http://www.rurdev.usda.gov/rbs/pub/may05/may05.pdf

"Cream of the CROPP: Demand Outstripping Supply as CROPP Organic Dairy Products Go Prime Time," by Dan Campbell. USDA, Rural Development, Rural Business-Cooperative Service. Rural Cooperatives (May/June 2005): 14-19. [NAL Call Number: aHD1491.U6R87] Full text online: http://www.rurdev.usda.gov/rbs/pub/may05/may05.pdf

World Agricultural Outlook Board (WAOB)

"Cultivating New Ground: Organic Agriculture Arrives at the Outlook," by Mark Lipson. USDA, World Agricultural Outlook Board. Agricultural Outlook Forum '98 Proceedings (1998): 159-160. Note: Meeting held February 23-24, 1998, in Washington, DC. [NAL Call Number: aHD1755.A376]

"The Evolving Organic Marketplace: Executive Summary," by Harvey Hartman. USDA, World Agricultural Outlook Board. Agricultural Outlook Forum '98 Proceedings (1998): 169-174. Note: Meeting held February 23-24, 1998, in Washington, DC. [NAL Call Number: aHD1755.A376]

"Organic Food Marketing Trends," by Gene Kahn. USDA, World Agricultural Outlook Board. Agricultural Outlook Forum '98 Proceedings (1998). Note: Meeting held February 23-24, 1998, in Washington, DC. PowerPoint file. [NAL Call Number: aHD1755.A376] Full text online: http://www.usda.gov/oce/waob/outlook98/98speeches.htm

"Direct Marketing Options: Farmers Markets, Restaurants, Community Supported Agriculture, and the Organic Alternative," by Steve Gilman. USDA, World Agricultural Outlook Board. Agricultural Outlook Forum 1999 Proceedings (1999): 118-121. Note: Meeting held February 22-23, 1999, in Arlington, VA. [NAL Call Number: aHD1755.A376] Full text online: http://www.usda.gov/agency/oce/waob/outlook99/99speeches.htm

"Organics...Yesterday, Today, and Tomorrow," by Larry Jacobs. USDA, World Agricultural Outlook Board. Agricultural Outlook Forum 2000 (2000). Note: Meeting held February 24-25, 2000, in Arlington, VA.
Full text online: http://www.usda.gov/oce/waob/Archives/2000/speeches/jacobs.txt

"Venture Capital in Organic Marketing," by Brooks Browne. USDA, World Agricultural Outlook Board. Agricultural Outlook Forum 2000 (2000). Note: Meeting held February 24-25, 2000, in Arlington, VA.
Full text online: http://www.usda.gov/oce/waob/Archives/2000/speeches/brooksbrowne.txt

"Establishing Protocols for Certified Organic and Natural Beef," by Diana S. Endicott. USDA, World Agricultural Outlook Board. Agricultural Outlook Forum 2003 (2003). Note: Meeting held February 21-22, 2003, in Arlington, VA. Large PowerPoint file.
Full text online: http://www.usda.gov/oce/forum/Archives/2003/Programs/program.htm

"Current Trends in the Retail Market for Organics: A Retailer's Perspective," by Mary Mulry. USDA, World Agricultural Outlook Board. Agricultural Outlook Forum 2004 (2004). Note: Meeting held February 19-20, 2004, in Arlington, VA.

"The Growth of the Organic Market: Producers' Perspectives," by Kevin Edberg. USDA, World Agricultural Outlook Board. Agricultural Outlook Forum 2004 (2004). Note: Meeting held February 19-20, 2004, in Arlington, VA.
Full text online: http://agecon.lib.umn.edu/cgi-bin/pdf_view.pl?paperid=12736&ftype=.pdf

"Organic Consumer Profile," by Laurie Demeritt. USDA, World Agricultural Outlook Board. Agricultural Outlook Forum 2004 (2004). Note: Meeting held February 19-20, 2004, in Arlington, VA.

"Recent Trends in Organic Production," by Catherine Greene. USDA, World Agricultural Outlook Board. Agricultural Outlook Forum 2004 (2004). Note: Meeting held February 19-20, 2004, in Arlington, VA.
Full text online: http://www.usda.gov/oce/forum/Archives/2004/speeches/catherinegreene.doc

Other USDA Publications

"How You Can Grow Food Organically," by W. Judkins and F. Smith. In Yearbook of Agriculture. USDA, 1978. 224-229p. [NAL Call Number: 1 AG84Y]
Report and Recommendations on Organic Farming. USDA, Study Team on Organic Farming, 1980. xiv, 94p. [NAL Call Number: aS605.5.U52]

"Cliff and Debbie Van Till: A California Family Finds out What 'Organic' Really Means," by J. Stumbos. In Yearbook of Agriculture, 28-30. (Americans in Agriculture: Portraits of Diversity, edited by D. Smith) USDA, 1990. [NAL Call Number: 1 Ag84y]

First International Conference on Kyusei Nature Farming: Proceedings of the Conference at Khon Kaen University, Khon Kaen, Thailand, October 17-21, 1989, by James F. Parr, S. B. Hornick, and C. E. Whitman. USDA, 1991. ix, 175p. [NAL Call Number: aS605.5.I56-1989]

About the Alternative Farming Systems Information Center at the National Agricultural Library

The Alternative Farming Systems Information Center (AFSIC), one of several centers at the National Agricultural Library (NAL) that provide in-depth coverage of specific agricultural subject areas, focuses on sustainable and alternative agricultural systems, crops and livestock.

AFSIC specializes in locating and accessing information related to many aspects of alternative agriculture: sustainable and organic crop and livestock farming systems;

renewable farm energy options; alternative marketing practices; crop and livestock diversification including aquaculture, exotic and heritage farm animals, alternative and specialty crops, new uses for traditional crops, and crops grown for industrial production; and small farm issues.

INDEX

A

B

C

D

G

H

N

O

P

T

U

V

W

Y

Z